OXFORD LOGIC GUIDES: 30

General Editors
DOV M. GABBAY
ANGUS MACINTYRE
JOHN SHEPHERDSON
DANA SCOTT

OXFORD LOGIC GUIDES

1. Jane Bridge: *Beginning model theory: the completeness theorem and some consequences*
2. Michael Dummett: *Elements of intuitionism*
3. A.S. Troelstr: *Choise sequences: a chapter of intuitionistic mathematics*
4. J.L. Bell: *Boolean-valued models and independence proofs in set theory* (1st edition)
5. Krister Segerberg: *Classical propositioned operators: and exercise in the foundations of logic*
6. G.C. Smith: *The Boole–De Morgan correspondence 1842–1864*
7. Alec Fisher: *Formal number theory and computability: a work book*
8. Anand Pillay: *An introduction to stability theory*
9. H.E. Rose: *Subrecursion: functions and hierachies*
10. Michael Hallett: *Cantorian set theory and limitation of size*
11. R. Mansfield and G. Weitkamp: *Recursive aspects of descriptive set theory*
12. J.L. Bell: *Boolean-valued models and independence proofs in set theory* (2nd edition)
13. Melvin Fitting: *Computability theory: semantics and logic programming*
14. J.L. Bell: *Toposes and local set theories: an introduction*
15. Richard Kaye: *Models of Peano arithmetic*
16. J. Chapman and F. Rowbottom: *Relative category theory and geometric morphisms*
17. S. Shapiro: *Foundations without foundationalism*
18. J.P. Cleave: *A study of logics*
19. R.M. Smullyan: *Gödel's incompleteness theorems*
20. T.E. Forster: *Set theory with a universal set*
21. C. McLarty: *Elementary categories, elementary toposes*
22. R.M. Smullyan: *Recursion theory for metamathematics*
23. Peter Clote and Jan Krajíček: *Arithmetic, proof theory, and computational complexity*
24. A. Tarski: *Introduction to logic and to the methodology of deductive sciences*
25. G. Malinowski: *Many valued logics*
26. Alexandre Borovik and Ali Nesin: *Groups of finite Morley rank*
27. Smullyan: *Diagonalization and self-reference*
28. Dov M. Gabbay, Ian Hodkinson, and Mark Reynolds: *Temporal logic: mathematical foundations and computational aspects* (Volume 1)
29. Saharan Shelah: *Cardinal arithmetic*
30. Sandewall: *Features and fluents: the representation of knowledge about dynamical systems*

Features and Fluents

The Representation of Knowledge about Dynamical Systems

Volume 1

ERIK SANDEWALL

Department of Computer and Information Science,
Linköping University, Linköping, Sweden

CLARENDON PRESS · OXFORD

1994

Oxford University Press, Walton Street, Oxford OX2 6DP
Oxford New York
Athens Auckland Bangkok Bombay
Calcutta Cape Town Dar es Salaam Delhi
Florence Hong Kong Istanbul Karachi
Kuala Lumpur Madras Madrid Melbourne
Mexico City Nairobi Paris Singapore
Taipei Tokyo Toronto
and associated companies in
Berlin Ibadan

Oxford is a trade mark of Oxford University Press

Published in the United States by
Oxford University Press Inc., New York

© E. Sandewall, 1994

All rights reserved. No part of this publication may be
reproduced, stored in a retrieval system, or transmitted, in any
form or by any means, without the prior permission in writing of Oxford
University Press. Within the UK, exceptions are allowed in respect of any
fair dealing for the purpose of research or private study, or criticism or
review, as permitted under the Copyright, Designs and Patents Act, 1988, or
in the case of reprographic reproduction in accordance with the terms of
licences issued by the Copyright Licensing Agency. Enquiries concerning
reproduction outside those terms and in other countries should be sent to
the Rights Department, Oxford University Press, at the address above.

This book is sold subject to the condition that it shall not,
by way of trade or otherwise, be lent, re-sold, hired out, or otherwise
circulated without the publisher's prior consent in any form of binding
or cover other than that in which it is published and without a similar
condition including this condition being imposed
on the subsequent purchaser.

A catalogue record for this book is available from the British Library

Library of Congress Cataloging in Publication Data

ISBN 0 19 853845 6

Typeset by the author using T_EX

Printed in Great Britain by
Bookcraft (Bath) Ltd
Midsomer Norton, Avon

Preface

This book is about logic-based methods for reasoning about actions and change. It has passed through several stages of evolution during its gestation period. At first, it was intended to be a synthesis and extension of a number of previous conference articles, a project which seemed to require a larger number of pages than would be possible, for example, in a journal article. At that time, I also intended to include various implementation aspects, ranging from software architecture to application examples.

While working on the book along those lines, one obvious step was to assemble a set of test examples for reasoning about actions and change, and to check out my own approaches and others that have been proposed with respect to whether they obtain the intended results for the test suite. This required a detailed review of several previously proposed approaches. Gradually, the focus of the book changed towards attempting to provide a precise and uniform account of the various approaches in this field, so that their properties could be compared more easily and concisely than has been the case before.

A number of chapters were written with this goal in mind. However, this work led in its turn to the insight that the intended purpose could never be achieved using test examples, even if these are combined with informal arguments in support of a chosen approach, no matter how convincing such arguments may seem. This problem will be discussed with more detail in Chapter 3. A second change of approach was necessary, therefore, and the revised goal was to obtain *assessments* of various nonmonotonic logics for actions and change, that is, theorems about the *range of correct applicability* of proposed logics for reasoning about actions and change.

The new goal turned out to be very challenging and constructive. It suggested a number of interesting questions, and was the basis for a sequence of nontrivial results. The only problem was that it opened up more questions than could be addressed at one time. In order to retain the focus, I have adopted the policy that whenever the material in the book seemed to branch off from the core too much, I have postponed the centrifugal material to the intended next volume. In this way, the contents of the present volume are coherent and tightly connected, although it does contain many references to the forthcoming Volume II.

If the original intention was to make a synthesis of some previous articles, the book became instead a replacement for articles: new results in my own research have been successively incorporated into the structure of the book, and have not always been presented in journal or conference articles. The reader of this book will therefore find a number of results which have not been published previously.

Many colleagues and friends have contributed during the four years that have been required for this process. The fundamental question 'for what class of problems is the method that we are discussing now guaranteed to work correctly?' was first asked to us by Lennart Ljung, Professor of automatic control in Linköping, during one of the many creative joint meetings between his group and ours. Bernhard Nebel's Ph.D. thesis [52] impressed me with its systematic approach, where the properties of a number of variants of a theory were analyzed with precision. It provided a pattern that I felt could be adapted to the present topic as well.

The discussions in our research group in Linköping, RKLLAB, have influenced each step of the work. Particular thanks go to Patrick Doherty, Thomas Drakengren, Witold Łukaszewicz, Christer Bäckström, Dimiter Driankov, Peter Jonsson, Lars Karlsson, Jacek Malec, Tommy Persson, and Hua Shu, who have given valuable and often detailed comments on successive versions of the manuscript.

I have also received valuable comments and advise from Michael Anderson, Hector Geffner, Matt Ginsberg, Joachim Hertzberg, Neelakantan Kartha, Vladimir Lifschitz, Pavlos Peppas, David Poole, Len Schubert, Yoav Shoham, and Lynn Stein, mentioning only the most detailed contributions. Some of their comments pertained to material which has now been postponed to Volume II; others have influenced already the present volume. I am equally indebted in either case.

Some of the results from this book have been used for conference and journal articles, and the reviews of those articles have influenced the book as well. Their contributions range from concrete suggestions for improvements to sharp disagreement with the whole approach. Both kinds of feedback have been useful in their respective ways.

This preface is being written at LAAS-CNRS in Toulouse, France, where I am spending a sabbatical which started half a year ago. I had been careful to finish the book before leaving, or so I thought, but the excellent research atmosphere at LAAS made it possible to develop additional results and to add them to the book. I am particularly grateful to Malik Ghallab, my host and friend at LAAS, who made my visit here possible.

The Department of Computer and Information Science (IDA) at Linköping University is my academic home. I am grateful to all my colleagues for their wholehearted contributions to our common research environment. The present book has benefitted from uncountable discussions with these colleagues at IDA, as well as from the interactions in the European cooperation projects – DRUMS II and Logic and change – where I also have the pleasure to participate.

The technical and administrative staffs at IDA in Linköping and at LAAS in Toulouse are equally efficient and pleasant to work with, and this book would probably never have been completed without them. Special thanks to our group's secretaries at IDA during successive years: Lillemor Wallgren, Ingrid Sandblom, Anne Eskilsson, and Lise-Lotte Svensson, and to Jackie Som at LAAS.

Our two sons, Anton and Örjan, have taken an active interest in this work. Almost every week for several years they have asked whether the book would soon be ready, and sometimes they have expressed their doubt that it would ever be completed. They have been disappointed when I could not explain in their terms what the book is about, and they have asked tough questions about how many people would read the book, how many would pay for one, and what part of the proceeds would benefit the family, questions which were good starting-points for talking about a researcher's priorities in life. Their alternative perspective has been a useful reminder of how this enterprise may look from outside the sphere of science.

Margareta, my wife, has always given her consistent support and encouragement to the work on this book. The greatest thanks of all goes to her.

Grants by a number of agencies have contributed to the research environment and the work that has resulted in this book, in particular (in Sweden) The Swedish Board of Technical Development, NUTEK, The Swedish Engineering Science Research Council, and (in France) the Centre National de Recherche Scientifique. Some of the Swedish support pertains to participation in the Esprit projects DRUMS II and Logic and change. All this support is gratefully acknowledged.

Contents

1 Inert and inhabited dynamical systems — 1
- 1.1 Topic — 1
  - 1.1.1 The two levels of system description — 3
  - 1.1.2 Perception — 4
  - 1.1.3 Inertia — 6
  - 1.1.4 Actions — 7
  - 1.1.5 Inhabited dynamical systems and AI — 8
- 1.2 The image level of description — 9
  - 1.2.1 Structure of an IDS — 9
  - 1.2.2 Fluents and occurrences — 10
  - 1.2.3 The monitor — 12
- 1.3 Ego/world interaction — 14
  - 1.3.1 Informal introduction to ego/world interaction — 15
  - 1.3.2 Finite, infinite, and complete developments — 17
  - 1.3.3 Egos and worlds — 19
  - 1.3.4 IDS Games — 21
  - 1.3.5 Formal IDS's — 22
  - 1.3.6 Scenarios — 23
- 1.4 The ontological taxonomy — 24
  - 1.4.1 Ontological designators — 24
  - 1.4.2 The ontological characteristics — 25
  - 1.4.3 Ontological subfamilies — 28
  - 1.4.4 Assignment of ontological family to a problem — 29
- 1.5 Summary — 29

2 Inference operations on scenario descriptions — 31
- 2.1 Scenario descriptions — 31
  - 2.1.1 The Stockholm delivery scenario — 32
  - 2.1.2 IDS scenarios vs. common-sense scenarios — 32
  - 2.1.3 Structure of scenario descriptions — 34
  - 2.1.4 Object domains — 37
  - 2.1.5 Timepoint and object constants — 37
  - 2.1.6 Chronicles — 38
- 2.2 Epistemological assumptions — 39

		2.2.1	Logics with implicit epistemological assumptions	39

x   *Contents*

      2.2.1 Logics with implicit epistemological assumptions   39
      2.2.2 Epistemological designators   41
      2.2.3 The repertoire of epistemological designators   43
  2.3 Formal definitions of scenario descriptions   43
      2.3.1 Scenario descriptions   43
      2.3.2 Model sets for scenario descriptions   46
  2.4 Entailment relations   47
      2.4.1 Entailment relations and weakened models   47
      2.4.2 Scenario completion   48
      2.4.3 Deduction and abduction with respect to chronicles   49
      2.4.4 Consistency-checking with respect to chronicles   51
      2.4.5 Chronicle completion as the pivotal operation   52
  2.5 Reasoning problems   53
      2.5.1 Definitions of reasoning problems   53
      2.5.2 Rephrasing logical operations as reasoning problems   54
      2.5.3 Discussion   57
  2.6 Terminology   58
      2.6.1 Domains and worlds   58
      2.6.2 Formulae, axioms, and premises   59
      2.6.3 Layers of logic and of logical language   59
      2.6.4 Realizations and implementations   61
  2.7 Summary   61

3 Underlying semantics for IDS worlds   63
  3.1 Some methodological aspects of common-sense reasoning   63
      3.1.1 The example-based methodology   63
      3.1.2 Systematic methodology used in this book   65
      3.1.3 Approaching common-sense reasoning with systematic methodology   67
      3.1.4 Combining the two methodologies   69
      3.1.5 The progression of more general ontologies   69
      3.1.6 Syntax based assessment methodology   71
      3.1.7 Validation using a chronicle description language   72
  3.2 Choice of underlying semantics   72
      3.2.1 Reasons for IDS world semantics   72

		3.2.2	Partial state-transition semantics	73
		3.2.3	Trajectory semantics	75
		3.2.4	Trajectory-semantics worlds	75
		3.2.5	Trajectory-semantics egos	78
		3.2.6	Trajectory semantics and ontological families	78
	3.3	Trajectory semantics in concrete terms		79
		3.3.1	The rainy-day shooting example	79
		3.3.2	The bus ride example	80
	3.4	Trajectories and the material system		80
	3.5	Summary		82
4	**Elementary feature logic and meta-logical concepts**			**83**
	4.1	Logical domains		83
		4.1.1	Object domain vs. value domains	84
		4.1.2	Lexical object domains	84
	4.2	Elementary feature logic		86
		4.2.1	Introduction	86
		4.2.2	Value domains and syntax	87
		4.2.3	Semantics	89
		4.2.4	Some axioms	89
	4.3	Meta-level concepts in the base logics		90
		4.3.1	Meta-level domains and the model set	90
		4.3.2	Quine quotes	90
		4.3.3	Meta-level operations	91
		4.3.4	Abbreviations	91
		4.3.5	Logical entailment relations	91
	4.4	Reduction of model sets		92
		4.4.1	Selection functions	93
		4.4.2	Alternatives to selection functions	94
		4.4.3	Semantic selection and selection functions	94
		4.4.4	Syntactic methods and premise integration	97
	4.5	Summary		99
5	**Lexical-domain object-feature logic**			**100**
	5.1	Ontology		100
		5.1.1	Object domains	100
		5.1.2	Features	101
	5.2	Basic definitions		101
		5.2.1	Object domains, similarity types, and vocabularies	101
		5.2.2	Syntax	102

		5.2.3	Denotations	103
		5.2.4	Valuation of formulae	104
		5.2.5	Model sets and entailment	106
	5.3	An example		106
		5.3.1	The assignment scenario	107
		5.3.2	The elementary codesignation algorithm	109
	5.4	Set-level approach		112
		5.4.1	Syntax and semantics	112
		5.4.2	Axioms	113
		5.4.3	The assignment scenario revisited	113
		5.4.4	Example 'mother of son'	114
	5.5	Summary		115
6	**Temporal feature logic for discrete time domains**			**116**
	6.1	Ontology for standard time		117
		6.1.1	Choice and use of time domain	117
		6.1.2	Rationale for standard time	118
		6.1.3	Timepoints	119
		6.1.4	States	120
		6.1.5	Situations	120
		6.1.6	Arguments in favor of non-metric time	121
		6.1.7	Intra-situation change and chronometric fluents	122
		6.1.8	Time structure	123
		6.1.9	Non-determinism in Herbrand time	124
	6.2	Reification of temporal formulae		125
	6.3	Main syntax		127
		6.3.1	Core of the main syntax for DFL-1	128
		6.3.2	Convenience abbreviations	130
	6.4	Semantics and axioms for the base logic		132
		6.4.1	Semantics	132
		6.4.2	Some axioms	134
	6.5	The side language for occurrences		134
		6.5.1	Approach	134
		6.5.2	The elementary occurrence language	135
		6.5.3	Reassignment formulae as abbreviations	137
	6.6	Branching time		138
		6.6.1	The linear and branching time domains	138
		6.6.2	Intervals and trees in branching time	140
		6.6.3	Inhabited dynamical systems with branching time	142
	6.7	Summary		146

7	**Chronicle completion in $\mathcal{K}$-IA**		147
	7.1 Chronicle completion		147
	7.2 Examples of chronicle completion in common-sense domains		150
		7.2.1 Yale shooting scenario (YSS)	151
		7.2.2 Hiding turkey scenario (HTS)	152
		7.2.3 Stanford murder mystery (SMM)	154
		7.2.4 Ferryboat connection scenario (FCS)	154
		7.2.5 Russian turkey scenario (RTS)	155
		7.2.6 Stolen car scenario (SCS)	156
		7.2.7 Ticketed car scenario (TCS)	157
		7.2.8 Furniture assembly scenario (FAS)	158
	7.3 The **IA** family of IDS worlds		159
		7.3.1 Definition of the **IA** ontological family	159
		7.3.2 Subcharacteristics for **I**	162
		7.3.3 Subcharacteristics for **A**	165
	7.4 The $\mathcal{K}$-**IA** family of chronicles		167
		7.4.1 Epistemological properties	167
		7.4.2 Classification of scenario examples	168
	7.5 Summary		168
8	**Intended models for chronicles in $\mathcal{K}$-IA**		169
	8.1 Full trajectory normal form for action laws		169
		8.1.1 An example	169
		8.1.2 Action laws for actions without arguments	171
		8.1.3 Action laws for actions with arguments	173
		8.1.4 Expressiveness of FTNF for infinite trajectory sets	174
	8.2 Chronicle structure in $\mathcal{K}$-**IA**		175
		8.2.1 The schedule in a chronicle	175
		8.2.2 The observation set in a chronicle	176
		8.2.3 The action laws in a chronicle	177
		8.2.4 Definition of $\mathcal{K}$-**IA** chronicles	178
		8.2.5 The labelled-formula layout	178
	8.3 Truth conditions for chronicles		180
		8.3.1 Auxiliary concepts	180
		8.3.2 Model set for an **IA** chronicle	182
		8.3.3 Intended model set for an **IA** chronicle	185
		8.3.4 Classical models for an **IA** chronicle	185
		8.3.5 Chronicle completion	186
	8.4 Models over restricted time domains		187
		8.4.1 Definitions and basic properties	187

	8.4.2	Progressive construction of restricted intended models	189
8.5	Adaptations for branching time		191
	8.5.1	Chronicle structure	191
	8.5.2	Truth conditions for chronicles	192
8.6	Summary		193

9 Entailment methods for $\mathcal{K}$-IA using DFL-1 — 194

9.1	Entailment methods		195
	9.1.1	Pretransformations and model selection	195
	9.1.2	Entailment methods formed using minimization	196
	9.1.3	Prototypical chronological minimization of change	198
	9.1.4	Original chronological minimization of change	199
	9.1.5	Prototypical global minimization of change	200
9.2	Assessments of applicability		201
	9.2.1	Time-restricted selection functions	202
	9.2.2	Assessment of PCM	203
	9.2.3	Assessment of OCM	206
	9.2.4	Assessment of PGM	208
	9.2.5	Discussion of the restrictions	209
9.3	Discussion of possible improvements		211
	9.3.1	Ambiguities	211
9.4	Entailment methods that use filtering, and their assessments		213
	9.4.1	Definitions	213
	9.4.2	Assessments	214
9.5	Branching time		215
9.6	Summary		215

10 Duration constraints — 217

10.1	Chronological assignment of valuation		217
	10.1.1	Discussion	217
	10.1.2	The model selection function for CAMC	218
	10.1.3	Reformulation functions	219
10.2	Intended models using executable schedules		220
	10.2.1	Executable schedules	221
	10.2.2	Auxiliary concepts	223
	10.2.3	Performing egos	224
	10.2.4	Finite intended model sets	226
10.3	Assessment of applicability for CAMC		228

	10.3.1	Auxiliary definitions	228
	10.3.2	Assessment	228
10.4	Generalization to branching time		230
10.5	Summary		231

11 Entailment methods for $\mathcal{K}$-IA using occlusion 232

- 11.1 Discussion 232
  - 11.1.1 Change incidence ambiguity 232
  - 11.1.2 Changetime ambiguity 235
- 11.2 Minimization of occlusion 237
  - 11.2.1 Syntax extensions for DFL-2 238
  - 11.2.2 Semantics extensions for DFL-2 238
  - 11.2.3 Correctness condition 239
- 11.3 Chronological entailment methods with occlusion 239
  - 11.3.1 Chronological minimization of occlusion and change 240
  - 11.3.2 Chronological assignment of valuation and minimization of occlusion and change 241
- 11.4 Syntactical approaches with occlusion 241
  - 11.4.1 Schema of nochange break premises 241
  - 11.4.2 Global minimizatpremisesion of occlusion with nochange premises 242
  - 11.4.3 Pointwise minimization of occlusion with nochange premises and filtering 243
- 11.5 Two-stage minimization of occlusion and change 244
- 11.6 Assessments of occlusion-based approaches 244
  - 11.6.1 Assessment of CAMOC 244
  - 11.6.2 Assessment of CMOC 246
  - 11.6.3 Assessments of methods that use nochange premises 247
  - 11.6.4 Assessments of a two-stage entailment method 250
  - 11.6.5 Summary 251
- 11.7 Oracle features 251
  - 11.7.1 The oracle reformulation function 251
  - 11.7.2 Entailment methods using oracles 253
  - 11.7.3 Discussion 254
- 11.8 A perspective on the chronicle description language $\mathcal{A}$ 255
- 11.9 Interpretation of assessment results 257
  - 11.9.1 General remarks 257
  - 11.9.2 The compatibility requirements 257
  - 11.9.3 The special case of single-step actions 258

12 Composite actions 261

12.1 Composite actions in schedules 261
    12.1.1 Syntax extension 262
    12.1.2 Intended models 262
    12.1.3 Direct translation of composite action statements 264
    12.1.4 World descriptions for composite action statements 266
    12.1.5 Aggregation 268
    12.1.6 Widened translation of composite action statements 269

12.2 Entailment methods for chronicles containing composite action statements 273
    12.2.1 Requirements for the generalized applicability of entailment methods 273
    12.2.2 Entailment methods using Π^{kw} 274
    12.2.3 Entailment methods using Π^{k} 275

12.3 Summary 275

13 Upper applicability bounds and assessment of soundness 276

13.1 Direct upper bounds 276

13.2 Principles for upper bounds for range of applicability 277
    13.2.1 Merely ontological constraints 277
    13.2.2 Ontological constraints in an epistemological range 278
    13.2.3 Upper bounds on the epistemological constraints 279
    13.2.4 Summary of the upper-bound results in this section 283

13.3 Additional upper bounds on range of applicability 283
    13.3.1 Original chronological minimization 283
    13.3.2 Prototypical global minimization 284
    13.3.3 Global minimization of occlusion with nochange premises 286

13.4 Summary of upper-bound results 287

13.5 Assessments of soundness 289

13.6 Summary of assessment criteria and assessment results 290

14 Future directions 292

			xvii
A	**Term index**		297
	A.1	Abbreviations	297
	A.2	Technical terms	299
B	**Notation**		311
C	**References to related work**		321

1
Inert and inhabited dynamical systems

1.1 Topic

This book addresses the knowledge representation and reasoning about inert and inhabited dynamical systems, where the agent(s) inhabiting the system are characterized by their ability to perform *actions*. A *system* is then defined in a quite general fashion as a collection of *objects* which have individual properties as well as relationships between each other. Color, weight, and velocity are examples of properties; adjacency, support, and mutual visibility are examples of relationships. The property assignments and relationships that hold in the system at a point in time constitute its *state*. A part of the state may be designated as the system's *input* at the present time.

A *dynamical system* is a system whose state changes over time, and where effects flow forwards in time so that the non-input part of the state at one time can depend only on its earlier states.

A *scenario* of a dynamical system is a set of possible developments in the system or, using another current term, runs of the system. A scenario is usually obtained using a *scenario description*, that is, a description of the system itself and of how its state has changed over time. For example, a scenario description may be a partial specification of the initial state of the system, combined with descriptions of some events that occurred during the run, and their timing.

The task of reasoning about such dynamical systems, and about scenarios in them, can be realized as follows. One needs a formal language whereby one can express statements about the system's properties, about the actions that may be performed there, and about system states. Given some statements of these kinds, the reasoning task is to derive additional statements that must necessarily follow, given what one knows in general about the properties of dynamical systems, and about the particular dynamical system at hand. Neither the premises nor the conclusions need to be exhaustive descriptions of a specific course of events; in most cases they are only partial descriptions. This is the basic reasoning task, and it is directly applicable for predicting the future states of the system. Planning and diagnosis can be viewed as derived reasoning tasks, and we shall show how they can be defined in terms of the basic task.

A dynamical system is *inhabited* iff it contains one or more *agents* which can influence the system's state at later times by performing *actions* or *activities*. Actions may be understood as time limited control regimes which are performed with a specific purpose, and which terminate when the purpose has been achieved. Activities are similar to actions but are terminated due to a separate decision by the agent.

In order to reason about inhabited dynamical systems, one needs the same resources as for reasoning about ordinary dynamical systems, but in addition one needs a formal language and a reasoning capability for actions and activities that are performed by the agents. Systematic methods for reasoning about inhabited dynamical systems have at least two applications: in order to verify that a given system has certain desirable properties, and as a part of the design for artificially designed agents, or 'robots' that inhabit the dynamical system and reason about their own actions.

A number of logics for reasoning about action and change have been proposed in recent years in Artificial Intelligence (AI) research, and are directly applicable to reasoning about dynamical systems. The purpose of this book is not to propose one more logic, but to present a framework where existing and new logics can be described, compared, and analyzed. I have two goals in mind when developing such a framework:

- To assess the range of applicability of existing and new logics for action and change. The main results of this book are theorems which specify, for each of a number of logics, the class of dynamical systems and scenario descriptions where the logic is guaranteed to obtain the correct set of conclusions. The logics that are defined and analyzed within the framework include both those previously proposed by others and those identified by myself. Until now, the proposals for the various logics have usually not been accompanied by assessments of this kind.

- To clarify whether and in what sense the logics of action and change are relevant for intelligent agents. The concept of an inhabited dynamical system is not only an underlying semantics for the various logics, it is also a 'model' (in the engineering sense of that word) for how the reasoning system in the intelligent agent is related to the world in which the agent operates. This clarification will hopefully be useful both for the design of intelligently reasoning agents, and as a basis for reasoning about the behavior of intelligent agents.

These goals are relevant for two specific topics within artificial intelligence, namely for cognitive robotics and for common-sense reasoning. Cognitive robotics is concerned with theory, tools, and techniques for the high-level or 'intelligent' control of robots, that is, for mechanical systems which are able to operate in the real world. A level of intelligence in such systems is essential for making the robot flexible and resilient so that it can recognize a wide range of possible problems that it may encounter, and find adequate ways of dealing with them.

Common-sense reasoning is concerned with methods whereby computers can understand a practical situation and reason about it in ways that are similar to the reasoning of humans. This topic overlaps with the cognitive robotics topic: they share the need to reason about common-world situations, but cognitive robotics emphasizes the role of sensors and actuators in the total system. Common-sense reasoning may be possible even without the direct contact with the world that is provided by sensors and actuators, but then it requires the use of natural language instead in order to communicate the common-sense information.

The logic-based methods for inhabited dynamical systems that will be presented here are intended to be equally relevant for cognitive robotics and for common-sense reasoning.

The underlying motivations and basic notions for reasoning about actions and change have not, in my opinion, been worked out with sufficient precision. One reason for this is that most authors emphasize the goal of common-sense reasoning, which is an inherently imprecise concept, at the expense of the robotic aspects. (The book [17] by Dean and Wellman is an important exception, but on the other hand, it addresses the logical aspects in a quite practical way). This is particularly unfortunate since the role of logic in knowledge representation is frequently debated. In the present book I felt it was necessary, therefore, to devote the introductory chapters to various aspects of the use of logic: what are the systems that we wish to describe with the logic, what types of reasoning operations are involved, what are the criteria of adequacy for a proposed logic, and so forth.

1.1.1 The two levels of system description

In order to analyze how one can reason about an inhabited dynamical system, it is convenient to assume the use of at least two levels of description for that system. In simple cases, the lower level uses continuous time and continuous-valued state variables that change according to the laws of physics, whereas the higher level uses discrete-valued state variables which characterize the system's present state in more global terms. I shall refer to the lower level as the *material* one, and to the higher level as the *image* level. This terminology is intended to suggest that the description of the system on the material level is closer to (or is even identified with) what the system really is, whereas the system description on the image level is secondary, and has been obtained from the material-level description by some kind of perception or classification process.

In fact, there are two different reasons for the occurrence of discrete descriptions. Some discrete components are the result of *perception*, where an intrinsically continuous state variable is divided into a number of intervals, and the momentary value of the state variable is replaced by a code for the interval in which the present value belongs. The value of the discretized state variable as a function of time is therefore piecewise constant. Other discrete components reflect the system's present *configuration*, that is, the structure of the constituent objects. The configuration may, for example, specify which objects are in physical contact with which others, and the laws governing the change in the state variables are then dependent on the configuration even on the material level of description. In the general case, therefore, the material level must be described in a mixed discrete/continuous fashion, whereas the image level uses a discrete description with few or no continuous components.

The term 'discrete' is also commonly used for a third purpose, namely to indicate that the time is 'sampled' rather than continuous. This distinction between discrete and continuous time is not of any particular interest here. Actually, we shall mostly use the non-negative integers as time domain, but there is no intrinsic reason why we could not have used the real numbers as the time domain even for the image level.

The logic-based methods for reasoning about the inhabited dynamical system operate on the image level, and it is perhaps not obvious that one needs to take the material level into account at all in order to develop those logics. Also, the analysis of the relationship between the two levels of description is a large topic of research in itself. For this book, although the study of logics using the image level of description is my main topic, I believe it is very important to also define the nature of the relationship between the two levels. Only in that way will it be possible to appreciate and evaluate the 'design decisions' that have been chosen in the logics and in the framework for analyzing the logics. Furthermore, any intelligent autonomous agent must include sensori-motoric capabilities whose very purpose it is to translate back and forth between the material and the image level of system description. Any reasonable account of such an agent must therefore be based on the recognition of (at least) these two levels.

1.1.2 Perception

At each point in time during its development, a dynamical system is in (or is perceived to be in) a particular state. Corresponding to the two levels of system description there are two distinct state domains. I shall represent the material-level state domain as $\mathcal{R}_M$ and the image-level state domain as $\mathcal{R}$. The following are some concrete ways in which image domains can be formed for a given material domain:

- States in $\mathcal{R}$ replace quantitative property values with qualitative ones such as 'high', 'medium', 'low'.
- States in $\mathcal{R}$ are partial interpretations (using the truth-values true, false, and unknown) with respect to relations between objects which are fully defined in $\mathcal{R}_M$.
- With a broader definition of state, one may choose states in $\mathcal{R}$ as sets of logical formulae, typically in a restricted logical language, which provide a partial characterization of states in $\mathcal{R}_M$.

The only constraint that one can impose in general on the relation between $\mathcal{R}_M$ and $\mathcal{R}$ is that the latter should provide less information than the former. For example, $\mathcal{R}_M$ may be of infinite size and $\mathcal{R}$ finite, or if they both have a finite number of members then $|\mathcal{R}| \leq |\mathcal{R}_M|$.

The operational mechanism for relating material- and image-level descriptions is by means of *perception*, that is, a transformation which accepts a material state as a function of time, and produces an image state also as a function of time. The domain of timepoints will be written as $\mathcal{T}$. In this chapter we take $\mathcal{T}$ to be either the set of all non-negative integers, or a finite set of integers $[0, n]$ for some n.

Definition. *Let $\mathcal{T}$ be a timepoint domain and $\mathcal{R}$ a state domain. A* **finite history** *for $\mathcal{R}$ is a mapping $[0, t] \to \mathcal{R}$ for some $t \in \mathcal{T}$. A* **complete history** *for $\mathcal{R}$ is a mapping $\mathcal{T} \to \mathcal{R}$.* ⋈

In other words, to each timepoint $s \in [0, t]$ the finite history assigns a state in $\mathcal{R}$. If $\mathcal{T}$ is finite then a complete history is a special kind of finite history; if $\mathcal{T}$ is infinite then no history can be both finite and complete.

Definition. *Let a material domain $\mathcal{R}_M$ and an image domain $\mathcal{R}$ be given. A* **perception function** *is a mapping* Perc *from finite histories for $\mathcal{R}_M$ to finite histories for $\mathcal{R}$ which satisfies the following assumptions:*
 1. *The argument and value are histories over the same interval.*
 2. *If R is a history over $[0, t]$, $s < t$, and $R_{0:s}$ is the restriction of R to the interval $[0, s]$, then* $\text{Perc}(R)_{0:s} = \text{Perc}(R_{0:s})$. ⋈

In other words, although the perception that is made during the additional time interval from s to t may be influenced by what has been perceived before time s, what has been perceived before time s is not reinterpreted because of what happens and is perceived later.

1.1.3 Inertia

Since the intended use of the image level is as the basis for reasoning about the dynamical system, computational properties are significant in the choice of the image domain and the perception function. A particular computational advantage can be obtained if the image-level description of the system is piecewise constant. This makes it possible to replace the continuous or quasi-continuous recalculation of derived quantities by an 'asynchronous' computation process that triggers when an image state variable changes from one value to another.

This standard method solves some problems, but it also introduces some new ones. In particular, continuous change on the material level may be more predictable. The resulting occasional changes of perceived discrete values on the image level are sometimes impossible to predict without access to the underlying, quantitative state variables.

There are some cases where the discrete description on the image level behaves in a particularly regular and predictable fashion. The simplest case, and the first one to be considered here, is when the image-level state variables are *inert*, which means that they do not change at all except when there is a specific reason why they must or may do so. The concept of inertia[1] is also of central importance in the AI literature about logics of actions and change.

The following definitions will be used. A dynamical system is *truly inert* iff its properties and relationships change only in well-understood ways, so that one can reason about developments in the system under the assumption that things do not change unless there is a specific reason why they must or may do so. It is *perceived inert* iff there is a (useful) perception function such that the image is truly inert.

In particular, every truly inert system is also perceived inert using $\mathcal{R}_M = \mathcal{R}$ and the identity function for perception. An example of a useless perception function is the one which maps all members of $\mathcal{R}_M$ to an image containing no information at all.

The following is an example of a dynamical system which is perceived inert although not truly inert. Consider a 'goldfish hotel' consisting of a number of goldfish bowls, each inhabited by one or a few fish (Figure 1.1). Each fish moves around continuously in its bowl, but it is not able to leave it. The states in $\mathcal{R}_M$ characterizing the material dynamic system will specify the exact spatial position of each fish at each point in time, and possibly a lot of other information as well. However, the image domain is chosen so that for each fish it only specifies which bowl it is in, and the

---

[1] The term *persistent* is often used in the AI literature as a synonym for inert, but since this also has other and more established meanings in logic I prefer to use the term 'inert'.

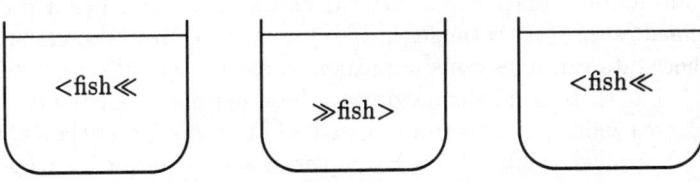

Fig. 1.1. Goldfish hotel

perception function maps material coordinates to the choice of bowl. In this case the system is not truly inert, since the fish do move around, but it is perceived as inert, since the bowl of each fish is constant over time.

If the states in the material domain are chosen so that they only contain the position and movement coordinates of each goldfish, then the material domain is also an abstraction, and one might consider a hierarchy of images that can contain many levels. However, for simplicity I will only be concerned with two levels of system states: the material level containing all the information that one cares to consider, and the image level that is the immediate basis for the reasoning system. There is no assumption that the material state would be completely knowable or finitely describable.

1.1.4 Actions

Inertia is only interesting if there are exceptions to it, otherwise the image-level system is completely static. A major source of inertia exceptions is through actions performed by agents. The presence of an agent which is capable of performing actions is therefore a crucial aspect of the systems under consideration here. Inhabited dynamical systems are those containing one or more agents. For example, the 'goldfish hotel' system can be extended into an inhabited dynamical system by introducing a fish-manager who can perform the action of moving a goldfish from one bowl to another. Such actions override the perceived inertia of the fish's position.

There is a relationship of complementarity between the related concepts of perception, images, and inertia on one hand, and actions that override inertia on the other hand. At the same time, perception and action are two essential capabilities of an agent. Another aspect of the complementarity is that the presence of inertia makes it much more feasible to reason about the effects of actions than it would otherwise be.

A (truly or perceived) inert system is said to be *strictly inert* if there is only one way in which the state can change, namely by the direct effect of an action. Each action type is assumed to influence a certain, well-defined set of state variables in the system state. It is also assumed that all other state variables remain unchanged even when an action is performed, and that there is no change during those periods when there is no action. Systems which enjoy inertia properties but with a larger class of exceptions will be called *weakly inert*. Examples of such additional exceptions are if inertia can be overridden by causal chains of events, by side-effects, or by sudden 'surprises' or 'miracles'.

1.1.5 Inhabited dynamical systems and AI

The intended application areas of knowledge representation for IDS include

- the principled design of autonomous intelligent agents;
- model-based diagnosis of dynamical systems, performed by autonomous or interactive diagnostic systems;
- automatic control of complex systems;
- co-pilot systems;
- artificial intelligence systems that understand or operate in the real world.

Reasoning about inert and inhabited systems has been a favorite subject of AI research since its inception in the late 1950's, and in that sense this book addresses one of the central topics of AI. However, the book is an attempt to change the perspective on the topic. Inertia is usually treated as a fact, an empirical or philosophical fact, which is thought to be characteristic of common-sense reasoning or the common-sense world. Here, instead, I begin by defining in precise terms a class of dynamical systems that have the property of perceived strict inertia, and then proceed to study the temporal logics that may be appropriate for characterizing such systems. Volume II of this work will extend the analysis to weakly inert systems.

Before these methods can be used in a practical application, one must of course determine whether the system one is actually interested in satisfies the inertia properties. The theory is, however, independent of that practical issue.

Several branches of AI have made use of the inertia assumption, in particular the areas of temporal reasoning, qualitative reasoning, knowledge-based planning (previously called problem-solving), and model-based diagnosis. Quite often these branches of research have been distinct activities,

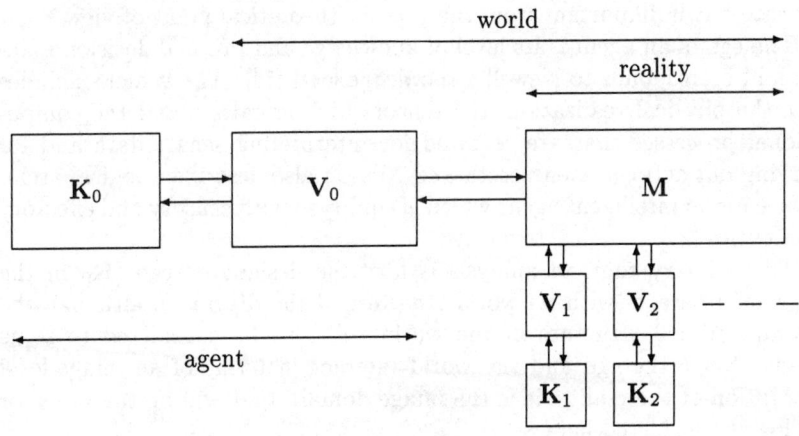

Fig. 1.2. Structure of an IDS

using different approaches and terminologies, but in recent years there has been a trend to pull them closer together and to develop their common formal basis [62, 1, 16, 17, 10]. It is in this context that the term cognitive robotics has been advanced. The work reported in this book is intended as a contribution to that development.

1.2 The image level of description

1.2.1 Structure of an IDS

Figure 1.2 describes the structure of the IDS that will be assumed in this book. An IDS consists of one or more *agents* which interact with a *material system* marked **M** in the diagram. Each agent consists of two parts, an *ego* K_i and a *vehicle* V_i. The vechicle is a system for perception and actuation. Relative to a given agent, the material system and the other agents are together called its *reality*[2]. Relative to one ego, the reality and its own vehicle are together called its *world*.

Although this block diagram might have been interpreted as a system architecture in a possible implementation, it is by no means intended as such, but only as a frame of reference for the formal definitions. Actual

---

[2]The 'reality' in this technical sense will if necessary be distinguished from the real reality by the prefix IDS: 'an IDS reality'. The same applies for the other technical terms defined here: 'IDS vehicle', 'IDS agent', etc.

implementations may differ in various ways which are significant for the software but unimportant from the present theoretical point of view.

The ego of an agent is its level of knowledge and rational decision making, and corresponds to Newell's knowledge level [54]. The vehicle includes both the physical realization, the sensors and actuators, and the computational processes that are required for interpreting sensor data and for carrying out actions. Genesereth and Nilsson also describe a logical architecture for an intelligent agent which is equipped with sensory and effectory capabilities [27].

The primary topic of analysis is how the designated ego ($\mathbf{K}_0$ in the diagram) interacts with its world (the rest of the diagram), although the decompositional structure of the world will also be considered to some extent. Since the ego and the world interact in terms of an image-level description of the reality, it is the image domain that will be the basis for the logics considered here.

1.2.2 Fluents and occurrences

As in many other block diagrams, the most significant aspect of Figure 1.2 is not the boxes but the interfaces. The interface between vehicle and material system is 'physical', and can be characterized in quantitative terms, for example by using differential equations, or their approximations in 'integer' time as difference equations. The interface between vehicle and ego uses an image domain, and is defined in terms of discrete-valued fluents and occurrences.

As in standard AI terminology, a *fluent* is a function from timepoints to corresponding values, usually taken from a finite set, and can be thought of as the trace over time of one of the state variables. Note that a fluent is considered here as a function and not as a function symbol. I shall use the term *feature* rather than, for example, 'state variable' for a characteristic aspect of an IDS which changes its value over time.

An IDS history is a function $R(f,t)$ which for each feature f and timepoint t assigns an appropriate value for that feature. If a specific feature f is chosen, then the resulting function of time is a fluent. On the other hand, if a specific timepoint t is chosen, then the resulting function from features to corresponding values is a state.

For example, the concept *parlsize(C)* for "the number of members in the legislative body of country C" could be a feature for each choice of C. A history R would then assign a value to $R(parlsize(C), t)$ for each choice of C and t. If C is chosen, for example as 'Sweden', one can restrict R to the specific feature *parlsize*(Sweden), and obtain a fluent $g(t)$ which for each year t specifies the size of the Swedish parliament in that year.

The value of a fluent is defined for each individual timepoint, distinguishing it from an *occurrence* whose truth-value is defined over intervals of time. There will be three types of occurrence: actions, activities, and events. An *action* happens over an interval of time through the intervention of an agent. "The robot fills the glass with water" (at a particular time) is an example of an action. (Robots are agents, of course). The general concept such as "a robot filling a glass with water at some time" is a class of similar actions: an *action class*. The definition of an action also specifies when it is to terminate, for example because the glass has been filled. An action is therefore started by a decision of the ego whereby the ego indicates to the vehicle that it shall start performing the action, and is terminated by the vehicle.

Note that the action is performed by the vehicle in the material system after an initial instruction by the ego. In particular, the termination condition for the action is specified in the quantitative terms of the material level and not solely in the qualitative terms of the image level.

It is appropriate to think of an action as a particular invocation of a program in a computer-based agent. The program is invoked at some point in time, and contains a specification for how to map sensor input to actuator output, for example for the purpose of performing a movement. The program is set up to accomplish a certain 'goal' or target condition, and terminates when that goal has been achieved. The state transition function defining how the world at time t depends on the world at earlier times (or how the rate of change in the world at time t depends on its state at the same time) is temporarily modified while an action is being performed.

An *event* happens in the IDS reality over an interval of time. If the event does not include any of the agents, it is called a *material event* and proceeds in accordance with 'natural' laws. "The cup falls to the floor" (or, more precisely: one particular occasion where a particular cup falls to the floor) would be an example of a material event. When a vehicle observes the reality, one of its ways of summarizing the observations is to state that an event is occurring or has occurred, and on this form the information can be delivered to the ego.

Unlike actions, events do not have an individual existence in the IDS reality; they are just part of the image. Events are identified by an observer (a part of the vehicle, or the monitor) in order to render the observations in compact and reasoning-friendly form for communication with, and use by, the ego. Their beginnings and endings are the results of circumstances in the IDS material state. For example, the event "the water in the puddle evaporates" starts when the puddle is formed, and goes on until all the water is gone. No agent or control mechanism is involved; the speed of the process and the timing of the event's end are due to the quantitative phenomena of nature.

An *activity*, finally, is similar to an action except that potentially it can go on for an indefinite time, and its termination as well as its start is decided by the ego. "To shower" would be an example of an activity: the ego decides when to go into the shower, and when to quit[3].

One important type of activity is by regulators which keep a fluent (state variable) at or near a prescribed value for as long as the activity is pursued.

Actions, events, and activities are together called occurrences. A structure consisting of several occurrences, sequential or concurrent, will be called a *process*. Sometimes it may be useful to consider a process consisting of actions as a higher-level or composite action. A *schedule* is a possibly parameterized description of a process expressed, for example, as a set of logical formulae. A *plan* is a schedule which satisfies certain constraints making it executable[4], and which is intended to be used for achieving a given goal or set of goals.

Many authors in AI treat the terms 'action' and 'event' as synonyms, and use them either for the general concept that is here called 'occurrence', or more specifically for what I called 'action'. I make a sharp distinction between actions and events. This volume will consider methods for reasoning about actions, excluding activities and events.

1.2.3 The monitor

Suppose that one of us owns an inhabited dynamical system, and that she has complete knowledge about all aspects of the system up to some point in time of its operation. Suppose also that the agents that inhabit the IDS are reasonably successful according to some chosen success criterion. This presumably means that there exists a useful image level of description, and a practical perception function for obtaining the image from the material level of the system. Therefore, again, when the owner of the system wants to reason about it, it would seem reasonable for her to use the same image level of description as the agents in the system are using.

Is it then appropriate to say that the owner of the IDS is simply one of its agents? In spite of its simplicity, such an assumption misses some important distinctions. The system owner is presumably only a perceiver and not an actor in the system. Also, the system owner may have more knowledge

---

[3] Some readers have objected that "to shower" is more like an action for them: showering has a well-defined termination criterion, namely that the body is cool and clean. Apparently, different people shower differently. This objection also illustrates that although the distinction between action and activity is well-defined in the formal system, its practical application may be less than well defined.

[4] A restricted case of the concept 'executable' will be strictly defined in Chapter 10, subsection 10.2.1.

about the current state of the system than do any of the inhabiting agents. Finally, the owner may be capable of operating on the system in other ways than an inhabitant can do, for example she may be able to run time backwards during a simulation.

It will therefore be useful to maintain the distinction between two types of user of the image level of description, namely the inhabiting agents, on the one hand, and the 'owner', which will henceforth be called the *monitor*, on the other hand. The monitor is to be understood as an idealized, omniscient, non-intervening observer/agent. The monitor's perception produces a *reference image* of the material history in the IDS.

One advantage of this view is that the reference image defines whether a belief which is held by one of the agents is correct or not. Without the assumption of the monitor, one would have to use a relativistic view of the correctness of the world image, for example that the image used by one of the agents is correct if and only if it supports adequate behavior in that agent, and (or) if it is retained and sufficiently stable under rational methods of observation and knowledge acquisition.

Although there may be good philosophical reasons for a position of this latter kind, I feel that for simple worlds with simple material domains and image domains, it is plausible to assume a monitor which 'is always right', and whose reference perceptions can be used (1) for the image level of world description that is used when 'we' (the system owner) reason about the IDS, and (2) as a litmus test for the truth or falsity of any belief held by one of the inhabitant agents.

Reasoning about the effects of actions can therefore be carried out from a *subjective* or an *objective* perspective, depending on whether the reasoning is performed by one of the agents or by the monitor. Objective reasoning is based on the reference image. Subjective reasoning occurs in inhabited dynamical systems with the following properties:

- Each agent makes use of a perception function providing it with an inert, subjective image of the surrounding system.
- The inhabitants are reasoning agents, which means that they use systematic reasoning methods for utilizing the information in the image.

Both the subjective and the objective case may be extended by methods of reasoning about the reasoning of an agent. In the subjective perspective, each agent is able to reason about the reasoning of other agents, possibly on several levels. In the objective perspective, the monitor is able to reason about the reasoning of the agents, but it is not able to perform actions, and the agents are not able to reason about the monitor. The task of reasoning about another agent's knowledge and reasoning processes is of course a topic of considerable difficulty. This book will not address that particular problem, and will be restricted to the simple objective case with a reasoning monitor and a single agent.

The least common denominator between the simple subjective and objective case is obtained by assuming that there is both a monitor and one or more agents whose perceptions equal the reference perception of the monitor. In other words, there is an objectively correct image of the material dynamical system, and this image is available not only to us (the monitor) but also to the agents as the basis for their reasoning and actions. In this case, the issue of limited knowledge or incorrect beliefs about the IDS reality on the part of the agents does not arise. This is the case that will be addressed in the present book.

The topic of this book can therefore be identified as knowledge representation for 'a reasoning monitor for perceived inert, dynamical systems which are inhabited by agents sharing the perceptions of the monitor', for which I shall use the shorter phrase 'inhabited dynamical systems' and the acronym IDS.

The chosen topic has aspects of both representation theory and systems design. I shall concentrate on the former issues, and not say much about software architecture or algorithms of the autonomous agent. Those topics are also very important, but there are a number of theoretical issues that need to be analyzed first.

1.3 Ego/world interaction

In order to define and analyse the behavior of the knowledge level or ego, it is natural to consider what the rest of the world looks like from the point of view of the ego. For simplicity we ignore the possible existence of other agents, and restrict outselves to the single-ego case that is illustrated in Figure 1.3. The ego has to deal with a 'world' consisting of the material system and its own vehicle or (using an anthropomorphic term) its 'body'. The relationship between ego and world is that the ego receives perceptions of the IDS reality which are produced by the sensory system in its vehicle, and, conversely, the ego invokes actions, that is, requests the vehicle to perform certain actions and activities. We shall analyze the ego/world interaction as a kind of game where ego and world make alternating moves.

The single-ego diagram does not formally exclude the possibility of there being additional agents within the IDS reality. However, the discussion of the vehicle/reality relationship in Chapter 3 will make assumptions about the IDS reality which are plausible only if the reality equals the material system. For all practical purposes this book will therefore only be concerned with single-agent IDS's.

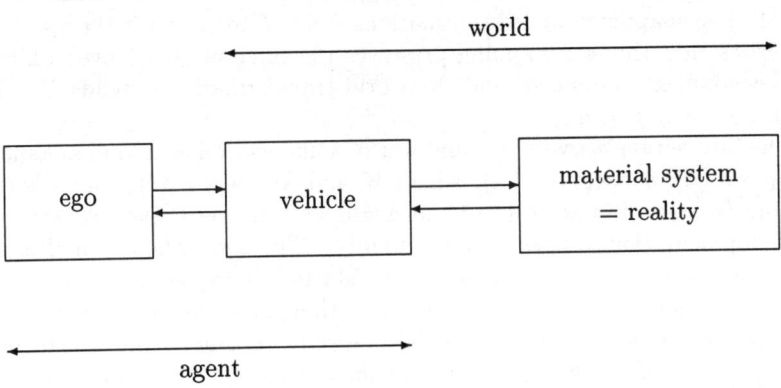

Fig. 1.3. Single-agent IDS

1.3.1 Informal introduction to ego/world interaction

The interaction between an ego and a world uses and gradually extends a finite history as defined above, but it also uses actions and other occurrences. The full and precise account of that interaction uses a formal construct that I shall call a *development*, two of whose components are (1) a history, which specifies the values of discrete-valued fluents at points in time, and (2) a set of occurrences (events, actions, or activities), which happen over specific intervals of time.

A development is called finite if it it is only defined over a finite interval $[0, n]$ of time. In particular, a finite development may contain both *past* occurrences which start and also end between 0 and n, and *current* occurrences, which start between 0 and n but which have not yet terminated at time n.

If occurrences are allowed to last for infinite time, then current occurrences can be part of infinite developments as well. However, that case will not be considered in the present text.

A finite development over $[0, n]$ constitutes a discrete image of what has happened in the world up to time n as the result of the interaction between the ego and that world. Reciprocally, such a finite development can be used by the ego as a means of understanding and influencing the world.

The domain of all possible finite developments will be written as $\mathcal{J}$.

Both the ego **K** and the world **W** will be defined in abstract terms (essentially) as mappings or transformations from $\mathcal{J}$ to $\mathcal{J}$, with the special properties that the ego mapping preserves the interval $[0, n]$ over which the development is defined, and the world transformation extends $[0, n]$ to $[0, n']$ where $n < n'$.

The interaction between ego and world is understood as a kind of game (compare [25], subsection 4.4), where **K** and **W** take turns, and where starting from an initial state of the world at time zero, they construct a development that ranges to plus infinity. The ego, when it makes a move, can specify that actions are to be started at the ending-time n of the present development. This time n is thought of as the *now* of the construction process. The ego can also specify that activities are to be started or terminated at time 'now'. In these decisions it is guided by the presently available development up to time n or now.

The world which makes the opposite moves is supposed to extrapolate what happens as a result of the current state of the world, the material laws, and the actions and activities that the ego has initiated. The world mapping extends the time interval, and consequently extends the definitions of feature-values over the additional timepoints, but preserves them for the previous interval (hence the system is dynamical). The world mapping also determines which events start and which actions and events terminate at the present now-time.

The intuitions underlying this definition should be obvious. The world **W** encapsulates the laws of the IDS material system that the agent interacts with, as well as the perception function whereby the agent interprets its environment and the motoric processes through which it acts in its environment. The material system will eventually be given a more fine-grained description. In particular for discrete time, the material system can be intuitively understood as operating sequentially from each timepoint to the next.

With this definition, only the world is able to determine the breakpoints, which are the points in time when interactions take place, since it is only the world mapping that extends to a new n. This may seem strange at first: shouldn't the ego also be able to define timepoints at which it wants its attention to be called, so that, for example, it can arrange to start an action when a certain condition has become true in the world? However, this effect can be achieved by starting an observation action[5] which terminates when the condition in question is satisfied.

The definitions in this book will be made in such a way that if no action or activity is invoked by the ego, then the world will advance by one single

---

[5] An action which does not have any effects on the world, but whose termination time is determined by the advent of some condition.

timestep while keeping all feature-values unchanged.

Nothing in this definition of an IDS world prevents it from also introducing discontinuities (value changes) at times between the breakpoints. At each interaction the world extends the range of the finite development from one breakpoint to the next, but discontinuities may occur both at breakpoints and between them.

The ego **K** can be thought of as an infinite state automaton[6] operating in real time. However, nothing is said about the computational process whereby the ego decides what actions or activities are to be started or terminated. It may be a simple tabular look-up process, but it may also involve arbitrarily complicated planning or other AI-style computation. Such an operational definition of the ego is an interesting research topic, but it is outside the scope of this book.

1.3.2 Finite, infinite, and complete developments

The following are the formal definitions of developments for the case of discrete time and discrete state variables. The definitions are written so that they apply both when $\mathcal{T}$ is the infinite time domain of integers in $[0, \infty]$ and when it is a finite time domain of integers in $[0, n]$.

Let a domain $\mathcal{R}$ of *states* and a domain $\mathcal{E}$ of *occurrence designators* be given. Each state $r \in \mathcal{R}$ assigns a discrete value to each of a number of *features*, for example "the current color shown by traffic light number 216 is Red", or "the temperature in boiler 8 is Low". The internal structures of $\mathcal{R}$ and $\mathcal{E}$ will be defined in Chapter 5, subsection 5.2.4, and in Chapter 6, subsection 6.5.2, respectively.

The members of $\mathcal{E}$ will be written E_1, E_2, etc. An *occurrence* is a tuple $\langle s, E, t \rangle \in \mathcal{T} \times \mathcal{E} \times \mathcal{T}$ where $s < t$. When the syntax of logic formulae is introduced in Chapter 6, subsection 6.5.2, we shall identify an occurrence $\langle s, E, t \rangle$ with the formula $[s, t]E$.

Finally, let a number of *temporal constant symbols* s_i and t_i and a number of *object constant symbols* o_i be given[7]. These constant symbols and the assignment of values to them will also be needed when developments are used as interpretations for a temporal logic.

Definition. *A* **finite development** *(over images) is a tuple*
$\langle \mathcal{B}, M, R, \mathcal{A}, \mathcal{C} \rangle$,

---

[6]It may require an infinite number of states since the decision of the ego at a point in time may depend on the entire previous history.

[7]The use of two separate symbols s and t is only for convenience reasons, and carries no formal significance. When we have a number of occurrences E_i, each of which has a starting time and a termination time, it is convenient to let the starting time be called s_i and the termination time be called t_i.

where

- $B \subseteq T$ is a finite subset of T, whose members will be called **breakpoints**. The largest member of B will be called **now** and will be written as n_B or (when the choice of B is clear from context) as n. The subset of T with all values between 0 and n_B inclusive will be written T_B.

- M, the **valuation**, is a mapping which assigns values to some or all of the temporal constant symbols t_i, and to all the object constant symbols o_i. The remaining temporal constants are left undefined[8].

- R, the **history**, is a mapping from T_B to $\mathcal{R}$. Thus it specifies the values of features at all times from zero until now.

- $\mathcal{A}$, the **past action set**, is a set of occurrences $\langle s, E, t \rangle$, where s and t are members of B.

- $\mathcal{C}$, the **current action set**, is a set of tuples $\langle s, E \rangle$, where s is a member of B. ⋈

$\mathcal{A}$ is understood as the set of all occurrences $\langle s, E, t \rangle$ which have occurred and which have terminated before or at time n_B. The s and t components specify the starting time and the termination time of the occurrence respectively. $\mathcal{C}$ represents the set of all occurrences which have been started but not yet terminated at time n_B.

Definition. A **complete development** (*over images*) is defined as a tuple
$$\langle B, M, R, \mathcal{A}, \mathcal{C} \rangle,$$
similar to a finite development, with the only differences that R is a mapping from T to $\mathcal{R}$, and that B may be either a finite or an infinite subset of T. An **infinite development** is a complete development over the time domain $[0, \infty]$. ⋈

The symbol $\mathcal{J}$ will represent the domain of finite developments, and $\mathcal{J}^*$ the domain of complete developments. Therefore $\mathcal{J}^* \subseteq \mathcal{J}$ for finite T and $\mathcal{J}^* \pitchfork \mathcal{J}$ (disjoint sets) for infinite T.

In a complete development the R component specifies the value for each combination of timepoint and feature. The present definition allows $\mathcal{C}$ to be non-empty even in the infinite developments, corresponding to actions of infinite duration. Later on, in Chapter 3, subsection 3.2.4, we shall

---

[8] Intuitively speaking, they are undefined because they are going to be assigned values greater than n, but history has not yet reached that far so the exact value is still indeterminate.

introduce a restriction so that $C = \emptyset$ in all infinite developments, that is, all actions have finite duration.

In the general case, the sets $\mathcal{A}$ and $\mathcal{C}$ will consist of arbitrary occurrences, and not only of actions. However, in the present book only actions will be considered.

1.3.3 Egos and worlds

It was described above how the construction of a finite development is performed by a kind of game, with the ego as one 'player' and the world as the other 'player'. The players and the game are formally defined as follows. The first definition is merely an auxiliary concept for the following definitions.

Definition. Let $J = \langle \mathcal{B}, M, R, \mathcal{A}, \mathcal{C} \rangle$ and $J' = \langle \mathcal{B}', M', R', \mathcal{A}', \mathcal{C}' \rangle$ be two finite developments, and let n and n' be the largest members of $\mathcal{B}$ and $\mathcal{B}'$ respectively. J' is said to be a **correct revision** of J iff the following conditions hold. It is required that $\mathcal{B} \subseteq \mathcal{B}'$, $M \subseteq M'$, $\mathcal{A} \subseteq \mathcal{A}'$, and that the restriction of R' to the time interval $[0, n]$ shall equal R. Every $b \in \mathcal{B}' - \mathcal{B}$ must satisfy $b > n$. For every $\langle s, E, t \rangle \in (\mathcal{A}' - \mathcal{A})$, it must be the case that $t = n'$, and that $\langle s, E \rangle$ is a member of $\mathcal{C}$ but not of $\mathcal{C}'$. For every $\langle s, E \rangle \in (\mathcal{C} - \mathcal{C}')$, it must be the case that $s = n$ or $\langle s, E, n' \rangle \in \mathcal{A}'$. Also, for every $\langle s, E \rangle \in (\mathcal{C}' - \mathcal{C})$, it must be the case that $s = n$. ⋈

The following changes can therefore be made to a finite development in a correct revision.

- To add members to $\mathcal{B}$ obtaining $\mathcal{B}'$. Any new member must be larger than the largest member of $\mathcal{B}$, and have the effect of extending the time interval of the finite development.

- To extend the time interval of the development, while defining the values of R' over the added interval but not changing it in the previous interval $[0, n]$.

- To add bindings to M, but not changing or removing previous bindings.

- To add elements of the form $\langle n, E \rangle$ to $\mathcal{C}$, meaning to start an action or activity.

- To remove an element of the form $\langle s, E \rangle$ where $s < n'$ from $\mathcal{C}$, and instead to add the element $\langle s, E, n' \rangle$ to $\mathcal{A}$, meaning to terminate the action or activity after a non-empty time interval.

- To remove a member of the form $\langle n, E \rangle$ from $\mathcal{C}$, without making a corresponding addition to $\mathcal{A}$. This represents that the world is unable to execute an action that has been invoked by the ego. The requested action is not recorded in $\mathcal{A}$.

There is no general restriction on how many such modifications may be performed in one move.

Definition. *A* **ego K** *is a mapping* $\mathcal{J} \to \mathcal{J}$ *where* $\mathbf{K}(J)$ *is a correct revision of* J, *and if*
$$\mathbf{K}(\langle \mathcal{B}, M, R, \mathcal{A}, \mathcal{C} \rangle) = \langle \mathcal{B}', M', R', \mathcal{A}', \mathcal{C}' \rangle,$$
it must be the case that $\mathcal{B} = \mathcal{B}'$. ⋈

Compared to the general concept of correct revision, an ego is not allowed to advance time. It can

- start an action or activity by adding $\langle n, E \rangle$ to $\mathcal{C}$;
- end an activity by removing $\langle s, E \rangle$ from $\mathcal{C}$ and making a corresponding addition to $\mathcal{A}$.

The ego in itself can also cancel an occurrence, but this case never arises in a game as defined below.

Definition. *A* **world W** *is a transformation from* $\mathcal{J}$ *to* $\mathcal{J} \cup \mathcal{J}^*$, *or in other words, a subset of* $\mathcal{J} \times (\mathcal{J} \cup \mathcal{J}^*)$, *such that* $\forall J \exists J' [\mathbf{W}(J, J')]$, *and if*
$$\mathbf{W}(\langle \mathcal{B}, M, R, \mathcal{A}, \mathcal{C} \rangle, \langle \mathcal{B}', M', R', \mathcal{A}', \mathcal{C}' \rangle),$$
then $\mathcal{C}' \subseteq \mathcal{C}$, *the second argument is a correct revision of the first argument, and if* n *is the largest member of* $\mathcal{B}$, *either of the following must hold:*

1. *there is some* $n' > n$ *such that* $\mathcal{B}' = \mathcal{B} \cup \{n'\}$;
2. $\mathcal{B} = \mathcal{B}'$, R' *is a mapping from* $\mathcal{T}$ *to* $\mathcal{R}$, *but its restriction to* $[0, n]$ *equals* R. ⋈

The world therefore has the following two options in its moves:

- In one alternative it adds exactly one more member n' to $\mathcal{B}$, and constructs a new finite development as a correct revision up to n'.
- In the other alternative the world leaves $\mathcal{B}$ unchanged, and extends the history R so that it becomes complete. This case can occur in two ways: for infinite time if the action continues over an infinite interval, and for finite time if the action continues to the end of time. In either case the 'game' is concluded, and the ego is not allowed to make any more moves.

The case of infinite actions will not be used in this book. The case of finite time will be of importance in the proofs in later chapters.

For both options, the world can only terminate occurrences that have been started by the ego, or in other words, actions. It can do so either successfully after the action has operated for a while, or reject the action immediately on its invocation.

This definition is adequate for actions and activities, but it has to be modified when events are to be accounted for as well. In that case the restriction $\mathcal{C}' \subseteq \mathcal{C}$ should be retracted, thereby allowing the world to initiate occurrences which is how events are obtained.

With this definition the world is considered as non-deterministic. The reason for this choice is because information is lost by the perception process, and non-determinism need not be intrinsic to the material system.

1.3.4 IDS Games

A finite (infinite) game is one with a finite (infinite) number of moves. Therefore, a finite game can extend over infinite time, but not vice versa.

Definition. *Let an ego* **K**, *a world* **W**, *an initial state* r_0, *and an initial valuation* M_0 *be given. A* **finite game** *for* **K**, **W**, r_0, *and* M_0 *is a finite sequence of developments*
$$J_0, J_0', J_1, J_1', ..., J_k,$$
where $J_k \in \mathcal{J}^*$, *and*
$$J_0 = \langle \{0\}, M_0, \{0 \mapsto r_0\}, \emptyset, \emptyset \rangle,$$
$$J_i' = \mathbf{K}(J_i) \text{ for } 0 \leq i < k,$$
$$\mathbf{W}(J_i', J_{i+1}) \text{ for } 0 \leq i < k.$$
If $J_k = \langle \mathcal{B}_k, M_k, R_k, \mathcal{A}_k, \mathcal{C}_k \rangle$ *and* $M_k \subseteq M$, *where* M *is a complete valuation, then* $\langle \mathcal{B}_k, M, R_k, \mathcal{A}_k, \mathcal{C}_k \rangle$ *is a* **resulting development** *from the finite game.* ⋈

The limit of an infinite sequence of sets is defined in a straightforward way by
$$\lim_{i \to \infty} S_i = \{x \mid \exists k \forall i [i > k \to x \in S_i]\},$$
which is used for the following definition.

Definition. *A* **infinite game** *for* **K**, **W**, r_0, *and* M_0 *(chosen as in the definition for a finite game) is an infinite sequence of developments*
$$J_0, J_0', J_1, J_1', ...,$$
where each J_i *and* J_i' *is a finite development, and*
$$J_0 = \langle \{0\}, M_0, \{0 \mapsto r_0\}, \emptyset, \emptyset \rangle,$$
$$J_i' = \mathbf{K}(J_i) \text{ for } i \geq 0,$$
$$\mathbf{W}(J_i', J_{i+1}) \text{ for } i \geq 0.$$
If for every i,
$$J_i = \langle \mathcal{B}_i, M_i, R_i, \mathcal{A}_i, \mathcal{C}_i \rangle,$$
then a **resulting development** *from the game is any development of the form*

$$\langle \lim_{i\to\infty} \mathcal{B}_i, M, \lim_{i\to\infty} R_i, \lim_{i\to\infty} \mathcal{A}_i, \lim_{i\to\infty} \mathcal{C}_i \rangle,$$
where, *for the purpose of taking the limit, each R_i is considered as a set of maplets of the form $(t \mapsto r)$, M is a complete valuation, and*
$$\lim_{i\to\infty} M_i \subseteq M. \quad \bowtie$$
Thus, with infinite time and for given **K**, **W**, r_0, and M_0, there is a set of resulting developments, each of which is infinite regardless of whether the (number of moves in the) game is finite or infinite. Notice that $\lim_{i\to\infty} \mathcal{C}_i$ may be the empty set even if all $\mathcal{C}_i \neq \emptyset$.

Informally, such a game proceeds as follows. Initially, the state r_0 of the world at time zero and an initial valuation M_0 are given. Define the initial finite development
$$\langle \{0\}, M_0, \{0 \mapsto r_0\}, \emptyset, \emptyset \rangle.$$
The ego performs the first move, deterministically transforming J_0 to $J_0' = \mathbf{K}(J_0)$. Then the world makes its first move by transforming J_0' to any J_1 such that $\mathbf{W}(J_0', J_1)$. Then the players alternate. After the game ends, arbitrary values are added to the remaining undefined temporal constant symbols.

It would be natural to also make a similar definition for the case of continuous time. However, for the purpose of the 'game' between ego and world where a finite development is gradually extended, it does not matter so much whether time is considered to be discrete or continuous, since it is sufficient to consider the timepoints in $\mathcal{B}$ where the world considers that 'something happens'. These timepoints, which are called breakpoints, serve the same role regardless of whether the time between them is discrete or continuous.

The set of resulting games can be viewed as a game tree, where the move of the ego is well defined at each step but the world has a choice between several moves. As time tends to infinity, either of two things can happen in each branch of the tree: either the world produces a member of $\mathcal{J}^*$ whereby the 'game' ends, or the 'game' continues with an infinite number of moves. Time then also goes to infinity since integer time is being used here. The resulting development is defined in the former case as the resulting member of $\mathcal{J}^*$, and in the latter case by taking the limit of the finite developments along the infinite branch of the tree.

1.3.5 Formal IDS's

The combination of an ego and a world constitutes an inhabited dynamical system, and for the purpose of this book we shall use the following formal definition.

Definition. *An* **inhabited dynamical system** *(IDS) is a pair $\langle \mathbf{W}, \mathbf{K} \rangle$, where $\mathbf{W}$ is a world and $\mathbf{K}$ is an ego, and where $\mathbf{W}$ and $\mathbf{K}$ use the same*

$\mathcal{R}$ and $\mathcal{E}$ domains. ⋈

For each combination of an IDS, an initial state r_0, and an initial valuation M_0 there is therefore a set of resulting developments. In Chapter 3 we shall consider the more detailed structure of the **W** component. It would also be an interesting topic of investigation to define a more detailed structure for the **K** component.

1.3.6 Scenarios

The concept of a scenario was mentioned at the beginning of this chapter. I shall make a distinction between a scenario and a scenario description, and define the former as follows.

Definition. *A* **scenario** *for an IDS world* **W** *is a set of games where* **W** *was the world player.* ⋈

The definition depends on the current choice of time domain. It is not necessary that the ego player is the same in all games in a scenario. Some of the games may be finite and some may be infinite. Scenarios arise in the following fashion. Suppose we have a fully specified world, for example because the material system obeys the laws of physics and the perception function is well defined[9]. At the same time we have only partial knowledge about how the ego operates, for example because it has been obtained as the result of evolution or learning. Some information about what has happened in the world is made available, either as statements about actions or other occurrences that have been performed, or as statements about the values of features at some points in time, or both of these. The available information, which will be called a *scenario description*, imposes additional constraints on the range of possible games that may have been played, besides the constraints obtained from knowledge about the ego. The scenario is then the set of games which can be played by the given world and which are also consistent with other parts of the given scenario description.

Scenario descriptions are significant both because they are used for reasoning about the actions of intelligent autonomous agents, and because they can be used within such agents which are to reason about the history and future of themselves and their environment.

In order to study scenarios and scenario descriptions systematically, it will be necessary to introduce a formal language of logic formulae for expressing statements about occurrences and feature-values, as well as for describing IDS egos and IDS worlds. The logics in later chapters of this book will be defined so that developments, with minor adjustments, can be used as interpretations in the logic.

---

[9]This does not preclude that the resulting IDS world is non-deterministic, of course.

1.4 The ontological taxonomy

The ego's understanding of its world is expressed in terms of (discrete-valued) features and occurrences. The understanding is intended to be used for prediction, postdiction (for example for diagnosis), generalization, planning, etc. In order to realize this understanding process in a correct and efficient fashion, it is useful to identify special cases with various restrictions on the structure of the world, over the general framework which has been defined so far.

1.4.1 Ontological designators

Table 1.1 lists a number of *ontological characteristics*, each of which may apply to an IDS world, for example whether concurrent actions are possible, whether there are actions which create and terminate objects, etc. Every such characteristic will be represented by a unique code letter.

Definition. *An* **ontological designator** *is a set of characteristic codes taken from the first column in Table* 1.1. ⋈

For example, the ontological designator **IAC** represents the assumptions that the world at hand has strict inertia (features do not change except as an immediate result of actions), concurrent actions are possible, and none of the other characteristics apply.

Ontological designators will actually serve a dual purpose. On the one hand, they will be used as symbols representing a particular set of ontological assumptions. These symbols may occur as elements in sequences and other set-theoretic constructs. On the other hand, each designator will be used as a name for the set of those IDS worlds which satisfy the ontological assumptions that the designator represents. Such a set of worlds will be called an *ontological family*.

One might consider expressing these ontological assumptions as logical formulae, and not merely as code letters such as the characteristic codes. However, such an approach would require the use of a quite powerful logic right from the start. My strategy will therefore be to first consider relatively narrow ontological families, formed from small ontological designators, and to define and analyze more specialized logics which are provably adequate for each ontological family at hand. In this way one may gradually advance in the direction of a powerful logic that can deal correctly with a larger number of characteristics.

Since each ontological family is defined in terms of a combination of characteristics, there is an obvious subsumption relation between families, and together they form an *ontological taxonomy*. This taxonomy will provide a structure to the assessment results which are obtained by analyzing the the ranges of applicability of various logics.

1.4.2 The ontological characteristics

Strictly inert dynamical systems with discrete time are chosen as the starting-point for the ontological taxonomy, and are represented by the designator **IA**. These are the systems where all perceived features have constant values unless there is an action that changes them; each action only changes some specific features and in specific ways; and there are no concurrent actions. This base case is the classical one in AI research since the time of the STRIPS system. However, **IA** also includes actions with extended duration, metric time, conditional effects, incomplete specification of the order of the actions, and non-determinism.

The table defines a smörgåsbord of possible exceptions or other complications relative to strict inertia. The code letters **I** and **R** represent inertia itself, in the case of discrete and continuous time respectively. The inertia can be either actual or perceived, so **I** does not require the IDS reality to stand still in the absence of actions; it only requires that features stay constant. The goldfish hotel is an example of **I**. It may then contain actions such as "put goldfish x in bowl number 2", provided that they are defined for any system state, including the case where x is already in bowl number 2, and provided also that they always have the same result on the affected features, namely that the position of x is bowl number 2 after the action.

If some of the actions have different effects depending on the state of the world when the action started, for example "put the goldfish x in the same bowl as y", we have a case of the alternative results characteristic **A**. The resulting ontological family is characterized as **IA**. In general, each family will be characterized by a combination of letters beginning with **I** or **R**, since inertia is the base assumption, and followed by the codes for a subset of the available characteristics.

Strict inertia is defined as the property shared by the **IA** and **RA** families. Non-deterministic actions will be considered as a case of **IA** rather than **I** since there are several possible outcomes of the action.

The terms 'conditional', 'non-deterministic', 'alternatives', and 'random' will be used as follows. The starting (ending) state of an action is the state of the world when the action starts (ends). An action is *unconditional* iff the set of possible ending states for the action is independent of the starting state, and it is *conditional* otherwise. An action is *non-deterministic* iff the set of possible ending states for some starting state has more than one member. An action *lacks alternatives* iff it is deterministic and unconditional; it *has alternatives* otherwise. In other words, if an action lacks alternatives then there is one single and the same ending state for the action, regardless of starting state.

Table 1.1. Ontological characteristics.

Code	Ontological characteristic	Absence
I	Context-free inertia (no alternatives, surprises, concurrency, etc.) and discrete 'integer' time. (None of the following arises).	Disorder
R	Like **I** but for continuous time.	Disorder
A	Alternative results of actions. (*Non-deterministic or conditional result*).	Context-free
U	Local ramification: dependencies between features of the objects directly involved in an action. (*Change in one feature implies possibility of immediate change in another feature*).	Local independence
D	Structural ramification - dependencies with features of objects which are only indirectly involved in an action. (*Change in one feature implies possibility of immediate change in another feature*).	Structural independence
L	Delayed effect [on features]. (*Changes due to an action may occur at a later time*).	Inertia
C	Concurrency [of actions]. (*Concurrency cannot be excluded*).	Sequentiality
S	Surprises [in features]. (*Additional changes, except those that are due to the previous characteristics, are possible but occur only infrequently*).	Predictability
N	Normality [in features]. (*Certain states or partial states of the world are stated to be more frequent than others. This may apply to the initial state or to the result of various actions or other occurrences*).	Equinormality
H	Hierarchical occurrences. (*Action/subaction, event/subevent*).	Flat occurrences
T	Timebound objects. (*Actions or events for the creation and/or termination of objects occur in the system*).	Timeless objects
M	Memory. (*Result of action may depend on feature values at earlier times than the start of the action*).	No memory

Finally, an action is *random* with respect to a particular feature f and starting state r iff for each possible value x of the feature f, there is some ending state of the action which may be obtained with r as the starting state, where f has the value x. For example, the action of throwing a die in an environment without undue constraints is non-deterministic, and it is random with respect to the display of the die and to arbitrary r.

Dependencies between features (**U** and **D**) can be obtained in several ways. One possibility is that there are dependencies between phenomena in the material system, and that those dependencies are reflected in the images as the values of features. Suppose, for example, that two water-pipes join, so that the rate of water flowing through the outgoing pipe C is the sum of the rate of water in the two ingoing pipes A and B. If one propositional feature specifies whether water is flowing through A or not, and similarly for B and for C, then the value of C is dependent on the values of A and B. Another possibility is that there are features on several levels, and that the value of a higher-level feature is defined to be a function of the values of some lower-level features.

The difference between local (**U**) and structural (**D**) dependencies is as follows. Consider an action which takes one or more objects as arguments, for example the action of lighting a candle using a match, or the action of moving an object to a certain location. If the action can affect features of objects other than those which occur as arguments of the action, or if its effects depend on some other features besides the starting-state features of those objects, then one has structural dependencies, and the characteristic **D** applies. Local dependencies according to the characteristic **U** are the ones which involve only features of the argument objects, for example if a scenario is modelled in such a way that if a lamp is unplugged then its light goes out but nothing else is affected.

An example of a delayed effect (**L**) is if there is an action to turn on the bathtub tap which changes the feature for 'tap open' from false to true, and which also causes the feature for 'floor wet' to become true a while later[10].

It is not necessary for the **D** or **U** characteristics that a dependent change occurs exactly at the same time as the independent change, but it must occur within the duration of the action. Suppose the material system contains two quantitative parameters (state variables) h_1 and h_2, which are characterized by discrete-valued features f_1 and f_2 in the interface between vehicle and ego, and where the value of h_2 depends momentarily on h_1. Even if f_1 and f_2 are inert, so that the value of f_2 outside actions is

---

[10]This applies for European bathtubs. American bathtubs usually have a safety drain which protects from overflow. (The common-sense world is full of exceptions and variation).

a function of the value of f_1, it is still possible that when an action changes the value of h_1 in such a way that it passes a threshold and also f_1 obtains a new value, the corresponding change in f_2 may occur a while earlier or a while later. However, necessarily in the case of dependency, the dependent change must occur within the period of the action affecting the independent feature. For delayed effects (**L**), on the other hand, the resulting change may well occur outside the period of the occurrence causing it.

The concurrency characteristic (**C**) refers to the case where concurrency is possible, although it is not required to arise. In scenarios with a number of actions which may or may not occur concurrently, one must reason about the possibility that there is concurrency, so such scenarios belong to a family with the **C** characteristic. In all cases, occurrences with the same occurrence designator are assumed to apply over quasi-disjoint time intervals, that is, time intervals that overlap in at most one point.

The classical 'qualification problem' in AI involves two characteristics, namely alternative results (**A**) and normality (**N**). It therefore arises in the **IAN** family and in all more general families, such as **IANC**. For example, the 'potato in tailpipe' scenario[11] invokes the alternative results characteristic: if the ignition key is turned, then [if there is a potato in the tailpipe then the car will not start otherwise it will start]. It also invokes the characteristic of normality: usually there is no potato in the tailpipe. A general treatment of normality must be able to account for *normality reversal*, that is, the fact that certain conditions may override a normality statement. For example, if a 'tailpipe marauder' is around then one cannot be so sure that the tailpipe is free. The intentions for the remaining characteristics should be clear from the text in the table.

This book, Volume I, addresses the case of the **IA** family. The characteristics **S**, **D**, **U**, and **N** will be addressed further in Volume II.

1.4.3 Ontological subfamilies

When we proceed to the analysis of proposed logics for each ontological family, it will actually be necessary to introduce more precise categories in terms of *subcharacteristics* and *subfamilies*. For example, the **I** family will have a subfamily **Is** for the case where all actions operate in a single timestep, and also a subfamily **Ib** for the case where all features are two-valued or 'binary'. Combinations of constraints such as **Ibs** are of course also possible. The **A** family for alternative results of actions will have one subfamily **Ad** for the case where all actions are deterministic, that is, the outcome of the action on the image level is completely determined by the state of the world when the action starts. The opposite case, **A** − **Ad**,

---

[11] This piece of AI folklore has been recorded, for example, in [29].

is obtained when some action is non-deterministic. Subcharacteristics will be defined in the chapter in which the respective main characteristic is introduced; subcharacteristics for **I** and **A** in Chapter 7, subsections 7.3.2 and 7.3.3.

1.4.4 Assignment of ontological family to a problem

Each combination of ontological characteristics defines an ontological family. For some practical examples it is fairly obvious which of these characteristics arise in the example. In other cases this depends on which abstraction (or 'model' in the engineering sense of the word)[12] one has selected. Consider an example where an action has the postcondition A for certain preconditions, and also occasionally has the delayed effect B. One possible abstraction will consider both effects on an equal basis, without expressing the frequency information. It will then be analyzed as a member of the **IAL** ontological family, since the action has alternative outcomes and a delayed effect. Another possible abstraction will express that B is a possible surprise in states where A holds, thereby capturing some of the frequency information. The same scenario will then be analyzed as a member of the **IAS** family since no delayed effects are recognized, but instead there are surprises.

When analyzing examples of common-sense reasoning and attempting to render them in logic, one can avoid a lot of confusion by first defining precisely which ontological phenomena are assumed to be present in the world. Only when this choice has been made is it meaningful to start looking for an appropriate logic for the examples. The ontological characteristics and taxonomy defined here are intended to help in this respect.

1.5 Summary

An inhabited dynamical system, which was introduced informally at the beginning of this chapter, has been defined as a pair $\langle \mathbf{W}, \mathbf{K} \rangle$, where $\mathbf{W}$ is an IDS world, and $\mathbf{K}$ is an IDS ego. A world is a binary relation on developments; an ego is a function of one argument over finite developments. A development is a tuple $\langle \mathcal{B}, M, R, \mathcal{A}, \mathcal{C} \rangle$, where $\mathcal{A}$ is a set of occurrences, for example actions. The R component is a history, that is, a function $R(f, t)$, which, for each combination of a feature f and a timepoint t, obtains the value of the feature f at time t. A state is a function from features to

---

[12] The word 'model' is used with one meaning in logic and another meaning in engineering. The two meanings are clearly different, but not different enough: it would be very confusing to use the word alternatingly for both meanings. Since the main approach of this book is logicist, I use the word 'model' as in logic, and use the word 'abstraction' for the engineering sense of model.

corresponding values, and can be obtained from a history by selecting a fixed value of t. Similarly, a fluent is a function from timepoints to values, and can be obtained from a history by selecting a fixed value of f.

A combination of an IDS, an initial state, and an initial valuation defines a set of games which is obtained by allowing the ego and the world components of the IDS to alternate their moves. More than one game may be obtained from such a combination, since the world may have more than one choice for each of its moves. For each game there is a resulting development which specifies the history and the occurrences during the game. A scenario is a set of games played by the same world. A scenario description is composed of statements which give a complete characterization of an IDS world, and a partial characterization of other aspects of the games.

An ontological taxonomy has been introduced as a tool for characterizing the properties of IDS worlds. It is defined in terms of ontological designators which are sets of code letters, each representing a characteristic, or in other words, an ontological complication.

Methods for reasoning on the basis of scenario descriptions are of interest both in order to verify properties of IDS's, and for the design of agents that inhabit an IDS. The purpose of this book is to assess the properties and the range of correct applicability of such reasoning methods. The range of applicability will often be expressed in terms of the ontological taxonomy: a particular reasoning method may be proved correct if the scenario world belongs to a certain ontological class in the taxonomy.

2
Inference operations on scenario descriptions

Reasoning about scenario descriptions may appear to be a straightforward application of formal logic, but it turns out to have a number of peculiar characteristics which have to do with how different types of premise are expressed. The present chapter attempts to express, in precise terms, both what the important issues are and how they can be broken down into more restricted subproblems.

2.1 Scenario descriptions

A number of logic-based methods for reasoning about scenarios in commonsense domains have been proposed in Knowledge Representation (KR) research within Artificial Intelligence. They have usually been illustrated with a number of naive examples, among which the 'Yale shooting scenario' (YSS) made the strongest impact. These examples are not an end in themselves, of course, nor has it ever been intended that the actual usage of the proposed logics should be restricted to tasks as trivial as those. Their major usage has been as counterexamples for some of the previously proposed approaches.

I shall argue emphatically in this book that simple examples, such as the YSS, are not an adequate basis for the development of logic-based knowledge representations. The roles of such examples will therefore be quite limited here, and in particular I will not use them as arguments for or against the various proposed logics. However, some simple examples may still be useful at the beginning of the text, in particular as an informal introduction to the precise structure for scenario descriptions that will be used throughout.

The following 'Stockholm delivery scenario' (SDS) is similar to the YSS, but it allows actions with extended duration in time, which are not really plausible for the YSS. In addition it is non-violent. Since actions with extended duration arise in many applications, and since they are addressed in the approach of this book, I will use the SDS by way of introduction.

2.1.1 The Stockholm delivery scenario

Consider a box, B, and a car, C, both of which are located in the city of Linköping at time 0, which represents the beginning of the scenario. The box is not in the car at time 0. Two action types are considered, namely to load a box into the trunk of a car, and to drive a car to a specified city. From time 8:15 [13] to time 8:20 the box is loaded into the car; from time 8:40 to time 11:15 the car is driven to Stockholm. Question: where is the box at time 13:00?

The common-sense answer that the box is in Stockholm at time 13:00 must be qualified by a number of additional assumptions: that the box was not unloaded from the car between time 8:20 and 8:40, that the box was not unloaded from the car somewhere en route, that the car was not driven further between time 11:15 and 13:00, and so on.

2.1.2 IDS scenarios vs. common-sense scenarios

Research on the logic of common-sense reasoning and on cognitive robotics attempts to identify systems of logic whereby scenario descriptions such as the SDS can be expressed as logical formulae, which will obtain the intended common-sense conclusions for each scenario. Its results are therefore potentially applicable to IDS scenarios as a special case.

Conversely, I shall argue that methods for the precise analysis of IDS scenarios and their descriptions may contribute to research about common-sense reasoning as well. One difficulty in such research is that it is not always clear what the common-sense conclusions for a given scenario description really are. Using IDS, the set of intended conclusions can be defined formally, and one can avoid direct reference to common sense[14]. The ego-world interaction which was defined in Chapter 1 will be used to define the intended developments and the intended conclusions of a given scenario description.

For example, the Stockholm delivery scenario will be viewed abstractly as an IDS scenario, that is, as a set of IDS games in the same IDS world.

---

[13] For many of the examples I shall use a notation that is compatible with conventional time, such as 8:15 or 0:12:45, and using a 24-hour clock. Formally, $h{:}m$ is defined to mean the integer $60h + m$, and $h{:}m{:}s$ is defined to mean the integer $3600h + 60m + s$. For example $8{:}15 = 495$. In this way all timepoints can be formally considered as integers.

[14] This does not preclude, of course, the fact that the formal definitions can be inspired by, and compared with their common-sense counterpart. The purpose of the proposed approach is to have a reliable basis for the theoretical work, and not to sever all links with common sense.

The positions of the car and of the box will be features in that world, and actions such as "driving the car to Stockholm" will be considered as actions in the strict sense of the previous chapter: occurrences which are initiated by the ego and executed by its vehicle[15]. The IDS of the Stockholm delivery scenario contains two action types, $Load(B, C)$ for "loading the box B into the car C", and $Drive(C, L)$ for "driving the car C to position L". If the ego initiates an action $Load(B, C)$ at a particular time n of the game between ego and world, then the world side will increment time to n' and extend the positional history of the objects B and C accordingly.

For example, suppose the load operation requires three steps: opening the trunk lid (30 seconds), putting in the box (0:3:30), and closing the trunk lid (1 minute). It is assumed that the box is loaded into the car from time (now using the seconds level as well) 8:15:0 to time 8:20:0. Assume also that the physical position of the box, as represented on the image level, only distinguishes whether the box is in the car or not, and which city it is in. Then in one of the possible developments the box is 'not in' the car at time 8:15:0 through 8:18:20 and 'in' the car at time 8:18:21 and onwards[16]. Therefore, if the move of the ego at time 8:15:0 is to add $\langle 8{:}15{:}0, Load(B, C)\rangle$ to $\mathcal{C}$, requesting the box to be loaded, then the move of the world may be to extend time from 8:15:0 to 8:20:0 and to extend the history component R from its previous range $[0, 8{:}15{:}0]$ to $[0, 8{:}20{:}0]$. Within the added interval $[8{:}15{:}01, 8{:}20{:}0]$ the values of the position features for the box and the car have to be updated in accordance with the definition of the action.

In general, I shall use an abstraction from the common-sense scenario to the formalized universe of IDS worlds and developments in them. This has the advantage of separating out any uncertainties as to what are the intended conclusions from a given scenario. Each scenario description defines an IDS world unambiguously. It will also be well defined which IDS games and, therefore, which IDS developments are correctly characterized by it. This is the key to the formal definition of 'intended conclusions' from a scenario description.

---

[15] The IDS vehicle represents the 'body' of the autonomous agent, and is of course not necessarily to be identified with the automobile in the SDS or any other driving scenario. The combination of IDS ego and IDS vehicle is more likely to be identified with the driver than with the car.

[16] Why doesn't the box have to be 'not in' the car until 8.19.0? This is because the ending condition of the action does not necessarily become true at the very end of the action period; it may become true earlier. For example, the action of putting the box into the car may involve moving it so that it is physically surrounded by the car, putting it down on the car floor, securing its position, and withdrawing one's arms. The box will then begin to be inside the car well before the action of putting it into the car has ended.

2.1.3 Structure of scenario descriptions

Besides formalizing the intended meaning in terms of developments in IDS worlds, it is also necessary to formalize the components of the scenario description itself. I will at least distinguish the following parts of such a formalized scenario description:

1. A description of the *world* where the scenario takes place. Remember that an IDS world was defined as a transformation (binary relation) on finite developments.

   Often the most natural description of the world uses a number of *action laws* which characterize the effect of each action type. In the Stockholm Delivery Scenario, there are two action laws. The action law for $Load(B, C)$ states that if B and C are in the same place when the action starts, then B will be contained in C at the end of the action. It must also be specified what happens if B and C are not in the same place when the action starts or is intended to start. The action law for $Drive(C, L)$ states that C is in L at the end of the action. Finally it must also be stated (using the action law or in some other fashion) that any object contained in C will likewise be in L at the end of the action.

   The choice of ontological family for the world at hand will often constrain the character of the actions, and will therefore impose rules for how the action laws are to be interpreted. The full world description will be represented as a tuple $\langle \Psi, \Gamma_1, ..., \Gamma_k \rangle$, where Ψ is an ontological descriptor. The subsequent Γ_i are sets of action laws or other formal entities, structured according to the choice of Ψ.

   The specification of the length of the timestep will be considered as a part of the description of the IDS world.

2. A statement about the *object domain*, that is, a set of objects which are involved in the scenario. The object domain contains only physical objects, not timepoints, numbers, names, and other similar, abstract entities. In the SDS only one box and one car are involved, and there are several ways of representing the locations where the journey starts and ends. One may choose to introduce objects for the two cities that are involved, or one may consider the two locations ('in Linköping', 'in Stockholm') as symbolic entities.

3. A statement of the *schedule* of the scenario, that is, which actions take place and at what time. In the SDS the schedule says '$Load(B, C)$ during the time interval $[8:15, 8:20]$' and '$Drive(C, \texttt{Stockholm})$ during the time interval $[8:40, 11:15]$'.

If there are actions which create or terminate objects, then the schedule will be expected to account for any changes of the object domain over time. The object domain specification of the previous item will then refer to the object domain at time 0.

4. A statement of the *observations* in the scenario, that is, statements about feature-values at distinct points in time. In the SDS the observations consist of statements about the state of the world at time 0. In other scenarios there may also be observations for later timepoints.

5. Finally, each scenario description is associated with an *epistemological assumption*. The Stockholm delivery scenario, like most scenarios that have been used as examples in knowledge representation research so far, makes an assumption of complete knowledge about the set of actions in the schedule. This particular assumption (which is more precisely defined below) will be represented using the epistemological designator $\mathcal{K}$. It is only by virtue of the $\mathcal{K}$ assumption that one can discount the alternatives that were mentioned above, for example that the box was unloaded sometime before or while it was being driven to Stockholm.

In summary, the scenario description consists of a number of collections of statements or other formal objects, plus two specific indications of assumptions: the designator of ontological assumptions, which must be part of the world description, and the designator of epistemological assumptions, which is a component of the whole scenario description.

For example, in understanding the description of the Stockholm delivery scenario, the epistemological assumption implies that all actions are known, that is, there are no additional actions besides those specified in the schedule. The ontological assumption says that in this world features do not change their values except within the duration of actions, and then only in those features which the action influences. Finally, another part of the epistemological assumption $\mathcal{K}$ is that one knows all of the features that an action may influence. (The observation that one must distinguish these assumptions was made by Morgenstern and Stein [51]).

Typically, the action laws and the corresponding world apply for arbitrary choices of object domain. The scenario description as a whole characterizes a set of infinite developments which can be obtained between the world characterized by the action laws and an arbitrary ego, and where the development has the given (initial) object domain and is characterized by the schedule and the observations.

The epistemological assumption is crucial to the possibility of understanding the scenario description. From the point of view of common-sense reasoning the $\mathcal{K}$ assumption seems quite questionable. How can one assume that there are no other actions in the real world? Of course there are also other actions that take place. If the assumption is that no other relevant actions take place, then what do we mean by a relevant action? However, this issue is brought under control by the formal view that a scenario description refers to developments in IDS worlds. By the $\mathcal{K}$ assumption one considers IDS worlds where there are no other actions besides those mentioned in the schedule. For a given practical situation it is up to the applier (the 'user') to decide whether a proposed abstraction, such as an IDS scenario, is an appropriate approximation of the common-sense scenario at hand.

The conventional practice in logic is to characterize all available knowledge as a set of logical formulae, and to analyze the correct conclusions from those formulae. In line with this practice, it would have been natural to represent the epistemological and ontological assumptions as additional logical formulae, and to investigate rules of inference for statements such as "I know about all the actions". I will differ from that practice in two ways: I encode the epistemological assumptions only as a separate assumption code, and I divide the encoded logical formulae into distinct groups, to be called *partitions*. Thus the SDS scenario would be represented as a fivetuple $\langle \mathcal{K}, \mathcal{O}, \text{Diw}, \text{SCD}, \text{OBS} \rangle$, where the $\mathcal{K}$ assumption is specified explicitly, $\mathcal{O}$ is the object domain, Diw is a description of the IDS world where the scenario is performed, and SCD and OBS are sets of propositions representing the schedule and the observations respectively.

The description of the world, Diw, will in this case have the structure $\langle \mathbf{IA}, \text{LAW} \rangle$, using the designator **IA** which was defined in Chapter 1. Here LAW is a set of action laws which are written in a form applicable for **IA** worlds. For more complex ontological families, it will be necessary to use more than one set of statements in order to characterize the world correctly. For example, **IAS** worlds are to be described by threetuples $\langle \mathbf{IAS}, \text{LAW}, \text{SPR} \rangle$, where SPR is a set of formulae characterizing possible surprises, similar to the 'fault model' in knowledge-based diagnostic reasoning. In each case Diw is a tuple whose first element is an ontological designator, determining the structure of the rest of the tuple.

The use of multiple partitions is closely tied to the method of semantic filtering, which was first proposed in [72] and which will be introduced in Chapter 9.

Beginning in Chapter 4 I shall introduce a formal-logical language for expressing the statements in the LAW, SCD, and OBS partitions. In this chapter I shall define the structures and operations which are built and used above the level of the individual statements. The members of LAW will actually be considered as syntactic transformation rules in some of the following chapters, but for the purpose of the present chapter one may simply assume that LAW is a set of logic formulae.

2.1.4 Object domains

This text will be restricted to domain-specific scenarios, that is, to scenarios where the object domain is finite, fixed for the duration of the scenario, and completely specified. There will be no actions for the creation or termination of objects, and no scenario descriptions of the form "there are either four or five cars". Therefore, the domain can be specified explicitly in each scenario. However, the same world and the same action laws can be used in several scenarios with different object domains. The choice of domain is therefore a component in each scenario description.

Formally, a *domain-specific scenario* for a set $\mathcal{O}$ of objects is a set of games with a world which does not have the ontological characteristic **T** that is, it does not create or terminate objects, and where all games use the same domain $\mathcal{O}$.

If there is just one object of each kind, for example just one box and one car as in the SDS, then one may omit the object domain altogether and consider the features of these objects as depending on time only.

The case of domain-specific scenarios is of particular interest, not only for its relative simplicity and the consequent logical properties, but also because many applications have this character. Most examples, both toy and non-toy examples, that have been discussed in the AI literature use domain-specific scenarios.

Even if a scenario is domain specific, the world in which it has been performed may very well allow different choices of object domain. Consider, for example, a blocks world where there are actions for moving the blocks around but not for creating or terminating blocks. In each pair $\langle J, J' \rangle$ of the formal definition of the world, the two components will use the same set of blocks, but two different pairs may use different block sets. The entire blocks world can therefore be divided into subsets such that the object domain (for example, the set of blocks) is held fixed within each subset. Each scenario in this world will only involve one of the domain-specific subsets of the world, but the action laws will describe the whole world in a domain-independent manner.

Domain-specific scenarios form the basis for the so-called 'closed world assumption' [60]. In our definition of a scenario description as, for example, a fivetuple $\langle \mathcal{K}, \mathcal{O}, \text{Diw}, \text{SCD}, \text{OBS} \rangle$, the explicit statement of the object domain as the $\mathcal{O}$ component expresses the closed-world assumption.

2.1.5 Timepoint and object constants

In the SDS the exact time interval was stated for both actions. In many other cases one has only partial information about the time of actions. This is formally expressed using temporal constant symbols. In one case where it was known that actions E_1 and E_2 were performed in that order, and with a pause of five time units between them, one could state, for example, 'E_1 takes place from time 240 to t_1' and 'E_2 takes place from time $t_1 + 5$ to $t_1 + 15$'. In this way it is possible to characterize cases where the timing, and even the temporal order, of the actions are incompletely known.

Similarly, it is useful to have object constant symbols for incompletely specified objects. One scenario might state, for example, "car o_1 drives box B_2 from L to M", and then "car o_1 drives box B_4 from M to N". Here it is clear that the same car is used for transportation of both boxes, but the identity of the car is left open.

The designation of the t_i and o_i as 'constants', rather than as 'variables' or 'parameters', is in accordance with the terminology of logic.

2.1.6 Chronicles

An important assumption, in the common-sense interpretation of the Stockholm delivery scenario and other similar scenarios, was that all actions are known with certainty. A special term will be introduced for such scenario descriptions: a **chronicle** is a scenario description which defines the set of actions (and other occurrences) in the corresponding games. (Recall that a scenario was defined as a set of games).

This concept will be made precise by the formal definition in subsection 2.3.1, and by the successive definitions of specific epistemological characteristics. Besides very simple cases, such as the Stockholm delivery scenario, it is intended to allow conditional actions which occur or don't occur depending on the current state of the world, as well as cycles, where actions of the same action type occur repeatedly. It is also intended to allow schedules containing statements of the form 'whenever condition ϕ holds, action E is invoked'.

Conversely, the concept of a chronicle is not intended to be used when actions or their effects are inferred in a modal fashion. Arguments such as "If A had done P, then B would have learned about it and consequently done Q, and since the effects of Q are not present we conclude that A did not do P" is an example of an action-based scenario that I would not call a chronicle. The present volume will only be concerned with a restricted class of chronicles, but many of the introductory definitions are written so that they apply to a wider class of chronicles, as well as to some non-chronicle cases. This is intended to put the present topic into perspective, and to facilitate future extensions.

The rest of the present chapter defines a framework for the precise expression and classification of chronicles, and for reasoning operations that apply to chronicles.

2.2 Epistemological assumptions

2.2.1 Logics with implicit epistemological assumptions

We have seen that the epistemological assumptions are essential for understanding a scenario description, and that the ontological assumptions are essential for understanding which world is intended by a given world description (for example, a set of action laws). There are several possible ways of dealing with these assumptions when one develops a logic.

One possibility is to consider the assumptions as logical propositions like all others. This requires the use of quite an expressive language for the logical formulae. Such an approach is theoretically appealing for its generality, but may be very costly in a practical implementation. The epistemological and ontological assumptions are not likely to change over time, and there does not seem to be much need to reason about them. Therefore, it may not be necessary to phrase them explicitly as propositions in the logic.

The other possibility is to make these assumptions on the meta-logical level, and to develop a logic that is adequate and tailor-made for a given set of assumptions. Most of the logics for reasoning about action and change that have been proposed in AI have that character. There are just two problems, as follows. The first problem is that the assumptions have not been made precise. The goal of the research is to develop a logic allowing fairly general modes of reasoning, that is, logics which make fairly weak assumptions. This is because one wants to deal with inertia, qualification, ramification, concurrency, and so on. However, the proposed logics do not in fact work in all such cases. They have only been verified by selected test examples, or formally verified for much more restricted cases than those ultimately intended.

The second problem is that if and when the assumptions are made precise, we are not going to have a single, canonical set of assumptions that everyone agrees about, but a reportoire of different possibilities with respect to those assumptions. To take a simple example: some applications require the use of non-deterministic actions; others do not. Similarly, some logics are correct even for worlds with non-deterministic actions and others are not.

In order to relate occurring applications to available logics, there is therefore a need for a taxonomy of scenario descriptions, where these descriptions are classified with respect to whether they allow (require) ramification, whether they allow concurrency, whether they allow non-deterministic actions, etc. In other words, the epistemological and ontological assumptions are used as the basis for the classification in the taxonomy. In terms of such a taxonomy it should be possible to identify, for each proposed logic, its range of applicability, that is, for which family of scenario descriptions in the taxonomy the logic obtains the correct conclusions.

The development of such a taxonomy was begun in Chapter 1 with the introduction of the ontological characteristics, and it will continue in this chapter with the addition of the epistemological part. However, since this is going to be a taxonomy of scenario descriptions, we must first clarify the formal structure of a scenario description. Consider the following two, similar but distinct, perspectives as applied to the example of the Stockholm delivery scenario:

1. A scenario description is a tuple consisting of an object domain $\mathcal{O}$, a set of action laws LAW, a schedule SCD, and a set of observations OBS. This tuple is to be understood in the context of the particular epistemological and ontological assumptions that were specified above. If the assumptions are changed then the tuple has a different meaning. However, in the context of other assumptions one may also have to change the structure of the tuple, for example by adding more elements.

2. A scenario description is a more extensive tuple which contains the four items that have just been mentioned, but which also contains an explicit symbol $\mathcal{K}$ for the epistemological assumptions, and an explicit symbol **IA** for the ontological assumptions. The meaning of the scenario description is unambiguously determined by the tuple. If other assumptions are to be used, then the assumption-code components of the tuple are replaced, and possibly the structure of the rest of the tuple also has to be changed.

In the first case the tuple may have the form $\langle \mathcal{O}, \text{LAW}, \text{SCD}, \text{OBS} \rangle$. In the second case it may have the form $\langle \mathcal{K}, \mathcal{O}, \langle \textbf{IA}, \text{LAW} \rangle, \text{SCD}, \text{OBS} \rangle$.

Clearly, the difference between these alternatives is of a technical nature. The first choice is close to the current practice in this research, where scenario descriptions are expressed as a set of logical formulae, with or without a subdivision into partitions. Under the first choice, one is led to address the question as to what assumptions does a given logic obtain the correct and intended conclusions. For example, it has been observed repeatedly that the principle of chronological minimization (without the use of filtering) is not correct for postdiction problems, that is, cases where there are observations for time later than zero. This is an example of an epistemological restriction on the range of applicability of the logic.

In this work, I shall use the second formulation initially, and then shift to the first formulation in later chapters. The reason for this choice is that the second formulation is more explicit. For example, it allows one to relate different scenario descriptions that make different assumptions. Later chapters of the book will address one choice of assumption at a time and analyze it in detail; it will be inconvenient then to repeat the same assumption code in all formulae throughout a chapter.

2.2.2 Epistemological designators

The basic definitions for the encoding of epistemological assumptions are very simple.

Definition. *An* **epistemological characteristic** *is one of the symbols (letter or pair of letters) listed in the first column of Table 2.1.* ⋈

Families of scenario descriptions can be further identified using epistemological properties, for example "all observations refer to time 0", or "the exact times of all the actions are known". Unlike the epistemological assumption that is encoded in the characteristics, the epistemological properties can be obtained by inspection of the formulae in the scenario description. Their purpose is therefore not to define how a scenario description is to be interpreted, but only as a classification which is useful for specifying the range of applicability of the various proposed logics. Epistemological properties will be denoted by lower-case sans-serif letters, for example p for "all observations refer to time 0".

Definition. *An* **epistemological property** *is one of the symbols listed in Table 2.2 (s or p), or one of those which will be added in later chapters. An* **epistemological designator** *is an epistemological characteristic, followed by a possibly empty set of epistemological properties.* ⋈

For example, $\mathcal{K}\{s, p\}$ is an epistemological designator. The interpunction will usually be omitted, like in $\mathcal{K}sp$.

In a similar way as for ontological designators, the epistemological designators will be used in two ways. An epistemological characteristic, for example $\mathcal{K}$, will be used as a symbol in the first position of a scenario description, and will then determine the structure of the remaining description. An epistemological designator will also be used to designate a family of scenario descriptions, namely those which have the epistemological characteristic as the initial element, and which also satisfy the epistemological properties in the designator.

For example, if the scenario description $\Upsilon = \langle \mathcal{K}, \mathcal{O}, \text{Diw}, \text{SCD}, \text{OBS} \rangle$ is formed so that all members of OBS are observations for time zero, then $\Upsilon \in \mathcal{K}p$. On the other hand, $\mathcal{K}p$ could not be the first element in the scenario description.

A code expression of the form $\mathcal{Z}\text{-}\Psi$, where $\mathcal{Z}$ is an epistemological designator and Ψ is an ontological designator, will be used to represent the

Table 2.1. Epistemological characteristics

Code	Epistemological characteristic	Absence
$\mathcal{N}$	No epistemological assumptions.	
$\mathcal{K}$	Chronicle with explicit, correct, and accurate knowledge. *No partiality, misperception, or illusions.*	Inaccurate
$\mathcal{Q}$	Chronicle with qualified action laws. *For some actions there are preconditions, such that the effect of the action is known when preconditions are satisfied, but unknown otherwise.*	Unqualified
$\mathcal{P}$	Chronicle with qualified action laws and composite actions. *Action sequences, conditional actions, and iteration are allowed in the schedule.*	Non-composite
$\mathcal{U}$a	Uninformed [about actions]. *Some of the instances of actions remain unspecified.*	Informed
$\mathcal{U}$l	Uninformed [about action laws]. *Do not always know the effects of actions, or even what features they affect.*	Informed
$\mathcal{U}$e	Uninformed [w.r.t. events]. *Sometimes fail to observe or infer an event.*	Informed
$\mathcal{M}$o	Misperception [w.r.t. observations]. *Sometimes observe the wrong feature-value.*	Observant
$\mathcal{M}$e	Misperception [w.r.t. events]. *Sometimes vehicle reports a non-occurring event.*	Observant
$\mathcal{I}$a	Illusion [of actions or activities]. *Sometimes vehicle believes that it performs an action that it actually does not perform.*	Illusion free

Table 2.2. Some epistemological properties for $\mathcal{K}$

Code	Epistemological property under $\mathcal{K}$
$\mathcal{K}$s	Complete knowledge about the initial state.
$\mathcal{K}$p	No information about any state later than the initial one (pure prediction problem).

set of all scenario descriptions which are members of $\mathcal{Z}$ and whose worlds are members of Ψ. For example,
$$\langle \mathcal{K}, \mathcal{O}, \langle \text{IA}, \text{LAW}\rangle, \text{SCD}, \text{OBS}\rangle \in \mathcal{K}\text{-IA},$$
and if all members of OBS are observations for time zero and all actions described by LAW are deterministic, then
$$\langle \mathcal{K}, \mathcal{O}, \langle \text{IA}, \text{LAW}\rangle, \text{SCD}, \text{OBS}\rangle \in \mathcal{K}\text{p-IAd}.$$

2.2.3 The repertoire of epistemological designators

The first four of the epistemological characteristics will be used in this book. $\mathcal{K}$ is the standard case which will also be at the center of interest in this volume, and $\mathcal{Q}$ is a slightly more general case. The difference between $\mathcal{K}$ and $\mathcal{Q}$ is that $\mathcal{Q}$ allows actions which are only defined as possible for some starting states, whereas in $\mathcal{K}$ for each combination of action and starting state the effects of the action must be defined, although they will possibly be non-deterministic.

Both $\mathcal{K}$ and $\mathcal{Q}$ are restricted to elementary actions, that is, a schedule specifies each action in terms of action symbol, arguments, starting time, and ending time. $\mathcal{P}$ is a generalization over $\mathcal{Q}$ allowing composite actions which are formed by the conventional operators of sequencing, conditional execution, and repetition.

$\mathcal{N}$, finally, is the trivial case and is only used for technical uniformity and to bind the object domain. If Γ is a set of logical formulae and $\mathcal{O}$ is a set of objects, then the model set of $\langle \mathcal{N}, \mathcal{O}, \Gamma\rangle$ consists of all those developments that satisfy the formulae in Γ and where $\mathcal{O}$ is the object domain.

As for the epistemological properties, their meanings were indicated in Table 2.2 and will be strictly defined in Chapter 7, subsection 7.4.1.

2.3 Formal definitions of scenario descriptions

The semiformal account of scenario descriptions in subsection 2.1.3 can now be made entirely precise on the syntactic level.

2.3.1 Scenario descriptions

Definition. *A* **scenario description** *is a tuple*
$$\langle \mathcal{Z}, \Gamma_1, \Gamma_2, ..., \Gamma_k\rangle,$$
where $\mathcal{Z}$ is an epistemological characteristic, and the following elements are sets of formulae or other formal objects. The number and type of the following elements is specific for each choice of $\mathcal{Z}$.

A **chronicle** *is a scenario description where the epistemological designator is one of* $\mathcal{K}$, $\mathcal{Q}$, *or* $\mathcal{P}$, *or a designator that is to be added later.* ⋈

The generic definition for scenario descriptions has to be complemented by specific definitions for the various choices of the epistemological characteristic. In particular, for $\mathcal{K}$ the specific definition is as follows.

Definition. *A scenario description in* $\mathcal{K}$ *is a tuple*
$$\langle \mathcal{K}, \mathcal{O}, \mathtt{Diw}, \mathrm{SCD}, \mathrm{OBS}\rangle,$$
where $\mathcal{O}$ *is an object domain,* $\mathtt{Diw}$ *is a description of an IDS world,* SCD *is a schedule specifying a set of actions and giving some information about their temporal order, and* OBS *is a set of observations of the values of features at some points in time.* ⋈

Let a scenario description Υ and a resulting development J be given. It will then be well defined (as a truth-value) whether Υ is a correct description of J. The correctness of the description requires partly that all statements in the various Γ_i are true[17], and partly that the epistemological assumptions expressed by $\mathcal{Z}$ are satisfied.

When writing out concrete examples it is inconvenient to write sets of sentences directly, and I will use a layout as in the following rendering of the Stockholm delivery scenario:

obs1	The box is not in the car at time 0
obs2	The box is in Linköping at time 0
obs3	The car is in Linköping at time 0
law1	If the box is loaded into the car from time s to time t, then the box is in the car at time t
law2	If the car is driven to L during the interval from time s to time t, then the car is in L at time t. If the box was in the car at time s, then the box will also be in L at time t.
scd1	The box is loaded into the car from time 8:15 to time 8:20
scd2	The car is driven to Stockholm from time 8:40 to time 11:15.

This is therefore a way of representing a particular scenario description $\langle \mathcal{K}, \mathcal{O}, \langle \mathbf{IA}, \mathrm{LAW}\rangle, \mathrm{SCD}, \mathrm{OBS}\rangle$, where the set $\mathcal{O}$ of objects is empty, the set LAW contains two propositions, namely those labelled law1 and law2 in the layout, and similarly for SCD and OBS. The choices of epistemological and ontological assumptions ($\mathcal{K}$ and $\mathbf{IA}$) and of the object domain are not included in the layout, and have to be indicated in the surrounding text.

There are a number of developments for which this chronicle is a correct description, differing as to exactly when the box changes from being out-

---

[17] Except for more advanced epistemological characteristics which may allow occasional misleading statements.

side to being inside the car, and exactly when the car switches between the successive discrete-valued locations: Linköping, the intermediate locations, and Stockholm. Notice, however, that this chronicle allows some developments which are presumably not intended, for example developments where the car and the box are temporarily in different cities during the trip from Linköping to Stockholm. A few additional statements are therefore needed in order to capture all the common-sense knowledge about the matter. However, the given chronicle will be adequate as an example for the continued discussion.

A development where the box is unloaded before or during the trip, either due to an additional 'unload' action or because a feature has changed its value without any causing action, are not considered to be correctly described by this scenario description. One aspect of the $\mathcal{K}$ epistemological assumption is that all actions have been specified explicitly. Also, one aspect of the **IA** ontological assumption is that all changes of feature-values must be due to actions. If a development is to be correctly described by this scenario description, its $\mathcal{A}$ component may not contain any occurrence designators other than those that have a counterpart in the scenario description, and its R component may not contain any changes other than those that are enabled by its $\mathcal{A}$ component.

We proceed now to the other types of scenario descriptions.

Definition. *A* **scenario description** *in $\mathcal{Q}$ is a tuple*

$\langle \mathcal{Q}, \mathcal{O}, \text{PRE}, \text{Diw}, \text{SCD}, \text{OBS} \rangle$,

where PRE *is a set of preconditions, that is, formulae which characterize whether the effects of an action are known,* Diw *is a description of an IDS world for the case where the effects of actions are known, and $\mathcal{O}$,* SCD*, and* OBS *are as for $\mathcal{K}$ scenarios.* ⋈

The difference between $\mathcal{K}$ and $\mathcal{Q}$ has already been discussed. This type of scenario description is particularly relevant for planning. If the effects of actions are only partially known, then temporal prediction is only possible if the schedule happens to be designed in such a way that all preconditions are satisfied. In the particular case of planning one can try to find a schedule, that is, a plan which satisfies this condition.

Scenario descriptions in $\mathcal{P}$ apply for the case where the schedule does not only contain statements of individual actions, but also composite statements that specify conditional actions (if-then-else expressions), sequences of actions, and repetition of actions.

Finally, we introduce the scenario type $\mathcal{N}$, where the scenario consists only of observations and a specification of the object domain.

Definition. *A* **scenario description** *in $\mathcal{N}$ is a tuple*

$\langle \mathcal{N}, \mathcal{O}, \text{OBS} \rangle$,

where $\mathcal{O}$ is an object domain and OBS is a set of observations. ⋈

The designator $\mathcal{N}$ represents the absence of any epistemological assumptions, so an $\mathcal{N}$ scenario description is true in any development where $\mathcal{O}$ is the object domain and all members of OBS are true. There is no schedule and no assumption of inertia.

According to the distinction in subsection 2.1.6, scenario descriptions in $\mathcal{K}$ and $\mathcal{Q}$ are chronicles, and those in $\mathcal{N}$ are not.

2.3.2 Model sets for scenario descriptions

Definition. *For a given scenario description Υ and time domain $\mathcal{T}$, $Mod(\Upsilon)$ denotes the set of complete developments which are correctly described by Υ.* ⋈

The formal definition of *Mod* is conditional on the choice of epistemological and ontological family in Υ. The precise definition of *Mod* in $\mathcal{K}$-**IA** chronicles will be made in Chapter 8, subsection 8.3.2. Basically, the idea is as follows. If $\Upsilon = \langle \mathcal{K}, \mathcal{O}, \langle \mathbf{IA}, \text{LAW} \rangle, \text{SCD}, \text{OBS} \rangle$, first construct the unique IDS world that is precisely specified by $\langle \mathbf{IA}, \text{LAW} \rangle$ and obtain an arbitrary IDS ego. Consider all possible games that can be played between them, and their resulting developments. Obtain the set of all those developments for arbitrary ego, and the subset consisting of those developments where the object domain is $\mathcal{O}$; where the set of actions $\mathcal{A}$ is exactly the set of actions specified by SCD; and where all the observations in OBS are satisfied. This subset is $Mod(\Upsilon)$.

The other epistemological characteristics require similar definitions, reflecting their particular structures for the scenario description. If $\Upsilon = \langle \mathcal{Q}, \mathcal{O}, \text{PRE}, \langle \mathbf{IA}, \text{LAW} \rangle, \text{SCD}, \text{OBS} \rangle$, one should first construct the unique IDS world that is precisely specified by $\langle \mathbf{IA}, \text{LAW} \rangle$, like in the case of $\mathcal{K}$. Then, however, this world is restricted in such a fashion that actions are only possible under the preconditions defined in PRE. In the opposite cases, the world quits the ego/world game, so that it is undefined whether a contribution to $Mod(\Upsilon)$ is obtained. The rest of the definition is like for $\mathcal{K}$.

If $\Upsilon = \langle \mathcal{N}, \mathcal{O}, \text{OBS} \rangle$, finally, $Mod(\Upsilon)$ is chosen as the set of all developments $\langle \mathcal{B}, M, R, \mathcal{A}, \mathcal{C} \rangle$, where $\langle M, R \rangle$ has been constructed using the object domain $\mathcal{O}$, and satisfies all formulae in OBS. There are no restrictions on the choice of actions, or their effects. Therefore, $Mod(\Upsilon)$ is simply the restriction of the set of classical models for OBS to those having $\mathcal{O}$ as their object domain.

2.4 Entailment relations

We shall define entailment relations in terms of $Mod(\Upsilon)$, and show how several common varieties of temporal reasoning can be expressed in terms of these entailment relations.

2.4.1 Entailment relations and weakened models

Definition. *Let Υ and Υ' be scenario descriptions. The following three* **entailment relations** *are defined:*
$\Upsilon \models \Upsilon' \Leftrightarrow Mod(\Upsilon) \subseteq Mod(\Upsilon')$,
$\Upsilon \cong \Upsilon' \Leftrightarrow Mod(\Upsilon) = Mod(\Upsilon')$,
$\Upsilon \parallel\diamond \Upsilon' \Leftrightarrow (Mod(\Upsilon) \cap Mod(\Upsilon')) \neq \emptyset$. ⋈

These relations may be pronounced 'entails', 'semantically equivalent', and 'consistent with', respectively. Notice that these entailment relations are quite conventional. For several types of reasoning operations, one is only interested in observation-type conclusions, such as: for a given schedule, what properties will hold in the world at what times? It is therefore convenient to introduce weakened entailment relations, based on a weakened model concept, as follows.

Definition. *The set $Mod(\Upsilon, \text{RM})$ of weakened models for Υ is defined as*
$\{\langle M, R \rangle \mid \langle \mathcal{B}, M, R, \mathcal{A}, \mathcal{C} \rangle \in Mod(\Upsilon) \text{ for some } \mathcal{B}, \mathcal{A}, \mathcal{C}\}$. ⋈

The following, analogous counterparts are obtained for the entailment relations.

Definition. *Let Υ and Υ' be scenario descriptions. The following three* **weakened entailment relations** *are defined:*
$\Upsilon \models_{\text{RM}} \Upsilon' \Leftrightarrow Mod(\Upsilon, \text{RM}) \subseteq Mod(\Upsilon', \text{RM})$,
$\Upsilon \cong_{\text{RM}} \Upsilon' \Leftrightarrow Mod(\Upsilon, \text{RM}) = Mod(\Upsilon', \text{RM})$,
$\Upsilon \parallel\diamond_{\text{RM}} \Upsilon' \Leftrightarrow (Mod(\Upsilon, \text{RM}) \cap Mod(\Upsilon', \text{RM})) \neq \emptyset$. ⋈

For the purpose of the weakened entailment relations, it is only the history (= assignments of feature-values over time) and the valuation of the constant symbols that counts, but not the choice and timing of occurrences. The following consequence is obtained.

Proposition 2.1. *If Υ is an arbitrary scenario description, and* OBS *is a set of statements that only refer to feature-values and not to occurrences, then*
$\Upsilon \models \langle \mathcal{N}, \mathcal{O}, \text{OBS} \rangle$
if and only if
$\Upsilon \models_{\text{RM}} \langle \mathcal{N}, \mathcal{O}, \text{OBS} \rangle$,
and similarly for $\parallel\diamond$.

The proof follows immediately from the definitions. We shall now proceed to discuss specific types of reasoning problems, which can be expressed in terms of $\cong_{RM}$, $\Vdash_{RM}$, and $\Vdash\!\diamond_{RM}$, respectively.

2.4.2 Scenario completion

Scenario completion is the operation of taking a scenario description and concluding as much as possible about the values of features at different points in time. In formal terms, therefore:

Definition. Scenario completion *for a scenario description* Υ *is the operation of finding a scenario description* $\Upsilon' = \langle \mathcal{N}, \mathcal{O}, \text{OBS} \rangle$ *such that*
$$\Upsilon \cong_{RM} \Upsilon',$$
or of determining the model set $Mod(\Upsilon, RM)$. ⋈

Notice that the two meanings of the term are equivalent from the semantic point of view that is taken here. Each solution is an exhaustive description of the possible histories that are compatible with Υ, but described only in terms of what features have what values at what point in time, and using a valuation for constant symbols that is consistent with Υ. There is no reference to actions or action laws in the solution Υ'.

If Υ is a chronicle then the problem is naturally called *chronicle completion*. Chronicle completion has been studied in the AI literature under the names of prediction and postdiction. Literally speaking, prediction could also refer to the prediction of future events which are consequences of current events and circumstances, but in the actual research literature, it is mostly the prediction of future facts from a given schedule, observations, and action laws that has been studied, and similarly for postdiction. The concept of scenario completion subsumes both prediction and postdiction, but allows as well for the case where a schedule (set of action statements) is given, and one deduces facts about the state of the world both before, during, and after the execution of those actions.

One can often distinguish a few main cases among the developments that are possible according to a chronicle Υ. For example, if one of the actions is nondeterministic, that is, it has several alternative outcomes, there may be significant differences between the continued histories of the world after that action. Also, if the schedule does not completely specify the temporal order of the actions, then different action orders may result in entirely different histories. On the other hand, each of these 'main cases' may contain several models which differ, for example, with respect to the exact timing of the actions, or with respect to the outcome of an action, in cases where the outcome does not make any difference for the continuation. Such a structure of 'main cases' can be captured by organizing the component OBS in the scenario completion as a disjunction, where each disjunct is a conjunction of elementary observation formulae. Then, each disjunct will be called an **extension** of the given chronicle Υ.

The definition of extension is intentionally somewhat imprecise at this point, since what qualifies as an extension depends on what qualifies as an elementary observation formula. For example, if alternative outcomes of an action are to be accommodated within one extension, then one must allow elementary observation formulae containing a disjunction between several possible values. The definition of chronicle syntax in later chapters will resolve the present ambiguity.

2.4.3 Deduction and abduction with respect to chronicles

Several reasoning operations can be expressed well in terms of $\models_{\text{RM}}$.

Prediction

The prediction operation consists of concluding whether a particular proposition p holds at the ending time t of the sequence of actions in a given scenario Υ. It can be formulated as follows. For given
$$\Upsilon = \langle \mathcal{K}, \mathcal{O}, \langle \mathbf{IA}, \text{LAW}\rangle, \text{SCD}, \text{OBS}\rangle,$$
for a constant symbol t which occurs in SCD and which in any model will obtain a value greater than or equal to the ending time of any of the actions there, and for a given property p, verify whether one or the other of the following conditions holds:
$$\Upsilon \models_{\text{RM}} \langle \mathcal{N}, \mathcal{O}, \{[t]p\}\rangle,$$
$$\Upsilon \models_{\text{RM}} \langle \mathcal{N}, \mathcal{O}, \{[t]\neg p\}\rangle.$$
Then answer 'Yes' (first condition), 'No' (second condition), or 'Do not know' (neither condition).

Plan construction in $\mathcal{K}$

Simple planning, or plan construction in $\mathcal{K}$, is the operation of finding a plan which will achieve a particular goal, expressed in terms of certain features which are to have certain values. This reasoning task is formulated as follows. For a given object domain $\mathcal{O}$, a given world description Diw, a given description of the initial state OBS, and a given description of the goal state GOAL, find a plan SCD such that
$$\langle \mathcal{K}, \mathcal{O}, \text{Diw}, \text{SCD}, \text{OBS}\rangle \models_{\text{RM}} \langle \mathcal{N}, \mathcal{O}, \text{GOAL}\rangle.$$

Coherence verification

Coherence verification is applicable for a world where the effects of actions are only known when certain preconditions are satisfied, that is, worlds with the epistemological descriptor $\mathcal{Q}$. The purpose of coherence verification is to check whether the preconditions of all actions in the schedule are necessarily satisfied. This operation is required for plan construction and plan verification.

Formally, coherence verification can be defined as follows. For a given object domain $\mathcal{O}$, a given world description Diw, a given set of preconditions PRE indicating restrictions on the world's ability to perform actions, a given a description of the initial state OBS, and a given schedule SCD, identify whether
$$\langle \mathcal{K}, \mathcal{O}, \text{Diw}, \text{SCD}, \text{OBS}\rangle \Vdash_{\text{RM}} \langle \mathcal{N}, \mathcal{O}, \text{PRE}[\text{SCD}]\rangle.$$
The partition PRE consists of rules which, for each action, specify the preconditions that have to be satisfied in order for the action to be performable, or at least for the action law in Diw to be applicable. If SCD is a schedule, then PRE[SCD] is a set of logical formulae which only refer to features (not to occurrences), and which express the preconditions for each of the actions in SCD. If the precondition is not satisfied, then it is impossible to perform the action or its effects are not known[18].

Plan construction in $\mathcal{Q}$

If the effects of actions are only partially known, like for coherence verification, then the definition of plan construction has to be revised: one is only interested in plans where each action is guaranteed to be performed in a situation where its preconditions are satisfied. This requires a combination of coherence verification and simple planning, and is defined as follows.

For a given object domain $\mathcal{O}$, a given world description Diw, a given set of precondition formulae PRE indicating restrictions on the world's ability to perform actions, a given description of the initial state OBS, and a given description of the goal state GOAL, find a plan SCD such that
$$\langle \mathcal{K}, \mathcal{O}, \text{Diw}, \text{SCD}, \text{OBS}\rangle \Vdash_{\text{RM}} \langle \mathcal{N}, \mathcal{O}, \text{PRE}[\text{SCD}] \cup \text{GOAL}\rangle.$$
This says that the plan SCD must both be coherent and achieve the goal. L. Karlsson [35] has shown how the TWEAK planning algorithm can be reconstructed in this framework.

Object identification

Object identification is the operation of finding all objects o that can safely be concluded to have a given property P within a given scenario $\Upsilon \in \mathcal{K}$. This example will use some notation that is defined in Chapter 5, and may therefore have to be skipped on a first reading of the book. With

---

[18] The reason why the case of unknown effects has to be given special treatment in this fashion, is because otherwise, under the inertia assumption, if no specific effects of the action are known, then it would follow that the action does not have any effects.

that caveat, object identification can be expressed as follows. Let $\Upsilon = \langle \mathcal{K}, \mathcal{O}, \text{Diw}, \text{SCD}, \text{OBS}\rangle$, and a property P of one argument be given in the problem statement. The operation consists of finding some object, or all those objects, #i for which

$$\langle \mathcal{K}, \mathcal{O}, \text{Diw}, \text{SCD}, \text{OBS}\rangle \models_{\text{RM}} \langle \mathcal{N}, \mathcal{O}, \{\ulcorner P(\text{\#i})\urcorner\}\rangle.$$

Most of the reasoning problems that have been defined here make sense only in the context of temporal reasoning. However, object identification also makes sense for static applications, in which case Υ is instead chosen as a member of $\mathcal{N}$ and without any aspects of time, action, and change.

Several classical operations in AI involve object identification. In particular, planning algorithms such as TWEAK [13] operate by successively assuming additional actions, and gradually constraining their temporal ordering and the object parameters (in an object domain which for many applications is held fixed) so that the choice of parameters for each action can be postponed for as long as possible. For example, in a blocks-world problem, where one action type is to move block x to the position (on top of block) y, one may wish to introduce an action for a given x, but postpone 'binding' y until additional constraints can be taken into account. Using object identification, one would introduce an object constant symbol o_k, to be used as the second parameter for the new action, and apply object identification methods on this o_k.

2.4.4 Consistency-checking with respect to chronicles

Finally, there are some reasoning operations which are the most faithfully expressed in terms of $\|\diamond_{\text{RM}}$, and we describe one of them.

Explanation

Explanation is a reasoning operation where observations are given for a later point in time, and one asks what the schedules (sets of actions) are that may possibly have led to these observations. The distinction between plan construction and explanation is particularly clear if non-deterministic actions are allowed. For example, given an observation that a person is standing in front of a gambling machine with a lot of coins in his hands, the explanation that he won them from the machine is adequate. However, playing the gambling machine is not an adequate plan for obtaining a lot of money.

The operation is defined as follows. For a given object domain $\mathcal{O}$, a given world description Diw, and two given sets of observations at arbitrary times OBS and OBS', where OBS' are to be explained and OBS do not need to be explained, a schedule SCD is a solution iff

$$\langle \mathcal{K}, \mathcal{O}, \text{Diw}, \text{SCD}, \text{OBS}\rangle \|\diamond_{\text{RM}} \langle \mathcal{N}, \mathcal{O}, \text{OBS}'\rangle.$$

One may wish to include a preference relation over possible answers in the problem specification, for example, to prefer short explanations over longer ones.

2.4.5 Chronicle completion as the pivotal operation

We have now shown how a number of temporal reasoning operations can be phrased in terms of $\Upsilon \rightsquigarrow \Upsilon'$, where $\rightsquigarrow$ is either of $\cong_{\text{RM}}$, $\models_{\text{RM}}$, or $\Diamond_{\text{RM}}$; Υ is a chronicle in $\mathcal{K}$, and Υ' is a chronicle in $\mathcal{N}$. The corresponding model-theoretic formulations are $Mod(\Upsilon, \text{RM}) \propto Mod(\Upsilon', \text{RM})$, where $\propto$ is either equality, $\subseteq$, or 'overlaps'.

Trivially, if $\Upsilon \cong_{\text{RM}} \Upsilon'$ and $\Upsilon' \rightsquigarrow \Upsilon''$, then $\Upsilon \rightsquigarrow \Upsilon''$, where $\rightsquigarrow$ is any of the three weakened entailment relations. The operation of chronicle completion is therefore of primary importance from a logical point of view: if chronicle completion is available, then all the other operations that were reviewed above can be reduced to an operation on two scenarios in $\mathcal{N}$, where there are no complications due to actions, inertia, and so on. The same does not necessarily apply from a computational point of view: it does not follow that all these operations are best computed by reducing them to a chronicle completion problem. In particular, Bäckström and Nebel have shown that there are some classes of problems where planning is tractable whereas prediction is not [53].

However, such computational considerations will not be addressed in the present book. We shall focus only on the entailment relations as such, and in this case the problem of chronicle completion is of pivotal importance. It can be viewed in terms of the $\cong_{\text{RM}}$ entailment relation, as above, or equivalently in terms of models: for a given $\Upsilon = \langle \mathcal{K}, \mathcal{O}, \langle \text{IA}, \text{LAW} \rangle, \text{SCD}, \text{OBS} \rangle$, how can $Mod(\Upsilon, \text{RM})$ be expressed in terms of the classical model sets for LAW, SCD and OBS, or variants of those sets?

The most obvious way to obtain $Mod(\Upsilon, \text{RM})$ is as a function of the conventional model sets for LAW, SCD, and OBS. We have already seen that the model set for LAW∪SCD∪OBS is not the right answer, or in other words, it is not correct to take the intersection of the model sets of the respective partitions. In fact, many of the non-monotonic logics which have been proposed for reasoning about actions and change amount essentially to a proposal for how to obtain $Mod(\Upsilon, \text{RM})$, for example, using the preference relation of chronological minimization to select a subset of $Mod(\text{LAW} \cup \text{SCD} \cup \text{OBS}, \text{RM})$, or adding additional 'explanation closure' axioms to achieve the same purpose.

The fact that we are only interested in the M and R components of models results in a particular problem. Rather than obtaining $Mod(\text{LAW})$, $Mod(\text{SCD})$, and $Mod(\text{OBS})$, combining them, and then extracting their $\langle M, R \rangle$ parts, we would prefer to start from $Mod(\text{LAW}, \text{RM})$, $Mod(\text{SCD}, \text{RM})$, and $Mod(\text{OBS}, \text{RM})$, and to operate further on those sets. Unfortunately, this does not work, because it would disable the connection between an action statement in SCD and the corresponding action law.

What we can consider, however, is to start from the two sets $Mod(\text{LAW} \cup \text{SCD}, \text{RM})$ and $Mod(\text{OBS}, \text{RM})$, and to study alternative ways of combining those two sets of restricted models of the form $\langle M, R \rangle$. As we shall see, this approach gives an expressivity that is adequate for many purposes, and it is the one that will be used in this volume.

2.5 Reasoning problems

The problem formulations in the previous section have a mathematical, rather than a logicist flavor: several sets of formulae are given, and the desired solution is yet another set of formulae, satisfying certain relations to the given sets. It is true that each set consists of logic formulae, and that the relations are expressed in terms of sets of models, so 'logic' (at least in the sense of syntax and model theory) is part of the details of the problem definition. However, its overall structure does not have the character that is usual in logic.

It is a matter of taste whether one considers this to be an advantage or a disadvantage. From the point of view of logic, including also the logic of common sense, one might argue that problems should best be stated in terms of formulae which specify 'what is the case', and from which one can derive 'what follows'. Already the increasing interest in abduction and the de-emphasis of deduction in contemporary research is arguably a step away from this logic-oriented way of stating problems, but the operations that were defined above seem to go further in the same direction.

One may ask, then, whether it would be possible to rephrase the problem statements for the various operations, so that they appear more logic-like. The present section explores how that could be done. It should be viewed as a thought-experiment; it is not something which I 'propose'; and it will not be used in the sequel. Later chapters in this book will be based on the concept of chronicle completion, which has already been introduced.

2.5.1 Definitions of reasoning problems

Definition. *A* **monadic reasoning problem** *is a fourtuple*
$$\langle \Upsilon, \Xi, \leadsto, \ll \rangle,$$
where Υ is a scenario description representing what is given, Ξ is a set of scenario descriptions wherein the solution is to be found, $\leadsto$ is a binary relation over scenario descriptions called an **inference operation**, *and $\ll$ is a preference relation over Ξ. A scenario $\Upsilon' \in \Xi$ is a* **solution** *to this*

reasoning problem iff $\Upsilon \leadsto \Upsilon'$, *and* Υ' *is* $\ll$-*minimal among those* $y \in \Xi$ *for which* $\Upsilon \leadsto y$. ⋈

Definition. *A k-adic reasoning problem is a fourtuple*
$$\langle\langle \Upsilon_1, \Upsilon_2, ..., \Upsilon_k \rangle, \Xi, \leadsto, \ll \rangle,$$
where all Υ_i *are scenario descriptions,* Ξ *is a set of scenario descriptions wherein the solution is to be found,* $\leadsto$ *is a* $(k+1)$-*ary relation over scenario descriptions written as*
$$\Upsilon_1, \Upsilon_2, ..., \Upsilon_k \leadsto \Upsilon,$$
and $\ll$ *is a preference relation over* Ξ. *A scenario* $\Upsilon' \in \Xi$ *is a* **solution** *to this reasoning problem iff*
$$\Upsilon_1, \Upsilon_2, ..., \Upsilon_k \leadsto \Upsilon',$$
and Υ' *is* $\ll$-*minimal among those* $y \in \Xi$ *for which* $\Upsilon_1, \Upsilon_2, ..., \Upsilon_k \leadsto y$. ⋈

Monadic reasoning problems are appropriate for reasoning about the world itself, whereas k-adic problems for $k > 1$ tend to be required for dealing with the beliefs, goals, and intentions of one or more egos.

The present definitions of reasoning problems suggest that one or more scenarios are given, a results scenario is computed, and the job is over. However, for the design of an autonomous agent one needs to consider the corresponding incremental problem, where the input scenario description(s) changes over time, and the corresponding output scenario description is to be updated correspondingly.

2.5.2 Rephrasing logical operations as reasoning problems

The logical operations that were defined in the previous section can be re-phrased as reasoning problems, as follows.

Prediction

Prediction, which is the problem of inferring whether a particular proposition p holds at the ending time **t** of the sequence of actions in a given scenario Υ, was formulated as follows. Let $\mathcal{O}$ be the object domain used in Υ. The answer to the question is '**Yes**' iff
$$\Upsilon \models \langle \mathcal{N}, \mathcal{O}, \{[\texttt{t}]p\} \rangle,$$
it is '**No**' iff
$$\Upsilon \models \langle \mathcal{N}, \mathcal{O}, \{[\texttt{t}]\neg p\} \rangle,$$
and it is '**Do not know**' otherwise. Here $[\texttt{t}]p$ stands for "p holds at time **t**". This formulation can be represented as a monadic reasoning problem with $\Xi = \langle \mathcal{N}, \mathcal{O}, \{[\texttt{t}]\alpha_i\} \rangle$, where α_i ranges over $\ulcorner p \urcorner$, $\ulcorner \neg p \urcorner$, and 'tautology'. The ordering relation $\ll$ is chosen so that the scenario containing 'tautology' is less preferred than the other two, and $\leadsto$ is chosen as $\models$. The three

members of Ξ correspond to the three answers 'Yes', 'No', and 'Do not know' respectively.

Plan construction in $\mathcal{K}$

Simple planning, or plan construction in $\mathcal{K}$, is the operation of finding a plan which will achieve a particular goal g. It is rendered as a dyadic reasoning problem where Υ_1 is a scenario description of the form
$$\langle \mathcal{K}, \mathcal{O}, \text{Diw}, \emptyset, \text{OBS}\rangle,$$
where in turn Diw is a world description and OBS is a description of the initial state. Υ_2 is a scenario description of the form $\langle \mathcal{N}, \mathcal{O}, \{g\}\rangle$, and Ξ is the set of scenario descriptions of the form
$$\langle \mathcal{K}, \mathcal{O}, \emptyset, \text{SCD}, \emptyset\rangle,$$
where SCD is an executable schedule, and $\ll$ is a preference order which for example may prefer shorter plans over longer ones. Finally, the inference operation $\leadsto$ is constructed so that
$$\langle \mathcal{K}, \mathcal{O}, \text{Diw}, \emptyset, \text{OBS}\rangle, \langle \mathcal{N}, \mathcal{O}, \text{GOAL}\rangle \leadsto \langle \mathcal{K}, \mathcal{O}, \emptyset, \text{SCD}, \emptyset\rangle$$
iff
$$\langle \mathcal{K}, \mathcal{O}, \text{Diw}, \text{SCD}, \text{OBS}\rangle \models \langle \mathcal{N}, \mathcal{O}, \text{GOAL}\rangle.$$
In other words, the problem is to identify a best possible, executable schedule which, when combined with the given action laws and the given initial state of the world (possibly incompletely specified), entails that the given goal will be achieved.

Coherence verification

The purpose of coherence verification is to check whether the preconditions of all actions in a given schedule are necessarily satisfied. It is defined as a monadic reasoning problem
$$\langle \Upsilon, \Xi, \leadsto, \ll\rangle,$$
where
$$\Upsilon = \langle \mathcal{Q}, \mathcal{O}, \text{PRE}, \text{Diw}, \text{SCD}, \text{OBS}\rangle,$$
Diw $= \langle \textbf{IA}, \text{LAW}\rangle$, and the set Ξ has two members representing 'Yes' and 'No'. 'Yes' is represented by the scenario description $\langle \mathcal{N}, \mathcal{O}, \text{PRE}[\text{SCD}]\rangle$, and 'No' by the tautology scenario description $\langle \mathcal{N}, \mathcal{O}, \textbf{T}\rangle$. $\ll$ prefers the former over the latter. Finally, $\Upsilon \leadsto y$ iff
$$\langle \mathcal{K}, \mathcal{O}, \text{Diw}, \text{SCD}, \text{OBS}\rangle \models y.$$
In other words, if one applies the postcondition rules for each of the action statements in the schedule, and one makes the completeness and inertia assumptions associated with $\mathcal{K}$, then the preconditions follow.

Plan construction in $\mathcal{Q}$

Plan construction in $\mathcal{K}$ has already been defined as a dyadic reasoning problem
$$\langle\langle \Upsilon_1, \Upsilon_2\rangle, \Xi, \leadsto, \ll\rangle,$$
where
$$\langle\mathcal{K},\mathcal{O},\text{Diw},\emptyset,\text{OBS}\rangle, \langle\mathcal{N},\mathcal{O},\text{GOAL}\rangle \leadsto \langle\mathcal{K},\mathcal{O},\emptyset,\text{SCD},\emptyset\rangle$$
iff
$$\langle\mathcal{K},\mathcal{O},\text{Diw},\text{SCD},\text{OBS}\rangle \Vdash \langle\mathcal{N},\mathcal{O},\text{GOAL}\rangle.$$
Plan construction in $\mathcal{Q}$ can then be defined as the dyadic reasoning problem
$$\langle\langle \Upsilon_1, \Upsilon_2\rangle, \Xi, \leadsto, \ll\rangle,$$
where the members of Ξ and the ordering $\ll$ are as for plan construction in $\mathcal{K}$, and
$$\langle\mathcal{Q},\mathcal{O},\text{PRE},\text{Diw},\emptyset,\text{OBS}\rangle, \langle\mathcal{N},\mathcal{O},\text{GOAL}\rangle \leadsto \langle\mathcal{K},\mathcal{O},\emptyset,\text{SCD},\emptyset\rangle$$
iff
$$\langle\mathcal{K},\mathcal{O},\text{Diw},\text{SCD},\text{OBS}\rangle \Vdash \langle\mathcal{N},\mathcal{O},\text{PRE}[\text{SCD}] \cup \text{GOAL}\rangle.$$

Explanation

Explanation can be rendered as a dyadic reasoning problem
$$\langle\langle \Upsilon_1, \Upsilon_2\rangle, \Xi, \leadsto, \ll\rangle,$$
where $\Upsilon_1 = \langle\mathcal{K},\mathcal{O},\text{Diw},\emptyset,\text{OBS}\rangle$ contains the world description and the given observations, that is, those observations that do not need to be explained. Also, $\Upsilon_2 = \langle\mathcal{N},\mathcal{O},\text{OBS}'\rangle$ characterizes those observations that are to be explained. Ξ is a set of chronicles characterizing schedules that may serve as explanations, on the form $\langle\mathcal{K},\mathcal{O},\emptyset,\text{SCD},\emptyset\rangle$. The inference operation $\leadsto$ is defined so that
$$\langle\mathcal{K},\mathcal{O},\text{Diw},\emptyset,\text{OBS}\rangle, \Upsilon_2 \leadsto \langle\mathcal{K},\mathcal{O},\emptyset,\text{SCD},\emptyset\rangle$$
iff
$$\langle\mathcal{K},\mathcal{O},\text{Diw},\text{SCD},\text{OBS}\rangle \Vdash \Upsilon_2.$$
In other words, any schedule which, when combined with the world description and given observations, is consistent with the explanendi, is a solution to the reasoning problem. The preference relation $\ll$ may be defined, for example, to prefer short explanations over longer ones, or to prefer more general explanations over more specific ones.

Object identification

Object identification is the operation of finding all objects o that can safely be concluded to have a given property P within a given scenario $\Upsilon_1 \in \mathcal{K}$. Using the same notation as above, object identification can be rendered as a dyadic reasoning problem
$$\langle\langle \Upsilon_1, \Upsilon_2\rangle, \Xi, \leadsto, \ll\rangle,$$

where $\Upsilon_1 \in \mathcal{K}$ is the given scenario, and $\Upsilon_2 = \langle \mathcal{N}, \mathcal{O}, \{\ulcorner P(o) \urcorner\}\rangle$ expresses the required property using an object constant symbol o that does not occur in Υ_1. Ξ is the set of possible solutions of the form
$$\langle \mathcal{N}, \mathcal{O}, \{\ulcorner o = \#\mathtt{i} \urcorner\}\rangle$$
for all choices of an object $\#\mathtt{i} \in \mathcal{O}$. The inference operation is defined so that
$$\langle \mathcal{K}, \mathcal{O}, \mathrm{Diw}, \mathrm{SCD}, \mathrm{OBS}\rangle, \Upsilon_2 \rightsquigarrow \langle \mathcal{N}, \mathcal{O}, \Gamma\rangle$$
iff
$$\langle \mathcal{K}, \mathcal{O}, \mathrm{Diw}, \mathrm{SCD}, \mathrm{OBS} \cup \Gamma\rangle \Vdash \Upsilon_2.$$
In other words, if the given scenario description Υ_1 is combined with the assumption that o is a particular object, then the conclusion $P(o)$ follows.

2.5.3 Discussion

The exercise in this section has shown, in an explorative fashion, that it is indeed possible to phrase the various logical operations in terms of the common 'reasoning problem' framework. This exercise has not been carried far enough to produce any stringent results, but one can still speculate whether it represents a useful way to proceed.

If one truly wishes to represent problems in terms of statement sets, where each statement expresses something which 'is the case', then one must first ask what different things 'are the case' in the operations that have now been reviewed. One assumes that everything that is given in the problem statement 'is the case' in one way or another, otherwise it could not be given. In a planning problem, for example, some formula φ describes the goal, and one could consider using a logical formula of the form Goal φ for "it is the case that φ is a goal at the present time". In an explanation problem, certain observations are made, and it is the case that some of them are to be explained, which could possibly be expressed by a formula of the form WonderWhy φ, and so forth.

The formulations in terms of reasoning problems can perhaps be considered as a way of writing such modal-looking formulae, namely if each of the partitions indicates the application of a modal operator such as Goal or WonderWhy on each of its members. In this case, the meaning of each modal operator would be obtained from the definition of $\rightsquigarrow$ in the corresponding reasoning problem, which would be a possible, although nonstandard, way of defining these modal operators. On the other hand, if such modal operators are needed then there may be other and more natural ways of defining them. Anyway, that would be outside the scope of the present book. We return from the present excursion to the pivotal operation of chronicle completion, and to the logic that is required for it.

2.6 Terminology

The material in this book is fairly complex and spans over a number of related issues and phenomena. As usual, it has been important to make precise definitions of all important concepts, but in this case the number of terms is fairly large. The fact that we are using a novel approach has further increased the need for explicit and precise definitions. The same applies for the notation that is used in formulae of various types.

Appendix A and B contain a term index and an index for the notation, in order to facilitate reference to previously introduced terms. The present section defines a few terms which are characteristic of our approach.

2.6.1 Domains and worlds

The KR and other AI literature uses the terms 'world' and 'domain' more or less interchangably for an area of knowledge, which may be either a toy microworld for a theoretical investigation, or a practical application area with greater conceptual richness. I use the word 'world' for that same purpose, and retain the term 'domain' for its mathematical meaning as the set of objects belonging to the same type. For example, the classical restacking problem of three blocks on a table [67] is a reasoning problem in the blocks world.

Within one world there may be several scenarios. For example, the blocks world is characterized by a set of action laws specifying the effects of 'Move', 'Clear', and a selection of other actions on blocks. Different scenarios in the blocks world share the use of the same set of action laws, but each scenario description is also associated with an object domain; a set of actions that have been performed; a set of observations, and so forth. In particular, this means that I will avoid using the common term 'closed-world assumption', since the assumption about a certain set of objects pertains to a specific scenario in the (blocks) world and not to the blocks world itself. One can easily consider a reasoning process which uses the action laws for the blocks world as its only premises, and which derives additional properties about the blocks world in general. The so-called closed-world assumption would of course not be applicable for such a reasoning task.

Formally, a world has already been defined as a transformation on finite developments, and as a 'player' in the game between ego and world. This is entirely consistent with the use of the word 'world' that is discussed here. Formally, a 'blocks world' is a set of pairs $\langle J, J' \rangle$, where J specifies what has happened in the world from time 0 up to the present now-time n, and also specifies what action (for example, a move action) the ego has initiated at the present time. In particular, J specifies the current state at time n. The corresponding J' extends J to the next relevant time, after the most recent action has been completed, and the current state at the now-time of J' defines the state of the same blocks world after the action has been completed. This is a complete (although somewhat elaborate) definition of what the blocks world is.

2.6.2 Formulae, axioms, and premises

I use the term *logical formula* instead of the usual term *well-formed formula* since the latter focuses needlessly on the syntactical aspect.

I shall use the word *axiom* in a restricted fashion for something which is obviously or intrinsically true, and I use the word *premise* rather than axiom for a formula that is used as an initial step in a formal proof. For example, in the operation of chronicle completion, the formulae in the LAW, SCD, and OBS partitions of the chronicle are considered as premises but not as axioms. Axioms are those formulae which either express mathematical truths, or express properties of the world according to the presently chosen ontology. For example, the formula
$$\forall t_1 \forall t_2 \forall t_3 [t_1 < t_2 \land t_2 < t_3 \to t_1 < t_3],$$
where the t_i vary over the domain of timepoints, would be called an axiom in the context of any ontology where the $<$ relation on timepoints is transitive.

2.6.3 Layers of logic and of logical language

The following terms relating to logic, language, and syntax will be used.

Base logic: a logic characterized in the ordinary way by its syntax and its conventional semantics. The situation calculus is one example of a base logic; first-order predicate calculus with explicit metric time as one of the sorts is another example.

Entailment method: A mapping from chronicles to sets of weakened models $\langle M, R \rangle$ which is usually defined using the following operations on the partitions of the chronicle:

- Obtaining the set of classical models of a partition (= set of formulae);
- Syntactic tranformations on formulae or sets of formulae;
- Mapping a set of models to a subset, in particular, the subset of its minimal members with respect to some preference relation;
- For a given set of formulae, obtaining the largest subset that satisfies a given property;
- Ordinary set-theoretic operations ($\cup$, $\cap$, etc) on sets of formulae and sets of models.

Entailment methods will be strictly defined in Chapter 9.

Logic: either a base logic alone, or a base logic together with an entailment method. For example, preferential non-monotonic logics are considered to consist of a base logic plus a preference relation on models. This preference relation is outside the base logic, and several non-monotonic logics may use the same base logic but different preference relations.

Main language vs. **side language:** for the specification of the syntax of the base logic, it will be convenient to divide the full language of the logic into a main language and a side language. The semantics is only defined for the main language. The meaning of the side language is formally defined using a translation function that translates side-language formulae into the main language. The formulae for expressing actions and other occurrences will be considered as a side language in the present volume.

Core syntax vs. **abbreviations:** the syntax of the main language is furthermore divided into a core syntax and definitions of abbreviations. Each abbreviation is formally defined just as a shorter way of writing something which, with less convenience, could have been written directly in core syntax.

The difference between side language and abbreviations is that the rewrite rules for abbreviations are held fixed and are specified in this book. The rewrite rules for a side language are 'left to the user', that is, they must be expressed in a partition of the scenario descriptions. In particular, action laws specifying the effects of actions will be viewed as translation rules, contributing to the definition of the translation of a side language.

Sublanguage: after the full language of the logic has been defined, sublanguages may be obtained by imposing syntactical restrictions. The partitions of scenario descriptions will be restricted to only allow certain sublanguages.

The use of these layers of logic and language is intended to serve the following two priority goals:

- Notational convenience. The logical language must make it practically possible to write concrete scenario examples in the logic, and to read them conveniently. This is, in particular, the reason for introducing a number of abbreviations along with the core syntax.

- Application orientation. The logic must be designed so that it fits the intended class of applications, both from a representation point of view and with realizations in mind. The use of partitions and of multiple languages has emerged as a way of assuring this goal.

2.6.4 Realizations and implementations

None of the concepts that were specified in the previous subsection has any direct connection with software implementation, but the indirect connection is significant. I view entailment methods as very high-level software specifications, since they specify what conclusions shall be inferred, or inferrable, but not how they are to be inferred. Therefore, for each logic in the sense that has just been defined, there may be one or several *implementations* consisting of data structures, specialized algorithms, general-purpose decision procedures, and so on.

The transition from an entailment method to actual software is often too complex to be made in one step. I will use the term *realization* for the intermediate level, which is closer to actual software than the entailment method is, but which anyway serves as a specification for the actual program. An algorithm for temporal reasoning is therefore one example of a realization. Another interesting class of realizations is obtained when an entailment method is realized as a translation from chronicles, or parts of chronicles, to equivalent formulations in terms of circumscription or logic programs. This implementation strategy may enable one to use the existing software technology for logic programming.

The present book does not address the questions of realization and implementation, for the simple reason that before we make the effort of realizing and implementing an entailment method we want to understand its logical properties. The present book is devoted to that analysis. However, other members of our research group have already presented results which are relevant for the realization of some of the entailment methods that are described in this book [23, 20].

2.7 Summary

Scenario descriptions and the special case of chronicles have been defined. Epistemological designators have been introduced in order to characterize assumptions that may accompany a chronicle. Model concepts and entailment relations for chronicles have been defined. A number of characteristic temporal reasoning problems can be expressed in terms of a relationship between the weakened model set for a chronicle in $\mathcal{K}$ and a chronicle in $\mathcal{N}$, the latter being essentially just a set of logical formulae for observations, combined with a specification of the choice of object domain.

Therefore, we shall need a base logic with the following characteristics. Its syntax must allow us to express action laws, schedules, and observations. Its semantics must be based on interpretations of the form $\langle M, R \rangle$, where M is a valuation for constant symbols, and R is a history, that is, an assignment of values to features as a function of time. Actions and other occurrences will not be directly represented in the semantics, therefore, and must be dealt with in some other way.

The purpose of the following chapters is to define such a base logic, and to extend it with entailment methods which obtain $Mod(\Upsilon, \text{RM})$ as a function of the base-logic model sets for the combination of LAW and SCD and for OBS. We shall see that this is possible, provided that some restrictions are imposed on the world in addition to the basic **IA** restriction. (For example, it can be done if all actions are deterministic, and are performed over a single step of time). After analyzing the shortcomings of the logic outside its range of correct applicability, we shall define and analyze several generalizations of the logic whereby the restrictions are overcome. That is, in brief, the agenda for this volume as a whole.

3
Underlying semantics for IDS worlds

The definition of IDS worlds given in Chapter 1 was quite broad. More specific classes of worlds can be obtained in two ways: first by imposing various constraints according to the ontological characteristics, and then by specifying exactly the effects of actions using action laws. Conversely, from the point of view of logic, IDS's provide a restriction on the classical set of models. Therefore, they can be used as an **underlying semantics** which defines the intended models or, equivalently, the intended conclusions from a given scenario description. In other words, there is a two-way relationship between the IDS 'model' and a logic of actions and change. Chapters 1 and 2 addressed this common topic from the IDS point of view; the present chapter will do it from the point of view of the logic.

3.1 Some methodological aspects of common-sense reasoning

Reasoning about actions and change is relevant for two distinct although overlapping purposes: for common-sense reasoning, for example in natural-language contexts, and for cognitive robotics. In this section, we shall discuss the methodology of research on common-sense reasoning about actions and change.

3.1.1 The example-based methodology

The present standard *example-based methodology* has been concisely described by Davis [16] as follows:

The basic approach used here, as in much of the research on automating commonsense reasoning, is to take a number of examples of commonsense inference in a commonsense domain, generally deductive inference; identify the general domain knowledge and the particular problem specification used in the inference; develop a formal language in which this knowledge can be expressed; and define the primitives of the language as carefully and precisely as possible.

This is true not only for reasoning about actions and change (where the 'Yale Shooting Scenario' is the most widespread example of common-sense inference), but also for inheritance reasoning where one has used examples such as the Nixon dilemma ('Nixon diamond'), the royal elephant, and the beer-drinking marine chaplain. Davis then proceeds to a discussion how the resulting formal language can be used in the design of actual AI programs.

The example-based methodology has been universally accepted, and there have also been proposals, for example by Lifschitz [41], to form a standard collection of representative examples of common sense reasoning, to be used as a test suite for proposed non-monotonic logics. This methodology has been instrumental for identifying a variety of different problems which come up in this type of reasoning ('the ramification problem' and 'the qualification problem', in particular), but it also has a number of problems:

- The example-based methodology assumes implicitly that there *is* such a thing as a logic of common-sense reasoning, which merely remains to be discovered. If it exists, then presumably it can be reconstructed from common-sense examples, but how do we know that it exists at all?

- There are sometimes 'clashes of intuitions' where different informers make incompatible statements as to the admissible common-sense inferences for a given example. This means that the empirical data for the research are unreliable. Also, since the informer is usually the researcher himself or herself, there is always a danger that the researcher will be influenced by the particular theory he has developed (Touretzky, [82]). A major reason for the uncertainty is that it is difficult to delimit the 'general domain knowledge' that has been used for a particular inference.

- As one goes beyond the trivial examples to more complicated ones, the potentially relevant general domain knowledge broadens, and the uncertainty as to what are the admissible common-sense inferences is therefore bound to increase. Therefore, it is difficult to develop a logic of common-sense reasoning based on such examples. Progress is hampered by the increasing uncertainty in the empirical data as soon as one goes beyond the simplest toy examples.

- The use of non-monotonic logics aggravates these problems. If classical logic is being used, then a proposed formalization can arguably be accepted even if it obtains too large a model set, because it could always be corrected by 'adding more knowledge', but it would definitely be rejected if it failed to include some intended models in the actual model set. However, with non-monotonic logics, *any* proposed formalization can possibly be salvaged by adding more premises whereby previously non-selected models regain the status of selected models.

3.1.2 Systematic methodology used in this book

The difficulties and limitations of the example-based methodology are the reason why an alternative methodology is needed. The *systematic methodology* that is used in this book has the following characteristic aspects:

- The use of an *underlying semantics* which characterizes inhabited dynamical systems in general.
- The use of the underlying semantics for a formal definition of the *intended models* for a given scenario description.
- An *ontological taxonomy* which is strictly defined in terms of the underlying semantics, and which characterizes classes of dynamical systems with specific restrictions on them.
- Explicit statement of *epistemological assumptions*.
- Formally proven *assessments* of the range of applicability of entailment methods, that is, the range of chronicles where the set of selected models according to the entailment method equals the set of intended models.

In this way one will consider a proposed non-monotonic logic as a function S from scenario descriptions to sets of models, and one is able to prove assessments of the form 'if a scenario description Υ belongs to a certain class $\mathcal{Z}$ of scenario descriptions, then $S(\Upsilon)$ equals exactly the set of intended models for Υ'. This set of intended models will essentially be defined as $Mod(\Upsilon, \text{RM})$ in the terms of the previous chapter. Notice that in order to obtain such results, it is necessary to have (1) a precise definition of the set of intended models $Mod(\Upsilon, \text{RM})$, and (2) a precise definition of the ontological taxonomy, which provides a reportoire of candidate classes $\mathcal{Z}$ that can be used for expressing the assessed range of applicability.

Referring to the subsection about realization and implementation (Chapter 2, subsection 2.6.4), we now have the following layers of expression:

- The underlying semantics, which defines the set of intended models for a given chronicle Υ in very basic terms.
- The entailment method, which is based on classical semantics and obtains the intended-models function using a number of semantical operations, for example, reducing a set of models to the set of its minimal members.

- The realization of a given entailment method, for example, a definition of the appropriate algorithms, or a translation to a known logical 'engine'.

- The actual implementation of the algorithm, the translator, and/or the 'engine' in terms of a programming language.

Corresponding to these four layers of expression there are three verification tasks. The verification of the correctness of an entailment method relative to a given underlying semantics will be referred to as *assessment*. The verification of the correctness of a realization for a given entailment method will be referred to as *validation*. The verification of the correctness of an actual implementation has already the name 'software verification'.

The underlying semantics is the fundamental instrument for defining both the function *Mod* and the various classes $\mathcal{Z}_i$ in the ontological taxonomy. The definition of the underlying semantics started in Chapter 1 through the general IDS structure consisting of ego, world, vehicle, etc., and with the definition of developments as the result of a game between ego and vehicle. The ontological taxonomy was also started in Chapter 1 by the introduction of the ontological designators, and continued in Chapter 2 with the epistemological designators. In the present chapter, one more aspect will be added to the underlying semantics, namely for each action type a definition of the possible ways of performing that action. This can be understood as a way of 'programming' the IDS world by defining its possible behaviors. This addition makes it possible to provide a precise and formal definition of the ontological characteristics, and to add precisely defined subcharacteristics which refine the classification.

If something is known about the general properties of the system, for example, that there is no possibility of concurrent actions, or that there is no possibility of surprises, then that extra knowledge will restrict the choice of IDS world, and thereby the set of intended models, increasing the set of possible conclusions. In a traditional view of the role of logic, one would use a logical system that is able to deal with all the ontological and epistemological characteristics that have been introduced, and to express knowledge about the system by additional statements in that logic. However, the use of such an ideal approach is not within reach, since we do not have a logic that covers such a broad range of system descriptions. Therefore, I use a progressive approach, and start with logics that are able to deal with a much more restricted family of scenario descriptions. Logics that are provably correct there provide a reliable basis for successive extensions to logics with a wider range of applicability.

It is in this perspective that we can perform an assessment of the range of applicability for each logic being considered. The ontological/ epistemological taxonomy serves as a given and fixed framework. The purpose

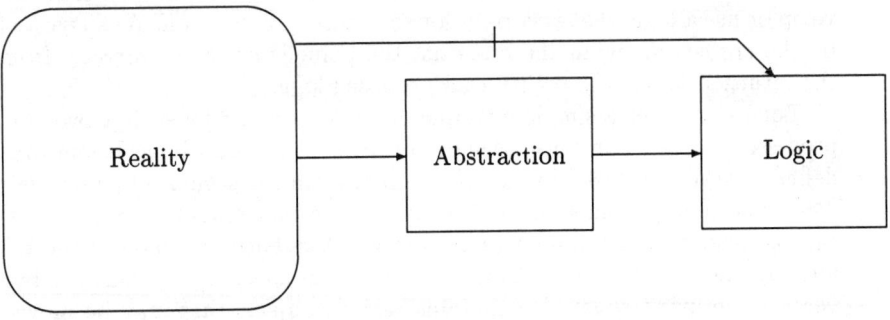

Fig. 3.1. Systematic methodology

of the assessment is to identify in a proven way, for each logic, the family of scenario descriptions for which it obtains exactly the intended set of models.

The central results in this book are assessment results. Chapter 9 and the following chapters contain formal proofs for the range of applicability of a number of logics for reasoning about action.

3.1.3 Approaching common-sense reasoning with systematic methodology

The systematic approach to common-sense reasoning is significantly different from the conventional methodology. Take, for example, the principle of 'minimization of change', which is often used in contemporary research on the representation of common-sense knowledge [28, 83]. In its simplest form, this principle states that in a transition from one state of the world to the next, and in the absence of knowledge to the contrary, one will prefer transitions which minimize the set of fluents that change their value. This principle is motivated by appeal to its intuitive plausibility, and to a (small) number of representative examples of common-sense reasoning where the principle obtains the intended conclusions. In a systematic methodology one would instead ask what the class of IDS systems is for which the principle of minimization of change obtains the intended conclusions.

The systematic methodology decomposes the transition from common-sense application to corresponding logic into two parts, as illustrated in Figure 3.1. The first step is to transform the application world and application scenario to an abstraction (or a 'model' in the engineering meaning of that word; see footnote in subsection 1.4.4). The second step is to develop or use a logic that is correct for the chosen abstraction. As suggested by the crossed arrow in the diagram, the point is not to go directly from the intuition about the reality to a proposed logic.

Let us consider a simple example of what is meant by such a two-step process. The common-sense scenario of 'Aunt Agatha's Living Room' was defined by Winslett [83] by modification of a similar scenario by Ginsberg [28]. The living room is heated using hot air that enters through vents, that is, openings in the floor (a system which is fairly common in the US although not so common in Europe), and in this specific case there are two vents. The living room also contains various objects that may be moved around, and that may in particular be used to block either of the vents. The room may or may not be stuffy. One of the laws of this world is that if both vents are blocked, the room is stuffy. (The time delay before it becomes stuffy is disregarded).

Consider now the situation where one vent is blocked, the other vent is not blocked, and the room is not stuffy. On the request of Aunt Agatha, the household robot performs an action that blocks the other vent as well. Winslett's reasoning system, which is based on minimization of change, allows two possible developments: either the room becomes stuffy, or the obstacle covering the first vent is concluded to be removed, and the room remains unstuffy. Both these two alternatives minimize the set of changes, under the constraints that a room with both vents blocked must always be stuffy, and that the present action cannot be immediately undone. It would seem that only the first model is intended by common sense.

Winslett discusses this apparent violation of what would be the common-sense conclusions, but argues that it is better to obtain too many preferred models than too few: then one can always constrain the model set by additional, domain-dependent premises. She also remarks that there may be circumstances where not only the first but also the second of these developments is admissible: when both vents are blocked, the resulting increased pressure in the hot-air system may remove the obstacle on the other vent. As a common-sense conclusion this one seems very contrived. One question, which is also mentioned by Winslett, is why the other vent cannot be cleared by the increased air pressure instead. Also, one may ask why the air is not able to clear both vents. But, more importantly, it seems that the rules of the game are changing while we play. One possible abstraction of the scenario is that the stuffiness of the room is a function of whether the vents are blocked or not, and the blockings are inert. In that case only the first development ought to be accepted. Another, and more elaborate, abstraction is that there is a possibility for causation of events, and the blocking of the vents leads to an increase in air pressure, which in turn leads to removal of all lightweight objects blocking the vents (regardless of which object was put there first).

A central idea in the systematic approach is that first one has to make up one's mind about which of these or any other abstractions one wishes to use. Then one chooses an appropriate logic from the toolbox of available logics (or develops a new one if necessary), and applies it to the scenario. As a necessary consequence of these steps, one obtains conclusions about the scenario which are guaranteed to hold within the framework of the chosen abstraction. If one is dissatisfied with the conclusions, then obviously one has chosen the wrong abstraction, or omitted some of the relevant facts. If the ground for dissatisfaction is that there is something more which was also relevant, and which one forgot to say, then that is one's own doing. A logic cannot read its user's mind, and rushing from one logic to another without first clarifying the intended semantics is literally a meaningless activity.

3.1.4 Combining the two methodologies

In practice, the systematic methodology and the example-based methodology have to be used together. The argument for a systematic methodology is not intended to imply that common-sense examples are useless, but only that they are not sufficient as a basis for developing a logic of common-sense reasoning. Representative examples of common-sense reasoning are still useful, both as a source of inspiration and challenge, and as a double-check that the logics being developed do their job.

In fact, one other problem with research in this area is that the body of common-sense examples has not been systematically used. Each author tends to use only a few of the examples that are in circulation. In later chapters of this book I try to improve on the situation by defining a test-suite of common-sense examples, and analyzing how a number of current approaches manage with them.

3.1.5 The progression of more general ontologies

Subsections 3.1.2 and 3.1.3 described the systematic methodology for the timescale of one project or one publication. In the longer range, my proposal is to use the taxonomy of scenario descriptions in order to structure the work and the results. Each family of scenario descriptions in the taxonomy is defined by its ontological and epistemological characteristics. It would be reasonable to begin with relatively narrow families, and after that, to address successively larger and more difficult families. At each step, one introduces an appropriate underlying semantics, and thereby a definition of the set of intended models for a given scenario description. This precise definition is used as the basis for assessment of entailment methods.

Such a strategy can be understood both from the point of view of cognitive robotics, and from the point of view of common sense reasoning. In cognitive robotics terms, the underlying semantics is an account for the type of worlds (in the sense of IDS worlds) where the robot will be supposed to operate. The strategy says that we first develop reliable methods for relatively simple types of worlds, and then proceed gradually to more difficult ones, which is a fairly obvious way to proceed.

From the point of view of common sense reasoning, one may worry about the relevance of any proposed underlying semantics. 'Why is your underlying semantics better than mine?' The question becomes even more burning as the strategy proposes the use of a progression of underlying semantics: which of them corresponds to common sense?

Well, first of all, what is common-sense reasoning? It might appear that, by its very nature, this concept should not need a definition. Of course everyone has common sense (every researcher, at least), and indeed the tradition in AI has been that we can simply go ahead and formalize the common sense that we all share. The tables of ontological and epistemological characteristics can then be seen as checklists: certainly a system for common-sense reasoning should be able to handle dynamical systems with all the ontological characteristics, and certainly it should be able to deal with the kinds of lack of information that are described in the table of epistemological characteristics. In such a perspective, the checklists also remind us of the magnitude of the undertaking, since those logics which have been proposed (let alone analyzed) to date are only able to deal with small subsets of all the characteristics.

Since full common sense reasoning requires all the epistemological and ontological characteristics which were listed above, as well as many which have not been listed, and since only a small part of these are well understood, my proposal is to consider full common sense reasoning as the asymptote which is only approached in the limit. The chronicle families, and the multiple underlying semantics, that are defined and used in the course of the proposed strategy, should be seen as successively better approximations of a phenomenon whose complete characterization remains a very distant goal indeed.

This systematic strategy has a very distinct advantage over the 'heads-on' strategy of attempting to deal with all aspects of common-sense reasoning at once. The systematic strategy makes it possible to obtain concrete and reliable results along each step of the way, and to use the results from earlier steps as the basis of the work in later steps. The present book will give specific examples of this incremental methodology.

3.1.6 Syntax based assessment methodology

This book uses an underlying semantics as the starting point for the systematic methodology. There have also been other approaches which allow the formal verification of systematic results, that is, those which are not limited to single examples or to informal support. One particular approach was first proposed by Winslett [83], namely to use a corresponding formalization in monotonic logic in order to evaluate the correctness of a proposed non-monotonic formalism:

I do, however, have faith that a given situation can be laboriously but correctly encoded in monotonic situation calculus, and that the PMA can be tested for correctness by comparison with the monotonic encoding.

Lifschitz [43] and Lin and Shoham [45] have later performed such formal justifications for different theories of action. Lin and Shoham define a notion of epistemological completeness for theories of action, and show that their formulation is adequate relative to this criterion. The major difference between their approach and mine is that they express the completeness criterion in terms of logical formulae with a particular syntax, together with a syntactically defined extension of the theory which is then treated monotonically. The corresponding definition in my case is expressed in terms of the underlying semantics.

Lin and Shoham obtain that simple chronological minimization (called PCM in this book) is correct relative to the monotonic theory that they use as reference. This may seem surprising since PCM is fairly restrictive, according to my assessment. However, the results are in basic agreement since Lin and Shoham's positive results are based on a number of assumptions namely (1) the use of situation calculus, and, therefore, single-timestep actions, (2) the restriction to deterministic actions, and (3) the restriction to 'causal theories' not containing observations. The analysis below verifies that this case is included in the range of applicability for PCM[19].

It is my intention that the use of underlying semantics as the basis for assessment will provide more intuitively convincing correctness criteria, a clearer way of identifying and working with the ontological hierarchy, and a wider spectrum of ontologies for the worlds in which the scenarios take place.

Lifschitz in [43] analyzes a new and relatively complex theory of action. The main result is that under certain assumptions on the domain formalization, the use of the proposed circumscription policy has the same effect as adding certain premises which, in particular, imply the conventional frame

---

[19] On the other hand, their approach also allows certain cases of so-called ramification, namely, those cases where the action laws and the domain constraints together determine the developments in the world completely.

axioms. Lifschitz's work is also based on the situation calculus, and his method of analysis does not seem to be applicable to previously existing logics with simpler entailment criteria.

The syntactic approach to the assessment of non-monotonic approaches to actions and change, has the disadvantage that it may focus the attention too much on those phenomena which can be expressed in standard logic, albeit in cumbersome ways. In order to address a wide variety of expressiveness issues it may be better to first characterize the required expressiveness in terms of the underlying semantics and the ontological taxonomy. This provides a specification, and consequently the task is to find a logic that meets the specification.

3.1.7 Validation using a chronicle description language

Gelfond and Lifschitz have introduced a separate chronicle description language $\mathcal{A}$ with its own syntax and semantics, and used it to define the intended conclusions for reasoning about actions [26]. Kartha [36] applied the same criterion to the theories previously proposed by Baker [6], Pednault [55], and Reiter [63]. Baral and Gelfond [7] generalized $\mathcal{A}$ to the case of concurrent single-timestep actions and used it for validations as well. Some of the later work using this chronicle description language is closely related to the entailment methods that will be introduced in later chapters of the present book. In terms of the distinctions that were introduced in subsection 3.1.2, their work may be understood as the realization of a given entailment method and the validation of that realization. This question will be discussed further in Chapter 11, Section 11.8.

3.2 Choice of underlying semantics

The previous section took the perspective of common-sense reasoning, and motivated the need for an underlying semantics as a basis for a systematic methodology, avoiding the weakness of the example-based methodology. This section will define the underlying semantics that is used in the present work, and give the reasons for this particular choice.

3.2.1 Reasons for IDS world semantics

Although the underlying semantics is introduced in order to have a precise basis for the assessment of the range of applicability of various logics, it can also be motivated with other familiar reasons:

- In order to understand better what the formal concepts really mean. In particular, for many applications we would like to think of the IDS

material system as a part of the real, physical world. In this case the
IDS image level corresponds to the discretized view of the physical
world that is traditionally used in qualitative reasoning research, and
the IDS vehicle performs the 'perception' transformation from actual
reality to the discretized image. A formal characterization of the IDS
material system and the IDS vehicle will render these concepts more
precise than before.

- In order to resolve some important special cases in the formalism.
  Here is one example. Suppose the effects of actions are characterized
  in terms of preconditions and postconditions, as is customary. Now
  consider a development where an action is started at a timepoint
  where the preconditions are not satisfied. How will such a case be
  handled? There are several possibilities, for example: it may be
  considered an impossible case, or the action may be considered to
  have no effects at all, or it may be assumed that the effects of the
  action are completely unpredictable. A semantics for actions in terms
  of how an action affects processes in the IDS reality can help to resolve
  such questions.

3.2.2 Partial state-transition semantics

Since the IDS world is described in terms of states and actions, where an
action causes the state of the world to change in a particular way, the most
obvious choice of underlying semantics might be a *state-transition semantics*, where each action is represented as a mapping from states to states
or, for non-deterministic actions, as a function $\texttt{Rstat}(E, r)$ whose value is
the set of all the posible resulting states r' when the action E is applied
to the starting state r. However, that choice fails to capture regularities
in how the state is modified. In particular, the essential characteristic of
actions in an inert world is that the action modifies a small number of
state components, and all the others are left constant. This regularity is
not captured in an underlying semantics that describes actions in terms of
input and output states.

A *partial state-transition semantics* is the simplest way to capture the
inertia property. In this case, an action is represented as a transformation
from states to partial states. The resulting partial states characterize features (= state components) that have changed; those features which are
not present in a resulting partial state are supposed to be unchanged. This
is a trivial change from the action's point of view, of course, but it does
capture the inertia property which is so important for the logic.

Actually, we are going to introduce another modification as well, but let
us first work out the formal details for partial state-transition semantics.

It serves as an introduction to the next step. The following notation will be used. Following the **Z** notation [80] I use $\oplus$ for the override operation on functions, so that $r \oplus r'$ is defined as the function r^* defined by $r^*(f) = r'(f)$ if the latter is defined, and $r^*(f) = r(f)$ for all other f. An IDS world is represented as a function $\texttt{Rstat}(E, r)$ whose value is a set of partial states. We shall assume that all members of $\texttt{Rstat}(E, r)$ are defined over the same set of features, for a given choice of E and r. In other words, they are 'equally partial'. It is convenient to assume that an IDS world is characterized as a pair $\langle \texttt{Infl}, \texttt{Rstat} \rangle$, where $\texttt{Infl}$ indicates the range of the partial states. Therefore, $\texttt{Infl}(E, r)$ and $\texttt{Rstat}(E, r)$ are functions which take an action designator and a system state as arguments. $\texttt{Infl}(E, r)$ is the set of those features which may be affected by the action, and $\texttt{Rstat}(E, r)$ is a set of partial states assigning values to the features in $\texttt{Infl}(E, r)$, but not to other features.

Accordingly, the set of possible resulting states, when the action E is performed from the initial state r, is $\{r \oplus r' \mid r' \in \texttt{Rstat}(E, r)\}$.

In order to apply such a world description $\langle \texttt{Infl}, \texttt{Rstat} \rangle$ for a chronicle, one uses the ego-world game. The world **W** is then defined as a player against the ego with the following behavior. If the ego has initiated the action E at time s, then **W** performs the action from the present time-point to the next. The present finite development is therefore extended by adding $s + 1$ to its $\mathcal{B}$ component, and $R(s + 1)$ is defined as $R(s) \oplus r'$ for an arbitrary $r' \in \texttt{Rstat}(E, R(s))$. In other words, all features which are not in $\texttt{Infl}(E, R(s))$ must have the same value in $R(s)$ and in $R(s + 1)$. On the other hand, if the ego does not perform any action at the present time, then the world's move is again to increment time by 1, but to choose $R(s + 1) = R(s)$.

This partial state-transition semantics is in principle adequate for almost all current work in AI on logics of actions and change, although it needs to be modified in order to deal, for example, with ramification or with concurrency. Notice, by the way, that concurrency gives another reason for using partial result states rather than total ones: if two actions operate on disjoint sets of features, then one may assume that concurrent execution is possible and results in the union of their separate effects. In [74], Sandewall and Rönnquist proposed one such approach to concurrency, where $\texttt{Rstat}$ is modified so that its value is a set of *pairs* $\langle r'', r' \rangle$ of partial states. Here r' still characterizes the postcondition, whereas r'' characterizes a **prevail** condition which is required to hold while the action is being performed in order for r' to be the result.

3.2.3 Trajectory semantics

The partial state-transition semantics has a limitation which is very undesirable for our purpose: it does not characterize intermediate states during the execution of an action. Since we defined an action as something that is performed by the lower levels of the agent's architecture and which takes place in the IDS reality, it is important to allow actions that are characterized by successive states during a time interval. Single-timestep actions are a special case of this more general notion. The ego-world game has already been defined so that it allows actions with extended duration.

Accordingly, the *trajectory semantics* is obtained by the following generalization of the partial state-transition semantics. An IDS world **W** is characterized by a pair $\langle \text{Infl}, \text{Trajs} \rangle$, where $\text{Infl}(E, r)$ is a set of features as for partial state-transition semantics. The function $\text{Trajs}(E, r)$ replaces Rstat, and has as value a set of finite sequences of partial states assigning values to the features in $\text{Infl}(E, r)$, but not to other features. Every such sequence will be called a trajectory. The formal definition is as follows.

Definition. *A* **trajectory** *for a set $F \subseteq \mathcal{F}$ of features is a finite nonempty sequence of partial states, defined exactly over F. A* **world description** *is a pair $\langle \text{Infl}, \text{Trajs} \rangle$, where Infl is a function from $\mathcal{E} \times \mathcal{R}$ to subsets of $\mathcal{F}$, and Trajs is a function from $\mathcal{E} \times \mathcal{R}$, such that $\text{Trajs}(E, r)$ is a set of trajectories for $\text{Infl}(E, r)$.* ◁

For given E and r, let r' be the restriction of r to the features in $\text{Infl}(E, r)$, and let $\langle r'_1, ..., r'_k \rangle$, where $k \geq 1$ is a member of $\text{Trajs}(E, r)$. This trajectory represents an execution of the action of the form $[s, s+k]E$, where the IDS system at times $s, s+1, ..., s+k$ is in states $r, r_1, ..., r_k$, where each r_i is obtained from r'_i by choosing, for all features outside $\text{Infl}(E, r)$, the same feature-value as in r. In this way, the trajectory $\langle r', r'_1, ... \, r'_k \rangle$ specifies the successive change in the members of $\text{Infl}(E, r)$, and all other features must be kept fixed. Different choices of initial state r may lead to different execution sequences, and for each r one or more execution sequences is possible.

3.2.4 Trajectory-semantics worlds

The basic notions of trajectory semantics are very simple, since they only involve two generalizations of the simple state-transition semantics: the use of partial result states and the use of trajectories. A trajectory is a sequence of partial result states that are applicable at successive timepoints for an action with extended duration.

The purpose of this semantics is to define the set of intended models for a chronicle. This requires a number of additional formal definitions, in order to render the obvious intuitive notions on a form that can be used in the subsequent analysis.

I use the following auxiliary notation for extending a finite history with a trajectory. Let R be a finite history over $[0,t]$, and let $v = \langle r'_1, r'_2, ..., r'_k \rangle$ be a trajectory. Then, $R \triangleright v$ is the finite history R' over $[0, t+k]$ which is defined by $R'(s) = R(s)$ if $s \leq t$ and $R'(s) = R(t) \oplus r'_i$ if $s = t + i > t$. In particular, $R \triangleright \langle \emptyset \rangle$ extends R by one timestep from t to $t+1$ such that $R(t+1) = R(t)$, that is, it is the characteristic extension when inertia applies and no action is starting or in progress.

Definition. *An IDS world* **W** *is the* **corresponding world** *of a world description* $\langle \mathtt{Infl}, \mathtt{Trajs} \rangle$ *iff it satisfies the following conditions. If* $J = \langle \mathcal{B}, M, R, \mathcal{A}, \mathcal{C} \rangle$, $0 \in \mathcal{B}$, $n_B = max(\mathcal{B})$, *and* R *is defined over* $[0, n_B]$, *then* $\mathbf{W}(J, J')$ *iff* $J' = \langle \mathcal{B} \cup \{n'\}, M, R', \mathcal{A}', \emptyset \rangle$ *where either of the following holds:*

- $\mathcal{C} = \emptyset$, $n' = n_B + 1$, $R' = R \triangleright \langle \emptyset \rangle$, *and* $\mathcal{A}' = \mathcal{A}$.

- *For some action designator* E *and for some trajectory* v *which is a member of* $\mathtt{Trajs}(E, R(n_B))$ *and has the form* $\langle r'_1, r'_2, ..., r'_k \rangle$, *it holds that* $\langle n_B, E \rangle \in \mathcal{C}$, $n' = n_B + k$, $R' = R \triangleright v$, *and* $\mathcal{A}' = \mathcal{A} \cup \{\langle n_B, E, n' \rangle\}$.

A **trajectory-semantics world** *is an IDS world which corresponds to some world description.* ⋈

Chapter 6, subsection 6.6.3, will generalize this definition to the case of branching time. Chapter 10, subsection 10.2.3, will generalize it to also allow M to be a partial valuation and to allow the world to add bindings to M. For the time being these amendments are not needed.

Since **W** is required to be an IDS world, it is also constrained by the general requirements in the formal definition of IDS worlds in Chapter 1. The pair $\langle \mathtt{Infl}, \mathtt{Trajs} \rangle$ characterizes precisely an IDS world **W** which is able to participate in the 'game' between ego and world. Given a partial development where the 'now'-time is n, there are only two possibilities for **W**:

1. The agent has decided not to perform any action at this time, so the set $\mathcal{C}$ in the partial development $\langle \mathcal{B}, M, R, \mathcal{A}, \mathcal{C} \rangle$ is empty. In this case **W** increases n by one, and the state for the new timepoint equals the state for the previous point in time.

2. The agent has decided to perform an action E, so the set $\mathcal{C}$ has the member $\langle n, E \rangle$ where $n = n_B$. Then the move of the world **W** is to modify the partial development as follows:

- Select some member $\langle r'_1, r'_2, ..., r'_k \rangle$ of $\mathtt{Trajs}(E, r)$.

- Let $n' = n + k$ be the now-time of the updated finite development, and let $R(n+i) = R(n) \oplus r'_i$ for $1 \leq i \leq k$.

- Add $\{\langle n, E, n'\rangle\}$ to $\mathcal{A}$, to represent that the action has been completed at the new now-time n'.

- Reset $\mathcal{C}$ to the empty set.

If the ego invokes several actions at the same time, then the world chooses one of them non-deterministically and discards the others.

The special case where $\text{Trajs}(E, r) = \emptyset$ remains to be defined. It may be taken to mean that the action cannot be performed for the given combination of E and r, or that the effect of the action is undefined or unknown. This detail is made precise in Chapter 7, subsection 7.3.1.

Several world descriptions may correspond to the same world, and one will often prefer to use the 'minimal' world description. The following concepts make this precise.

Definition. *Two world descriptions $\langle \text{Infl}, \text{Trajs}\rangle$ and $\langle \text{Infl}', \text{Trajs}'\rangle$ are* **equivalent** *iff*
$$\{\langle r\rangle \triangleright v \mid v \in \text{Trajs}(E, r)\} = \{\langle r\rangle \triangleright v \mid v \in \text{Trajs}'(E, r)\},$$
for every E and r. ⋈

It follows from these definitions that equivalent world descriptions correspond to the same world. If a world description $\langle \text{Infl}, \text{Trajs}\rangle$ is given, and there is some $f \notin \text{Infl}(E, r)$ for some E and r, then an equivalent world description $\langle \text{Infl}', \text{Trajs}'\rangle$ can be constructed, where $\text{Infl}'(E, r) = \text{Infl}(E, r) \cup \{f\}$ and all trajectories in $\text{Trajs}(E, r)$ have been extended by explicitly specifying that the value of f is $r(f)$ in all elements of the trajectory. Thus, the new world description specifies explicitly that f is kept constant when this particular E is executed with this particular r as the starting state. Formally, we define this notion as follows.

Definition. *A world description $\langle \text{Infl}', \text{Trajs}'\rangle$ is a* **widening** *of a world description $\langle \text{Infl}, \text{Trajs}\rangle$ iff they are equivalent, and $\text{Infl}(E, r) \subseteq \text{Infl}'(E, r)$ for all E and r.* ⋈

It follows that in any equivalence set of world descriptions, there is one which is minimal with respect to widening. Minimal widening is obtained in the world description where, for every E and r and for every $f \in \text{Infl}(E, r)$, there is some trajectory $\langle r'_1, ..., r'_k\rangle$ in $\text{Trajs}(E, r)$ where the values $r[f]$, $r'_1[f], r'_2[f], ..., r'_k[f]$ are not all equal.

Definition. *A world description $\langle \text{Infl}, \text{Trajs}\rangle$ is* **uniform** *iff $\text{Infl}(E, r)$ is equal to $\text{Infl}(E, r')$ for all E, r, and r'.* ⋈

It follows that every world has some uniform description, for example, one where $\text{Infl}(E, r) = \mathcal{R}$ for all E and r, as well as a widening-minimal uniform description.

3.2.5 Trajectory-semantics egos

A trajectory-semantics world **W** was defined so that it will necessarily execute actions in sequence, even for an ego **K** which invokes several actions at the same time. If only sequential actions are going to be considered, one can impose restrictions on the ego, therefore, without any loss of generality. The following definition is chosen.

Definition. *A* **trajectory-semantics ego** *is an IDS ego* **K** *such that if*
$$\mathbf{K}(\langle \mathcal{B}, M, R, \mathcal{A}, \mathcal{C} \rangle) = \langle \mathcal{B}', M', R', \mathcal{A}', \mathcal{C}' \rangle$$
then $\mathcal{A} = \mathcal{A}'$, $\mathcal{C} \subseteq \mathcal{C}'$, *and* $\mathcal{C}' - \mathcal{C}$ *has at most one member.* ⋈

A trajectory-semantics ego is only able to invoke actions, therefore, and not activities, since it is not able to stop the execution of an occurrence. Also, it can only invoke one action at a time. The following observation follows immediately:

Proposition 3.1. *If an IDS* $\langle \mathbf{W}, \mathbf{K} \rangle$ *has a resulting development J and* **W** *is a trajectory-semantics world, then there is some trajectory-semantics ego* **K**′ *such that* $\langle \mathbf{W}, \mathbf{K}' \rangle$ *has J as a resulting development.*

The proof is trivial: for a given J one simply constructs an ego **K**′ that invokes the actions in J at the appropriate time if possible. This proposition allows us to simplify the definition of $Mod(\Upsilon)$ for a chronicle description Υ: without any restriction we can assume that all egos used in the definition are trajectory-semantics egos. This simplification is possible because $Mod(\Upsilon)$ was defined to be the set of developments that can be obtained in games between a specific and specified world defined by Υ and arbitrary egos, and that satisfy the conditions of the scenario description. According to Proposition 3.1 it is sufficient to consider games that have been played with trajectory-semantics egos.

3.2.6 Trajectory semantics and ontological families

Trajectory semantics is sufficient for the **IAD** ontological family, and in fact **IAD** will be defined in Chapter 7 as the class of those IDS worlds that are characterized by a trajectory semantics but with a minor restriction. Then, its subfamilies **IA** and **I** are defined by restrictions on **IAD**. It seems possible to make a generalized definition, still based on the notion of trajectories, that spans additional ontological characteristics such as **C** (concurrent actions) or **L** (delayed effects). Some of the other characteristics such as surprises (**S**) and normality (**N**) require extensions that are more or less orthogonal to the trajectory concept and its extensions. Several of these extensions will be addressed in Volume II.

Table 3.1. The world description $\langle \texttt{Infl}, \texttt{Trajs} \rangle$

Starting state r	$\texttt{Infl}(Fire, r)$	Member of $\texttt{Trajs}(Fire, r)$
$\{a:\text{T}, l:\text{T}, p:\text{T}\}$	$\{a, l\}$	$\langle \{a:\text{T}, l:\text{F}\}, \{a:\text{F}, l:\text{F}\} \rangle$
$\{a:\text{T}, l:\text{T}, p:\text{F}\}$	$\{a, l\}$	$\langle \{a:\text{T}, l:\text{F}\}, \{a:\text{F}, l:\text{F}\} \rangle$
$\{a:\text{T}, l:\text{F}, p:\text{T}\}$	$\emptyset$	$\langle \emptyset \rangle$
$\{a:\text{T}, l:\text{F}, p:\text{F}\}$	$\emptyset$	$\langle \emptyset \rangle$
$\{a:\text{F}, l:\text{T}, p:\text{T}\}$	$\{l\}$	$\langle \{l:\text{F}\} \rangle$
$\{a:\text{F}, l:\text{T}, p:\text{F}\}$	$\{l\}$	$\langle \{l:\text{F}\} \rangle$
$\{a:\text{F}, l:\text{F}, p:\text{T}\}$	$\emptyset$	$\langle \emptyset \rangle$
$\{a:\text{F}, l:\text{F}, p:\text{F}\}$	$\emptyset$	$\langle \emptyset \rangle$

3.3 Trajectory semantics in concrete terms

We shall illustrate the trajectory semantics by two examples, and then summarize the assumptions that are being made.

3.3.1 The rainy-day shooting example

Consider a world similar to the Yale shooting scenario (see Chapter 7) with three propositional features, namely a ('Fred is alive'), l ('the gun is loaded'), and p ('it is raining'). There is also one action $Fire$, which has the effect of unloading the gun, if it was loaded, and of killing Fred iff the gun was loaded. The rain does not affect these outcomes, and the rain itself is not affected. Furthermore, for the sake of the example, assume that in the case where Fred is alive and the gun is loaded the action takes two time units, where the gun becomes unloaded during the first timestep, and Fred dies in the second timestep. In all other cases it takes one timestep.

Under these assumptions the set $\texttt{Trajs}(Fire, r)$ will have exactly one member for each choice of r. Table 3.1 specifies, for each choice of r, the value of $\texttt{Infl}(Fire, r)$ and the single member of $\texttt{Trajs}(Fire, r)$. Notice that the set $\texttt{Infl}(Fire, r)$ has two members a and l in those cases where both the loadedness of the gun and Fred's aliveness are affected by the action. It is the empty set in all four cases where the gun was unloaded at the start of firing, since in that case no feature at all is affected. The chosen description is minimal with respect to widening.

Now consider the move of the world in a situation where the current finite development is
$$\langle \mathcal{B}, M, R, \mathcal{A}, \mathcal{C} \rangle$$
if $\mathcal{C} = \{\langle 16, Fire \rangle\}$, 16 is the largest member of $\mathcal{B}$, and
$R(16) = \{a:\text{T}, l:\text{T}, p:\text{F}\}$.
The world is to select one of the member trajectories in $\texttt{Trajs}(Fire, R(16))$ and it has only one to choose from, namely,

$\langle \{a:T, l:F\}, \{a:F, l:F\}\rangle$.
Since the length of this trajectory is 2, the number 18 will be added to $\mathcal{B}$. Also, the history R will be extended so that
$R(17) = \{a:T, l:F, p:F\}$,
$R(18) = \{a:F, l:F, p:F\}$,
where $R(17)$ and $R(18)$ have been obtained from $R(16)$ by overriding it with the first and second element of the trajectory respectively. The element $\mathcal{C}$ is reset to the empty set, and $\langle 16, Fire, 18\rangle$ is added to $\mathcal{A}$.

In some lines of this table, the trajectory which is the single member of $\texttt{Trajs}(Fire, r)$ is $\langle \emptyset \rangle$, meaning that if firing occurs at time n, then the current history of the world is extended by one timestep, and $R(n+1) = R(n)$ since $r \oplus \emptyset = r$.

3.3.2 The bus ride example

The following is an example where the trajectories are slightly less trivial. Consider a common-sense world where 'taking the bus home' is one of the actions. States in this world are constructed using the following features: Inbus for being inside the bus, Athome for being at home (or at least, at the bus stop near home), and Paid for having paid the bus fare. Each of the features has the truth-value T or F, in the obvious fashion. We abbreviate states in such a way that, for example, FTF means {Inbus:F, Athome:T, Paid:F}. The natural starting state for the action 'taking the bus home' is of course FFF, and normal trajectories for this action have to form
TFF, TFT, TFT, ..., TFT, TTT, FTT.
Such a trajectory represents the common-sense changes of first entering the bus (Inbus becomes true), then paying the fare (Paid becomes true), and then staying in the bus until Athome is true, and which point one causes Inbus to become false again. Different trajectories for this action differ with respect to the duration of the action, and probably also with respect to the time that is required for entering the bus, paying the fare, and leaving the bus.

In more realistic examples, the world state may contain a large number of features, and a number of them may be involved in an action. The set of trajectories is large in such cases. It is not the intention to enumerate the members of the trajectory set in an explicit fashion. Instead, the purpose of using a logic is to find a much more compact representation for expressing the information in trajectory sets.

3.4 Trajectories and the material system

Where do trajectories come from? In the perspective of common-sense reasoning, it is common to consider verbally expressed knowledge as the

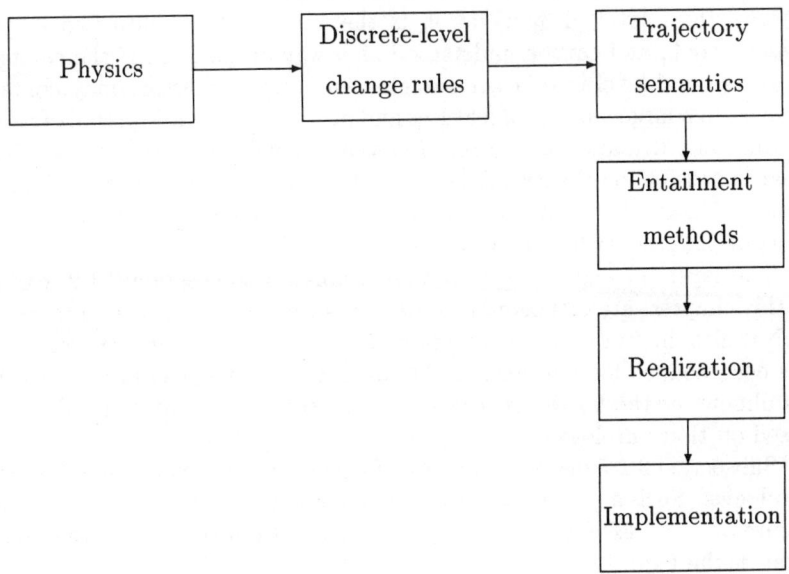

Fig. 3.2. Levels of aggregation and layers of expression

primary source of information. To the extent that this verbal knowledge can be expressed as logical formulae, the role of the trajectory semantics will be to provide a hypothetical semantical infrastructure: one chooses to assume that the logical formulae in question refer to developments in a trajectory-semantics IDS world. This choice of assumption provides a guideline for the development of an appropriate logic and reasoning system.

In the perspective of cognitive robotics, trajectories can instead be viewed as a compact description which in principle, although not necessarily in practice, has been derived from a more detailed description of the material system. Figure 3.2 describes my present view of how underlying semantics may be chosen and used. The three horizontally aligned boxes represent successive levels of aggregation, where the second level is in principle derived from the first one, and the third level from the second one. The primary level is offered by physics, and is as close to the world 'in itself' as we can get. The second level is already discrete in character and uses image-level states, but changes are expressed in a local and unstructured fashion. It may for example use low-level rules of cause and effect, which state that a change of value in one feature results, after a certain delay, in a consequent change of value in another feature.

The third level of aggregation is the one which is addressed in the present book, and can be understood as a way of 'packaging' the changes on the second level into larger parcels. An action trajectory may contain changes in a large number of features and over a considerable period of time. In this way, it allows one to reason about changes in larger chunks than what is possible on the second level. State-transition semantics (without a notion of metric time) may be considered as a fourth level or, alternatively, as a special case within the third level.

The layers of expression that were defined in subsection 3.1.2, earlier in this chapter, are orthogonal to the levels that have now been proposed. This is also illustrated in the diagram. Thus, once the trajectory semantics has been defined for the third level of aggregation, one may proceed to the entailment methods, the realizations, and the implementations that are based on that ontology.

Subsection 3.1.5 discussed the use of a progression towards more general ontologies. Such a progression runs orthogonally to the two dimensions in Figure 3.2, and extends in the third dimension from the box on the upper right in the figure.

The blank space in the diagram is not so blank in reality: the first and second level have, of course, their counterparts for logics, realizations, and implementations. The establishment of horizontal cross-connections in the lower layers of the diagram is likely to become important eventually.

3.5 Summary

An assessment of a non-monotonic method S is a proven statement to the effect that for every scenario description in a certain family $\mathcal{Z}$, the set of intended models $Mod(\Upsilon, \text{RM})$ equals $S(\Upsilon)$. In order to formulate and prove such assessments, it is necessary to have strict definitions of the function Mod and of the family $\mathcal{Z}$. Different S may be assessed using different $\mathcal{Z}$, so Mod should be defined over a broader domain, containing (as far as possible) the various ranges of applicability that one is likely to encounter.

The present volume will use the trajectory semantics, which is expressive enough to characterize the ontological family **IAD** that was introduced in Chapter 1. The trajectory semantics has a completely precise definition from a formal point of view. On the other hand, the interpretation of this semantics in a broader context depends on whether one adopts a perspective of common-sense reasoning or of cognitive robotics.

4
Elementary feature logic and meta-logical concepts

We proceed now to the introduction of the base logic, using a combination of two perspectives. One perspective is from the point of view of inhabited dynamical systems: given the definitions of Chapters 1–3, what logic can properly express scenario descriptions and obtain the intended conclusions according to the underlying semantics? The other perspective is in relation to previous research on common-sense reasoning about actions and time: how does the present material build on and extend what has been published before? I am going to alternate between those two perspectives, both in the present chapter and in those that follow.

This chapter will introduce, as a first step, a simple logic called EFL, for elementary feature logic. EFL is 'propositional', in the sense that it does not allow any object domain $\mathcal{O}$ which was used in developments, and it does not account for time. Objects and time will be introduced by two successive generalizations, in Chapters 5 and 6. The present chapter will also define meta-logical concepts and notation, and a framework for non-monotonic logics that can be used in later chapters.

4.1 Logical domains

Conventional logic is 'flat': there is one language of logic formulae, and deduction takes one unstructured set of premises and obtains a set of conclusions. However, knowledge representation tasks have often more structure than that. Their additional structure should, in the end, benefit the theory as well as the computational performance.

I shall use several specific and non-standard techniques which rely on the structure of the knowledge representation application. One technique which has already been mentioned is the use of premise partitions distinguishing, for example, between action laws, schedules, and observations, and the use of those partitions in the entailment methods.

4.1.1 Object domain vs. value domains

Another technique being used here is a distinction between *value domains* and *object domains*. The object domain consists of the elements or components of the IDS system; value domains contain those items which are used for characterizing objects. The set of real numbers is a value domain. If objects are characterized in terms of color then a set of color names {Red, Green, Blue, ...} is another value domain, and so on.

Typically, an object domain is specific for a scenario, whereas the same value domain often applies throughout a given world. The use of value domains can be understood in the context of image-level descriptions of the IDS: as a perception function transforms a quantitative material-level description of the system to a qualitative description on the image level, it is often appropriate to use features with more than two possible values for the features in the image.

Some value domains, such as the colors, are just a set of possible values without additional structure. In the 'goldfish hotel' example, fish could have a position feature with (codes for) the various fish bowls as the value domain. However, there are also domains which have considerable structure, such as the real numbers, the domain of time intervals, and domains with a graph structure. Since these domains are usually quite well understood and good computational methods have been developed for them, there is no point in reducing them to an axiomatization in the logic.

For these reasons, I will make a distinction beetween a domain of 'objects', which may vary between different models, and domains of 'feature values', which are specified in the vocabulary (= signature) that is chosen for a particular usage of the logic. Consequently, there will be no need for axioms characterizing the functions on value domains.

Time will be introduced in Chapter 6, and will have the same formal character as a value domain, although it is treated differently syntactically, and has a different ontological status.

4.1.2 Lexical object domains

Yet another technique will be to use *lexical domains* for objects in the description of specific systems or scenarios. This comes naturally from the IDS point of view, since each chronicle specifies its object domain explicitly, and thereby constrains the set of intended models with respect to their object domains. However, the use of lexical object domains is also a sensible move from the common-sense reasoning point of view, and I shall discuss it as such here.

Domain closure

Most application examples that have been considered in the KR literature about logics for action and change use a fixed, known, and finite object domain. This assumption is usually formulated in logical terms as a 'unique names hypothesis' and a 'domain closure assumption'. Formally, one introduces one constant symbol a_i for each object, and uses premise schemas characterizing the known domain, namely a *domain closure premise* [60] of the form
$$\forall x[x = a_1 \vee x = a_2 \vee ... x = a_n]$$
and a *unique name premise* as the conjunction for each i and j where $i \neq j$ of
$$a_i \neq a_j.$$
These schemas are only to be used for scenario-specific reasoning, not for obtaining conclusions that hold in all scenarios of a given world.

This standard approach is inherently uneconomical: it means that one first introduces full predicate logic, and then constrains it to a special case which is intertranslatable with propositional logic. The lexical domain approach is a way of building the known-domain assumption into the syntax and the semantics of the logic itself.

Object names

The following approach will be used. The logic will contain a special kind of symbol for objects called *object names*, together with the customary constant symbols and variable symbols. Object names are unique; there is a one-to-one correspondence between them and the objects themselves, and formally they may be considered as being the objects themselves. The set of all the object names is then the object domain. However, unlike standard predicate calculus semantics where the object domain is part of each interpretation, the lexical object domain consisting of object names is associated with the similarity type. Interpretations in a lexical domain logic specify only the denotation of functions and relations in the object domain, but not the domain itself.

From a conventional (KR) point of view, one may think of object names as a special kind of constant symbols, which are chosen in such a way that axioms for 'closed worlds' and for 'unique names' apply over the set of object names, but not for other constant symbols. Furthermore, since this has the effect of fixing the object domain modulo the equality relation to the same size as the set of object names, I choose for convenience to identify objects and object names. Finally the term 'constant symbol' will only be used for the other constant symbols besides the object names.

For example, if one uses the object names #1, #2, and #3 and the (other)

object constants o_1 and o_2, then in all models
$$\#1 \neq \#2 \wedge \#1 \neq \#3 \wedge \#2 \neq \#3,$$
and
$$\forall o[o = \#1 \vee o = \#2 \vee o = \#3],$$
from which it follows, for example, that
$$o_2 = \#1 \vee o_2 = \#2 \vee o_2 = \#3.$$

Applicability

How wide is the applicability of the known-domain assumption and lexical-domain semantics? Many classical AI toy and non-toy scenarios have been formulated for a fixed number of objects, so lexical object domains are sufficient for their description. See, for example, [62] or [27]. Also, and perhaps more importantly, the first reasoning task for an intelligent autonomous agent is to reason about the objects that it 'sees' around itself. The existence and the identity of objects is then determined by relatively low-level, sensory processes, and the reasoning system can take the set of current objects as given. More advanced reasoning systems must of course also be able to reason about an unlimited number of other objects which are not seen by the agent. However, that may come as a next step. The use of lexical domains is, therefore, a very reasonable restriction from a cognitive robotics point of view.

If the object domain and the value domains are finite, then it follows immediately that each predicate logic with lexical domains can be translated to a propositional logic. From an abstract point of view such lexical-domain logics thus have no independent interest. However, from an application point of view the situation is reversed. It is often stated that propositional logic is not expressive enough for knowledge representation tasks (see, for example, [8][20] or [16][21]), but lexical-domain predicate logic does in fact suffice for most of the scenarios that are commonly addressed in AI research.

4.2 Elementary feature logic

4.2.1 Introduction

The language of elementary feature logic (EFL) contains logic formulae such as
$$f_1 \hat{=} \mathrm{GY} \rightarrow f_2 \hat{=} \mathrm{D},$$

---

[20] Page 163.
[21] Page 33.

Elementary feature logic

with the intended reading 'if f_1 has the value G (for green) or Y (for yellow), then f_2 has the value D (for drive)'. Thus, the logic formulae are constructed using the usual propositional connectives, from atomic formulae of the form
$$f \triangleq \mathcal{X},$$
where f is a *feature symbol* and $\mathcal{X}$ is a subset of the value domain for the feature f [22]. For each feature f there is a feature domain with a finite number of members, for example G, Y, and R. The symbol $\triangleq$ connects a feature and its explicitly specified value, or possible range of values. A variable or a composite expression is not allowed on the right side of this symbol. The reason for allowing multiple items in the right-hand side of such expressions is in order to allow multiple outcomes of an action, as was already discussed in Chapter 2, subsection 2.4.2.

Clearly, by a simple transformation one can translate such a feature logic into a corresponding propositional logic, simply by mapping each feature symbol into one or more proposition symbols depending on the size of the corresponding value domain[23]. However, EFL is a useful starting point for the presentation, for the following reasons:

- In many cases it is a convenient representation of an application.

- The generalization to temporal feature logic, or discrete-valued fluents, is quite useful. In particular, it is very appropriate for 'planning' in the AI sense of the word.

- The use of multi-valued features give a significant advantage in the analysis of the complexity of algorithms for planning and temporal prediction [4].

- In another direction than is taken in the present book, EFL can be generalized in a natural way into a terminological language.

Some of these generalizations will follow in later chapters, along with the generalization to admit an object domain also. The syntactic and other machinery from the present chapter will then be used as a starting point.

4.2.2 Value domains and syntax

First it is necessary to define the value structure that is to be used in the feature logic formulae, along the lines of subsection 4.1.1.

A *similarity type* for EFL is a mapping[24]

---

[22]The key distinction is between 'feature symbol' and corresponding 'feature value'. The word 'feature' itself will be assigned a precise meaning in Chapter 5, subsection 5.2.2. Every feature symbol in EFL is a feature.

[23]Conversely, any propositional logic can also be translated to EFL by viewing each proposition symbol as a feature symbol with the domain {T, F}.

[24]See Appendix B for notation.

$$\{f_1 : \mathcal{D}_1, f_2 : \mathcal{D}_2, ..., f_n : \mathcal{D}_n\},$$
where $n \geq 1$, each of $f_1, f_2,...$ is a *feature symbol*, and each of $\mathcal{D}_1, \mathcal{D}_2,..., \mathcal{D}_n$ is a corresponding non-empty *value domain* consisting of a finite number of elements called *items*. In other words, the value domains are integrated into the similarity type. Items will be written as single capital letters, for example G, or occasionally as a capitalized word, like in Green. A set of items will often be written in abbreviated form by immediate concatenation of the items, for example GY is the same as $\{G, Y\}$. Thus, even in the formula $f \hateq D$, the expression to the right of the $\hateq$ sign actually represents $\{D\}$.

The feature domains are assumed to be disjoint, and if the same letter is used in several of the domains it is still assumed to represent several different items. If necessary the distinction can be made explicit by calling the items in domain $\mathcal{D}_i$ for A_i, B_i, etc.

If $\mathcal{X}$ is a set of feature items, that is, a subset of $\mathcal{D}_i$ for some feature domain $\mathcal{D}_i$, then $\overline{\mathcal{X}}$ is the complement set of feature items, $\mathcal{D}_i - \mathcal{X}$. The empty set of feature items will be denoted $\emptyset$ in logical formulae.

An *elementary formula* in EFL with similarity type σ is an expression of the form
$$f_i \hateq \mathcal{X},$$
where f_i is a feature symbol defined in σ, and $\mathcal{X}$ is a subset of the domain assigned to f_i by σ.

A *logic formula* in EFL with similarity type σ is either an elementary formula as just defined, or it is composed from logic formulae using the logical connectives T, F, $\neg$, $\wedge$, $\vee$, $\rightarrow$, or $\leftrightarrow$. Of these, T and F have zero arguments, $\neg$ has one argument, and the others are infix operators with two arguments, as usual.

Feature symbols will usually be chosen as f_i for some i, except when the domain consists of the truth-values $\{T_i, F_i\}$, in which case the feature symbol will be chosen as p_i for some i. In the latter case the formula
$$p_i \hateq T_i$$
will be abbreviated simply as p_i.

Example. The following formula
$$f_1 \hateq GY \rightarrow f_2 \hateq D \wedge p_3$$
or, equivalently,
$$f_1 \hateq G_1 Y_1 \rightarrow f_2 \hateq D_2 \wedge p_3$$
is correctly formed with respect to the similarity type
$$\{f_1 : \{G_1, R_1, Y_1\}, f_2 : \{D_2, S_2\}, p_3 : \{T_3, F_3\}\},$$
where the indices on the feature items can be omitted since no ambiguity can arise. ⋈

4.2.3 Semantics

Let
$$\sigma = \{f_1 : \mathcal{D}_1, f_2 : \mathcal{D}_2, ..., f_n : \mathcal{D}_n\}$$
be a similarity type for EFL. An *interpretation* for this similarity type is an assignment
$$I = \{f_1 : d_1, f_2 : d_2, ..., f_n : d_n\},$$
where $d_k \in \mathcal{D}_k$ for each k.

Let σ be a similarity type for EFL, let α and β be logic formulae for σ, and let I be an interpretation for σ. The value of α in I is written $val[\alpha, I]$, and is defined in the natural fashion as either T or F as follows:

$val[f_i \hat{=} \mathcal{X}, I] = $ T iff $I[f_i] \in \mathcal{X}$,

$val[\text{T}, I] = \text{T}$,

$val[\text{F}, I] = \text{F}$,

$val[\neg\alpha, I] = neg(val[\alpha, I])$,

$val[\alpha \wedge \beta, I] = min(val[\alpha, I], val[\beta, I])$,

$val[\alpha \vee \beta, I] = max(val[\alpha, I], val[\beta, I])$,

$val[\alpha \rightarrow \beta, I] = max(neg(val[\alpha, I]), val[\beta, I])$,

$val[\alpha \leftrightarrow \beta, I] = $ T iff $val[\alpha, I] = val[\beta, I]$,

where $neg(\text{T}) = \text{F}$, $neg(\text{F}) = \text{T}$, and max and min are defined with respect to the ordering $\prec$ over $\{\text{T}, \text{F}\}$ where $\text{F} \prec \text{T}$.

It follows that a formula of the form
$$f \hat{=} \emptyset$$
represents a contradiction for any fluent symbol f. It also follows that the value of $\neg p$ is always the same as the value of $p \hat{=} \text{F}$.

The symbols T and F will therefore be used both as (parts of) formulae and as values of formulae. A similar policy will be adopted for object names in the next chapter. A third use of the symbols T and F arises when the indices are omitted from the members T_i and F_i of a propositional feature domain.

4.2.4 Some axioms

The following schemata for logical axioms[25] are evident for arbitrary f and corresponding $\mathcal{X}$ and $\mathcal{Y}$:

$f \hat{=} \mathcal{Y} \leftrightarrow f \hat{=} \mathcal{X} \vee f \hat{=} \mathcal{X}'$, where $\mathcal{Y} = \mathcal{X} \cup \mathcal{X}'$,

$f \hat{=} \mathcal{Y} \leftrightarrow f \hat{=} \mathcal{X} \wedge f \hat{=} \mathcal{X}'$, where $\mathcal{Y} = \mathcal{X} \cap \mathcal{X}'$,

$f \hat{=} \mathcal{Y} \leftrightarrow \neg(f \hat{=} \mathcal{X})$, where $\mathcal{Y} = \overline{\mathcal{X}}$,

---

[25] The logical axioms are of relatively marginal importance since we use semantics but disregard inference rules and proofs. They are included mostly as a set of corollaries of the defined semantics.

$f \stackrel{\scriptscriptstyle\triangle}{=} \mathcal{X}$, where $\mathcal{X} = \sigma[f]$ for the current similarity type σ,
$\neg (f \stackrel{\scriptscriptstyle\triangle}{=} \emptyset)$,
$f \stackrel{\scriptscriptstyle\triangle}{=} \mathcal{X} \to f \stackrel{\scriptscriptstyle\triangle}{=} \mathcal{X}'$, provided that $\mathcal{X} \subseteq \mathcal{X}'$.

Notice that the set $\mathcal{X}$ must be a fixed set of items each time an axiom is used, so the formulae above are axiom schemata rather than specific axioms. The language of EFL does not allow variables or constant symbols denoting items or item sets.

4.3 Meta-level concepts in the base logics

A non-monotonic logic is often obtained by starting with a monotonic base logic and adding a preference relation on models, or by some other means of modifying the set of models for a given set of premises. The following meta-level notions will be used for the base logics. Some of these notions are analogous to the function *Mod* and the entailment relations $\models$, $\cong$, and $\Vdash\!\diamond$ which were defined over scenario descriptions in Chapter 2, subsection 2.4.1.

4.3.1 Meta-level domains and the model set

The set of logic formulae for a given similarity type σ will be written $\mathcal{L}_\sigma$. The set of interpretations for σ will be written $\mathcal{S}_\sigma$. The set of features for σ will be written $\mathcal{F}_\sigma$.

In the context of a defined similarity type σ, the *model set* $[\![\alpha]\!]$ for a given formula $\alpha \in \mathcal{L}_\sigma$ is a subset of $\mathcal{S}_\sigma$, and is defined as usual as $\{I \in \mathcal{S}_\sigma \mid val[\alpha, I] = \mathsf{T}\}$. If Γ is a set of formulae for σ then $[\![\Gamma]\!]$ is defined by

$$[\![\Gamma]\!] = \bigcap_{\alpha \in \Gamma} [\![\alpha]\!],$$

as usual. If ζ is a function whose argument is a set of interpretations, then $\zeta[\![\Gamma]\!]$ will be an abbreviation for $\zeta([\![\Gamma]\!])$. In particular, $\zeta[\![\Gamma]\!] \cap W$ means $\zeta([\![\Gamma]\!]) \cap W$.

The notation $[\![\Gamma]\!]$ will have to be modified slightly in the next chapter when models also contain an object domain.

4.3.2 Quine quotes

I use Quine quotes [58] $\ulcorner ... \urcorner$ for quoting formulae. These quotes are transparent for meta-variables, which are often written as lower-case Greek letters[26].

---

[26] Lower-case Latin letters with a numerical index are always object-level symbols, for example p_1. Lower-case Latin letters with a variable index or prime, and lower-case Greek letters, are always meta-level symbols, for example p_i, p', and α. Single lower-case Latin letters, for example p, may be either, depending on context.

For example, if $\alpha = \ulcorner p \wedge q \urcorner$, then $\ulcorner \alpha \to s \urcorner = \ulcorner p \wedge q \to s \urcorner$.

If a function or relation takes quoted formulae as its argument in at least one argument position, it is a meta-function or meta-relation. In such cases the round parentheses that usually surround the list of arguments, can be replaced by square brackets and the Quine quotes can be omitted. For example, $val[p, s]$ is an abbreviation for $val(\ulcorner p \urcorner, s)$. The context defines which of the arguments is implicitly Quine-quoted. This convention has already been used in the definition of *val* earlier in this chapter.

The infix operation $\equiv$ is equality with implicit Quine-quoting on both arguments. For example, $\alpha = \ulcorner p \wedge q \urcorner$ can equivalently be written as $\alpha \equiv p \wedge q$. Note that $\ulcorner \alpha \urcorner$ is the same as α, since Quine quotes around a single metavariable are redundant.

Similarly, a maplet (an element of a mapping) which is written $p : v$ is an abbreviation for $\ulcorner p \urcorner \mapsto v$. For example, $\{p_1 : T, p_2 : F\}$ is an interpretation which could also be written as $\{\ulcorner p_1 \urcorner \mapsto T, \ulcorner p_2 \urcorner \mapsto F\}$ and which maps the symbol $\ulcorner p_1 \urcorner$ to true and $\ulcorner p_2 \urcorner$ to false.

Application of functions inside Quine quotes will be indicated by underlining that function. For example, a part of the definition of substitution in a formula could be

$$subst[\alpha \wedge \beta, x, \tau] = \ulcorner \underline{subst}[\alpha, x, \tau] \wedge \underline{subst}[\beta, x, \tau] \urcorner,$$

saying that in order to substitute τ for x in $\alpha \wedge \beta$, one should substitute separately in α and in β and form the conjunction of the results.

Quine quotes will often be omitted when their meaning is obvious. In particular, in a set or sequence of logic formulae they can be omitted, since there would be little interest in the set (sequence) of the truth-values of the participating formulae.

4.3.3 Meta-level operations

If $\Gamma = \{\gamma_1, \gamma_2, ..., \gamma_k\}$ then $\bigwedge \Gamma \equiv \gamma_1 \wedge \gamma_2 \wedge ... \wedge \gamma_k$. (This operation is written $Cnj(\Gamma)$ by some authors). The case of $k = 0$ is defined by $\bigwedge \emptyset \equiv T$.

4.3.4 Abbreviations

Definitions of syntactic abbreviations will be written using $\equiv$. For example, if $\to$ were not part of the core syntax, it could have been introduced as an abbreviation defined by

$\alpha \to \beta \equiv \neg \alpha \vee \beta$.

4.3.5 Logical entailment relations

If W is a set of interpretations then I write
$W \models \alpha$ iff α is true in every member of W,
$W \mathop{|\diamond} \alpha$ iff α is true in some member of W.

The former notation is standard; the symbol $|\diamond$ should be read 'is consistent with'. If Γ is a set of formulae then 'Γ entails α' can therefore be written as
$$[\![\Gamma]\!] \models \alpha$$
and will be abbreviated to the usual notation
$$\Gamma \models \alpha.$$
This is of course equivalent to $[\![\Gamma]\!] \subseteq [\![\alpha]\!]$. The symbol $|\diamond$ is generalized in the analogous fashion.

If α and β are formulae then $\alpha \simeq \beta$ means $[\![\alpha]\!] = [\![\beta]\!]$, that is, each of them entails the other.

The following are some simple examples of the use of these symbols. Just as $\Gamma \models F$ reads 'Γ is inconsistent', so $\Gamma \mid \diamond T$ reads 'Γ is consistent'. Therefore,
$$\Gamma \mid \diamond T \iff \Gamma \not\models F.$$
Also, just as
$$\Gamma, \neg \alpha \models F \iff \Gamma \models \alpha,$$
so
$$\Gamma, \alpha \mid \diamond T \iff \Gamma \mid \diamond \alpha.$$
Furthermore,
$$\Gamma, \neg \alpha \models \beta \iff \Gamma, \neg \beta \models \alpha$$
and
$$\Gamma, \alpha \mid \diamond \beta \iff \Gamma, \beta \mid \diamond \alpha.$$

4.4 Reduction of model sets

In the context of logics for action and change, there will be a use for the following two aspects of non-monotonic logics:

- *Reduction of the model set.* If the set of intended models is a subset of the set of classical models for a given set of premises, then techniques are needed for reducing the model set to the intended one.

- *Recovery of the model set.* In some situations one obtains a model set which fails to include some of the intended models, and in particular one may obtain an empty model set even when some models were intended. We shall be particularly concerned with the case where an operator for reduction of the model set, as in the first item, reduces the model set to nil. In such cases one needs an operation that recovers a non-empty model set and obtains the intended one.

The present volume will only be concerned with the former problem. Recovery of model sets is needed for dealing with qualification and surprises, which are topics for Volume II.

In the context of classical logic, if a scenario is given and a proposed axiomatization of that scenario suffers because the set of actual models is a superset of the set of intended models, then obviously there are some facts that are part of one's intentions but which have not been properly included in the axiomatization. The natural way of proceeding, then, is to add the missing premises to the axiomatization, so that actual models and intended models agree. Problems arise, however, if the omitted facts can not easily be expressed in the logic at hand. This is exactly the case for reasoning about actions and change in worlds with strict inertia. The statement that there are no changes in the world, besides those implied by the action laws, is not directly expressible in a first-order logic where the chronicle itself has been represented in a straight-forward way.

A drastical solution for that problem is to abandon first-order logic, but then one loses the reportoire of methods and results that are associated with that standard. The basic idea with non-monotonic logics, from the present perspective, is that they offer alternative ways of reducing the model set so that it agrees with the intended one. The description of entailment methods in Chapter 2, subsection 2.6.3, captures the approach: the scenario description is divided into partitions, and one uses a range of syntactic and semantic operations on partitions and model sets in order to obtain the intended models.

In this section we shall briefly discuss how major kinds of non-monotonic logics fit into the perspective that has just been described.

4.4.1 Selection functions

Many non-monotonic logics can be characterized by selection functions of the following kinds:

Definition. *A* **selection function** *is a function ζ from sets of interpretations to sets of interpretations, which satisfies $\zeta(W) \subseteq W$. Selective entailment $\models_\zeta$ is defined so that $W \models_\zeta \alpha$ iff $\zeta(W) \models \alpha$. Similarly, if $\Gamma \subseteq \mathcal{L}_\sigma$, $\Gamma \models_\zeta \alpha$ iff $\zeta(\llbracket \Gamma \rrbracket) \models \alpha$.* ⋈

The subscript ζ on $\models$ will be omitted whenever it is clear by context.

The main purpose of a selection function is to reduce the classical model set of a given set of premises. This is a well-known technique in non-monotonic logics, where it allows one to obtain conclusions that would otherwise not have resulted. Non-monotonicity of the set of conclusions follows since there is no requirement for ζ itself to be monotone. Often one chooses a selection function of the form $\zeta(W) = Min(\ll, W)$, that is, the set of members of W which are minimal with respect to the preference relation $\ll$.

From the IDS point of view, model selection is required since we have

observed that
$$Mod(\langle \mathcal{K}, \mathcal{O}, \langle \mathbf{IA}, \text{LAW} \rangle, \text{SCD}, \text{OBS} \rangle) \subseteq Mod(\langle \mathcal{N}, \mathcal{O}, \text{LAW} \cup \text{SCD} \cup \text{OBS} \rangle)$$
with a strict $\subset$ relation in most cases. Some kind of model selection is needed for reducing the set on the right-hand side to the set on the left-hand side.

If ζ is a model reduction function and W is a set of models, then the members of $W - \zeta(W)$ are called the *removed* models, and the members of $\zeta(W)$ the *retained* models with respect to ζ.

4.4.2 Alternatives to selection functions

The use of selection functions is not the only possible way to achieve model reduction. The following are the main alternatives:

- Axiomatic methods: add extra, chronicle-independent premises which express the ontological and epistemological assumptions. This method is the most explicit and declarative one, and as such is closest to the spirit of logic, but for many of the purposes addressed in this book it is impossible or unknown how axiomatic methods could be used.

- Syntactic methods, using a premise integration function which maps a set of formulae Γ to a modified set Γ' such that $\Gamma' \models \Gamma$. The classical model set of Γ' is used as the reduced model set for Γ. Circumscription as originally defined is an example of a syntactic method for model reduction.

- Semantic methods, using a selection function which maps any set of interpretations to one of its subsets.

Both the use of selection functions and premise integration methods can be used for the purpose of chronicle completion.

4.4.3 Semantic selection and selection functions

Selection functions as defined above or similar notions have previously been used by Rychlik [65], Bell [9], Lindström [46], and Rott [64]. Since sets of models correspond directly to theories, that is, sets of formulae which are closed under logical inference[27], selection functions represent a straightforward reformulation of the C operator for non-monotonic consequence as used, for example, by Makinson [47]. In particular, the characteristic properties of the C operator on theories, such as cumulativity and cautious monotonicity, can be routinely transferred to selection functions ζ on model sets. I prefer the formulation in terms of models rather than theories since

---

[27] At least in the propositional case.

it relates more easily to preference relations on models, and to the set of resulting developments $Mod(\Upsilon)$ for a given scenario description.

There are a few standard ways of defining such selection functions:

Preference relations

Let $\ll$ be a strict partial order over interpretations. The selection function ζ *corresponding to* $\ll$ is defined such that $\zeta(W)$ is the set of $\ll$-minimal members of W, that is,
$$\zeta(W) = Min(\ll, W) = \{I \in W \mid \neg \exists I'[I' \in W \land I' \ll I]\}.$$
I write $W \mathrel{\approx_\ll} \alpha$ iff $W \mathrel{\approx_\zeta} \alpha$ for the selection function ζ corresponding to $\ll$, and similarly for $\Gamma \mathrel{\approx_\ll} \alpha$.

The specification of preference relations can in turn be performed in a number of ways:

- for the logic as a whole, for example by specifying that a certain abnormality predicate Ab is always to be minimized, as in classical circumscription [39, 11];

- specific for the application area and specified outside the logic, for example the use of chronological minimization of change [38] or chronological maximization of ignorance [78];

- specific for the particular usage of the logic and specified using logical formulae, for example using an explicit modal operator for defaults [19, 22, 21].

Only the second one of these alternatives will be used in the present work.

Default inference relations

A *default inference relation* (d.i.r.) δ is a binary relation on sets of interpretations such that if $\delta(W, W')$ then $W' \subseteq W$. Informally, this relation says that if W is our current set of models, that is, one of the members of W encodes the actual truth, then it is probably found in the subset W'. However, unlike a selection function ζ, the d.i.r. does not proceed directly to the fully reduced set of models, it just performs one step of inference.

Let $W = W_0, W_1, W_2,...$ be a finite or infinite sequence of sets of interpretations such that $\delta(W_i, W_{i+1})$ for each i. The last element or the limit $\lim_{i \to \infty} W_i$ is called an *extension candidate* of W. An extension candidate for W which is minimal, that is, no other extension candidate is a subset of it, is called an *extension* of W. If every W has a unique extension for δ, then the selection function *corresponding to* δ is the function which maps every W to its extension.

Default inference relations are customarily defined using *default rules* [68, 61]. A default rule written as

$$\frac{\alpha \;:\beta}{\gamma}$$

defines a d.i.r. δ such that $\delta(W, W')$ iff $W \models \alpha$, $W \mathrel{|\!\diamond} \beta$, and $W' = W \cap [\![\gamma]\!]$. A set of default rules defines a d.i.r. formed as the union of the d.i.r's of each of the rules.

There are also other ways of defining a selection function from a default inference relation, for example as the intersection of all non-empty extensions, or by using a fixpoint definition [61]. One may also wish to require from δ that it does not have any empty extension. Methods for non-monotonic temporal reasoning based on default logic and formulated in terms of d.i.r. will be addressed in Volume II.

Semantic filters

A *filter* is simply a predicate $F(I)$ on interpretations. The selection function ζ *corresponding to* a filter is defined by $\zeta(W) = \{I \in W \mid F(I)\}$.

Multiple extensions

If a default inference relation has several extensions then the definition of the corresponding selection function is not entirely obvious. The same problem also arises in some other approaches. In order to deal uniformly with those problems, it is convenient to define an *extension relation Z* as a binary relation on sets of interpretations, usually chosen such that if $Z(W, W')$ then $W' \subseteq W$. W' is called an *extension of W* (for Z) iff $Z(W, W')$. If δ is a d.i.r., then the *corresponding* extension relation Z is defined so that $Z(W, W')$ iff W' is an extension of W according to δ, that is, it is obtained from W by repeated application of δ, and is minimal.

The *skeptical selection function* ζ corresponding to an extension relation Z is then defined as the union of all W' such that $Z(W, W')$.

Maximal consistency selection

A set Δ of formulae used as *default formulae* defines an extension relation Z where $Z(W, W')$ iff $W' = W \cap [\![D]\!]$, where D is a maximal subset of Δ for which $W \cap [\![D]\!] \neq \emptyset$. The corresponding skeptical selection function is defined as above.

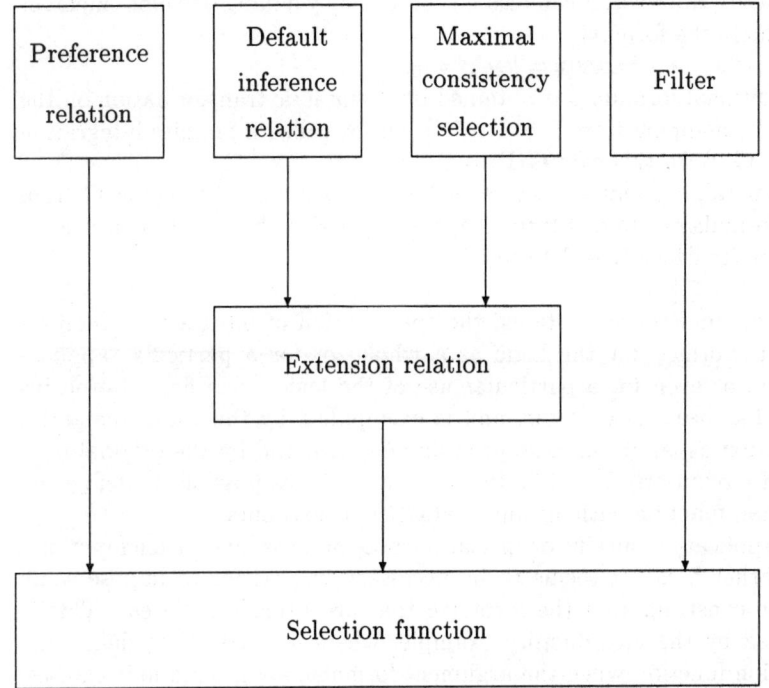

Fig. 4.1. Mechanisms for defining non-monotonic logics

Overview

Figure 4.1 contains an overview of how the different concepts that have been defined in this subsection build on each other. Many varieties of non-monotonic logics can be expressed in these terms. Notice that even default logic fits in nicely. The following section describes a method that can be understood only on the syntactical level.

4.4.4 Syntactic methods and premise integration

Suppose it is known that the objects a, b, and c all have the property P. This knowledge can be expressed as follows in standard logic, provided that ⌜a⌝, ⌜b⌝, and ⌜c⌝ are understood as constant symbols, and ⌜P⌝ as a predicate symbol:

$$\{P(a), P(b), P(c)\}.$$

Let Γ be this set of formulae. The additional knowledge that these are the only objects having the property P can be expressed by model selection,

for example using a preference relation that minimizes the extent of the predicate P. However, it can also be expressed by adding one more member to Γ namely the formula
$$\forall x[P(x) \rightarrow x = a \lor x = b \lor x = c].$$
This additional formula β is obtained by a syntactic transformation on the set Γ. The mapping from Γ to $\Gamma \cup \{\beta\}$ will be called a premise integration function $\mathbf{G}$, so in this case $\mathbf{G}(\Gamma) = \Gamma \cup \{\beta\}$.

In general, a *premise integration function* is a partial function $\mathbf{G}$ from sets of formulae to sets of formulae, which satisfies the case that if Γ is an argument for which it is defined, then
$$[\mathbf{G}(\Gamma)] \subseteq [\Gamma].$$
Just as for preference relations, the specification of integration functions can be performed for the logic as a whole, or for a particular application area, or even for a particular use of the logic. The first alternative is what has been used so far, and is exemplified by the Clark semantics for negation-as-failure in logic programming [14] and by the original definition of circumscription [48]. In those cases it was possible to define the integration function without any syntactical constraints.

For application-specific or special-purpose premise integration functions, on the other hand, it seems to be necessary in practice to impose some syntactic constraints on the formulae that are being integrated. This is illustrated by the introductory example, where it is trivial to define the integration function when the argument formulae are ground unit clauses, but hardly possible if arbitrary first-order formulae are allowed.

The selected set of models for a given set Γ of formulae is then defined as $[\mathbf{G}(\Gamma)]$. There is a possibility but naturally no guarantee of monotonicity: $\mathbf{G}$ is said to be *monotonic* iff for arbitrary Γ and Γ', $\Gamma \subseteq \Gamma'$ implies that $[\mathbf{G}(\Gamma')] \subseteq [\mathbf{G}(\Gamma)]$.

An integration function $\mathbf{G}$ is said to be *syntax neutral* iff it satisfies the condition that for all Γ and Γ' where it is defined,
$$[\Gamma] = [\Gamma'] \rightarrow [\mathbf{G}(\Gamma)] = [\mathbf{G}(\Gamma')].$$
Syntax-neutral integration functions can in some cases have a counterpart on the semantic level, but in general there does not seem to be any general way of transforming premise integration to a semantic formulation, in particular because of the syntactic constraints.

Axiom schemata

Premise integration is not to be confused with the use of *axiom schemata* such as the schema for the equality relation in standard first-order logic. This schema includes
$$\forall x \forall y[x = y \rightarrow f(x) = f(y)]$$

for every function f of one argument, and similarly for functions of several arguments and for relations. The set of formulae that are generated by an axiom schema depends only on the present similarity type, and not on the present application-dependent premises. Unlike premise integration, an axiom schema does not have to be reconsidered if additional premises are added describing more knowledge about the application, and an axiom schema cannot be a source of non-monotonicity. Both premise integration and axiom schemata will be used in this book.

4.5 Summary

This chapter has defined elementary feature logic as a first step towards the base logic for expressing statements about scenario descriptions. Meta-level concepts and notation, and concepts for non-monotonic logics have been introduced.

5
Lexical-domain object-feature logic

We now proceed along the lines defined in Chapter 4, subsection 4.1.2 and introduce a lexical object domain alongside, but distinct from, the value domains for the features. This extension allows the logic to refer to object domains in the $\mathcal{O}$ component of scenario descriptions. I will refer to the resulting logic as *lexical-domain object-feature logic*, or LFL. Section 5.4 will outline how a terminological language can be defined on top of the LFL semantics, and therefore indicates a bridge to KL-ONE-type languages.

LFL does not contain a notion of time, so it can only be used for describing static or momentary phenomena. However, within LFL we define a concept of state which will be used extensively to represent the 'snapshots' or momentary states in later chapters.

5.1 Ontology

5.1.1 Object domains

No distinction is made in this ontology between different types of objects within one world. The object domain will be written $\mathcal{O}$, as in IDS developments, and members of the object domain will be chosen as #1, #2, etc. These members will be variably referred to as *objects* and as *object names*, so no formal distinction is made between the object and the object's unique name. In particular, an object name such as #3 can occur in a formula, but at the same time a term in the logic language can have an object (object name) as value. Since there is one-to-one correspondence between objects and object names, the equality relation is defined so that $\#i = \#j$ iff $i = j$.

This identification of entities in the language and values of formulae is contrary to established practice in logic, but it is harmless and quite convenient for our purposes. The dual use of the symbols T and F as components of formulae and as truth-values is another manifestation of the same viewpoint.

The universe of all possible objects $\{\#1, \#2, ...\}$ will be written Ω. Thus in all cases the object domain $\mathcal{O}$ will be chosen as a subset of Ω. From the IDS perspective, Ω will be the universe of objects from the point of view of the IDS egos and worlds. The finite developments which are generated by egos and worlds use subsets of Ω, and each scenario description selects a particular subset $\mathcal{O}$ of Ω.

One might consider the extension to a more expressive logic, including multiple object domains and the use of object types. Although potentially very useful, such extensions will not be considered here.

5.1.2 Features

Each feature symbol will be characterized by its *arity* ≥ 0 and its *value domain* ($=$ co-domain). The value domain may be either a finite set of feature values, as in EFL, or the domain $\mathcal{O}$ of objects. An n-ary feature symbol names a function from $\mathcal{O}^n$ to the value domain of the feature symbol.

Features serve the roles of both functions and relations in predicate logic, therefore. Otherwise put, we confine the logic to a small number of relations such as equality, and define those relations once and for all. All world-specific and scenario-specific concepts are expressed using features, which from the point of view of predicate logic are functions with specific value domains.

It is straightforward to embed a terminological language in this ontology. Classes of objects or 'concepts' can be represented by features of one argument with the binary value domain $\{T, F\}$. Binary relations between objects ('roles' in terminological logic) are represented as features of two arguments with the same binary value domain. This representation would also facilitate a later generalization to fuzzy concepts and fuzzy valued roles.

Partial functions are formally not allowed, but can be obtained for object-valued features by adding one more object #0 to the object domain for representing the unknown, and an unknown item ¶ to each domain of feature-values.

5.2 Basic definitions

5.2.1 Object domains, similarity types, and vocabularies

Each chronicle description identifies a specific choice of object domain $\mathcal{O}$, so reasoning about the chronicle can be done in the context of a fixed domain. However, one may also wish to reason about an IDS world in general and keep the choice of object domain open. Therefore, we need formal constructs that can accommodate both of those cases, which I shall refer to as *lexical-domain* and *free-domain* semantics respectively.

Note that only the lexical-domain case is computationally crucial. The computer-based agent needs to reason mechanically about specific scenarios, whereas reasoning about a world in general is not so urgently needed to have on line.

An *object domain* $\mathcal{O}$ is a finite non-empty set of *objects*, whose members will be written #1, #2,..., #m, where m $= |\mathcal{O}|$.

A *similarity type* σ is a mapping
$$\{f_1^{i_1}: \mathcal{D}_1, f_2^{i_2}: \mathcal{D}_2, ..., f_n^{i_n}: \mathcal{D}_n\},$$
where $i_k \geq 0$ for each k, and each $f_k^{i_k}$ is a feature symbol for a feature with i_k arguments. Here $\mathcal{D}_k$ is either a corresponding non-empty and finite domain for feature-values, or the integer 1 indicating that the feature has objects as values. (In a possible generalization to multiple object domains, they can be called $\mathcal{O}_1$, $\mathcal{O}_2$, $\mathcal{O}_3$, and so on, and can be indicated by the corresponding number in the similarity type).

Let σ be a similarity type. A *vocabulary* for σ is a pair $\langle \sigma, \mathcal{O} \rangle$, where $\mathcal{O}$ is an object domain.

The *k-th domain* of a vocabulary ν is written as $Dom_k(\nu)$ or $Dom_k(\sigma, \mathcal{O})$, and is defined as $\mathcal{D}_k$ in σ if that is a set, and as $\mathcal{O}$ if $\mathcal{D}_k$ is the number 1.

An *interpretation*, which is defined below, assigns values to each combination of a feature symbol and its arguments. A *structure* will be defined as a pair $\langle \mathcal{O}, I \rangle$, where I is an interpretation. These definitions accommodate both lexical domain logics using a vocabulary $\langle \sigma, \mathcal{O} \rangle$ and an interpretation I, and free domain logics using a similarity type σ and a structure $\langle \mathcal{O}, I \rangle$.

5.2.2 Syntax

Let $\nu = \langle \sigma, \mathcal{O} \rangle$ be a vocabulary as defined above.

An *object expression* is either an object name (member of $\mathcal{O}$, for example #4), an object constant symbol, or an object variable. Object constants will be written o_k and object variables as o_k for positive integers k. One may think of object names as a particular kind of object constant, where two different objects cannot have the same name and every object has a name.

A *feature expression* is either of the following:
f_k^0,
$f_k^i(\omega_1, \omega_2, ..., \omega_i)$,
where all ω with an index are object expressions, and the f with an index are feature symbols in σ. The *domain* of such a feature expression is $Dom_k(\nu)$. The superscript for arity will be omitted when it can be derived from context, or when it is irrelevant.

An *elementary formula* is either of the following:
$f \hat{=} \mathcal{X}$,
$\omega = \omega'$,

where f is a feature expression, $\mathcal{X}$ is a subset of the domain of f, and ω and ω' are object expressions. Note that $\mathcal{X}$ may be either a set of items or a set of objects. In the latter case it may be written by direct concatentation of the object names, for example #4#6#7 is a set of three objects[28].

A *logic formula* is defined by finite composition of elementary formulae using logical connectives as in the previous chapter, or as $\forall o[\phi]$ or $\exists o[\phi]$, where o is an object variable and ϕ is a logic formula.

A feature expression is *rigid* iff it is either f_k^0, or $f_k^i(\omega_1, \omega_2, ..., \omega_i)$, where all arguments are object names. For example, $f_3^2(\#2, \#4)$ is a rigid feature expression, but $f_3^2(o_2, o_4)$ is not. Rigid feature expressions will also be referred to as *features* for short.

An object constant plays the same roles in LFL as a feature of zero arguments with an object value. However, constants and features will be treated differently in the next chapter, where time is introduced into the notation, since features will then be changeable over time, and constants will not.

5.2.3 Denotations

For the purpose of the semantics, for each feature expression f we define its *denotation* $den[f, s]$ in a structure $s = \langle \mathcal{O}, I \rangle$ as a rigid feature expression, that is, a feature. For example, in a context where the value of the object constant o_2 is #6, the denotation of the feature expression $f_3^1(o_2)$ is the rigid feature expression $f_3^1(\#6)$. The domain of rigid feature expressions for $\nu = \langle \sigma, \mathcal{O} \rangle$ will be written $\mathcal{F}_\nu$.

In general, for a given interpretation and a given formula we distinguish two different operations, namely obtaining the *value* of the formula and obtaining the *denotation* of the formula. The value is an object or a feature-value (including a truth-value) as usual. The denotation is a modified formula where the top-level operator remains, but each argument has been replaced by its corresponding value, that is, by a rigid expression.

Interpretations are defined so that one of their components is a mapping from denotations to corresponding values. The definitions of value and denotation are interdependent, therefore: the calculation of a denotation sometimes requires calculation of a value, and vice versa.

For example, the value of the feature expression $f_3^1(o_2)$ in an interpretation $\langle M, r \rangle$ is obtained as follows:

---

[28]Several aspects of this syntax are motivated by the forthcoming generalization to a temporal logic in Chapter 6. The reason for the construct $f \stackrel{.}{=} D$ is because it will be matched by assignment statements where f changes its value at a point in time. The syntax does not allow nested terms, such as $f(f'(\omega))$.

- Identify the denotation of $f_3^1(o_2)$. To do so, first evaluate the object symbol o_2.

- The value of o_2 is obtained as $M[o_2]$, where M is a part of the interpretation at hand, and was assumed to be #6.

- The denotation of $f_3^1(o_2)$ is the expression $f_3^1(\#6)$, therefore.

- The value of $f_3^1(o_2)$ is obtained as $r[f_3^1(\#6)]$, where r is the other component of the interpretation at hand, and is a mapping from denotations to corresponding values.

The real use of denotations arises when temporal logic is introduced in Chapter 6. Then, interpretations will specify the mapping from denotations to corresponding values as a function of time. Denotations will also be convenient for defining the set of changes from one state to another. For the sake of uniformity the denotations are introduced already here.

Notice that the use of denotations for defining the semantics has no relationship to the use of Herbrand models, except in the superficial sense that the value or denotation of a formula is another formula.

5.2.4 Valuation of formulae

Let $\nu = \langle \sigma, \mathcal{O} \rangle$ be a vocabulary as defined above. A *feature assignment* or *state* (synonyms) for ν is a mapping from rigid feature expressions in ν, to members (items or objects) in their respective domains. Feature assignments will be denoted by the letter r. The domain $\mathcal{R}$ of image-level states that was introduced in Chapter 1, subsections 1.1.2 and 1.3.2, is now identified with the domain of feature assignments.

Example. Let
$$\nu = \langle \{f_1^0 : \{\text{A}, \text{B}\}, f_2^1 : \{\text{C}, \text{D}, \text{E}\}, f_3^1 : 1\}, \{\#1, \#2, \#3\} \rangle$$
be the vocabulary. The following is an example of a feature assignment for ν:
$\{f_1^0 : \text{A}, \quad f_2^1(\#1) : \text{D}, \quad f_2^1(\#2) : \text{C}, \quad f_2^1(\#3) : \text{D},$

$f_3^1(\#1) : \#2, \quad f_3^1(\#2) : \#1, \quad f_3^1(\#3) : \#3\}.$

If this feature assignment is called r, we would accordingly write, for example, $r[f_2^1(\#1)] = \text{D}$ on the meta-level. Therefore, the LFL formula $f_2^1(\#1) \doteq \text{D}$ has the value T in r.

Remember that the maplet[29] f_1^0 : A is the same as $\ulcorner f_1^0 \urcorner \mapsto$ A, and $r[f_2^1(\#1)]$ abbreviates the Quine-quote-using expression $r(\ulcorner f_2^1(\#1) \urcorner)$. ⋈

---

[29] Maplets are the elements that mappings are made of.

Basic definitions 105

A *valuation* for ν is a mapping from object constant symbols to $\mathcal{O}$. An *interpretation* for ν is a pair
$$\langle M, r \rangle,$$
where M is a valuation and r is a feature assignment for ν. Interpretations will be denoted by the letter I.

Let σ be a similarity type. A *structure* for σ is a pair $s = \langle \mathcal{O}, I \rangle$, where $\mathcal{O}$ is a finite non-empty set of objects, and I is an interpretation for $\langle \sigma, \mathcal{O} \rangle$. Structures will be denoted by the letter s.

The *value* of a closed logic formula[30] α is defined for a given structure $s = \langle \mathcal{O}, I \rangle$ in order to accommodate both lexical and free semantics, and is defined using the following steps:

1. Each feature expression f has a *denotation* $den[f, s]$, which is a rigid feature expression, that is, a feature.

2. Each object expression ω has a *value* $val[\omega, s]$, which is an object name, that is, a member of $\mathcal{O}$.

3. Each closed logic formula α has a value $val[\alpha, s]$, which is defined to be either T or F.

The functions *val* and *den* are defined as follows, using $s = \langle \mathcal{O}, \langle M, r \rangle \rangle$.

If ω is an object name, then $val[\omega, s] = \omega$.

If ω is an object constant symbol, then $val[\omega, s] = M[\omega]$.

The denotation of a feature expression is defined by
$$den[f_k^0, s] = f_k^0,$$
$$den[f_k^i(\omega_1, \omega_2, ..., \omega_i), s] = \ulcorner f_k^i(\underline{val}[\omega_1, s], \underline{val}[\omega_2, s], ..., \underline{val}[\omega_i, s])\urcorner.$$

The value of a variable free elementary logic formula is either T or F and is defined by
$$val[f \hat{=} \mathcal{X}, s] = \text{T iff } r(den[f, s]) \in \mathcal{X},$$
$$val[\omega_k = \omega_j, s] = \text{T iff } val[\omega_k, s] = val[\omega_j, s].$$

The value of a logic formula obtained using propositional connectives is defined as in Chapter 4, subsection 4.2.3, and for quantified formulae as follows:
$$val[\forall o[\alpha], s] = \min_{n \in \mathcal{O}} val[\alpha_o^n, s],$$
$$val[\exists o[\alpha], s] = \max_{n \in \mathcal{O}} val[\alpha_o^n, s],$$

where α_o^n is the result of substituting n for o throughout α (that is, for all free occurrences of o in α), and *min* and *max* are defined as in Chapter 4, subsection 4.2.3. The value of a single variable, or of any other expression containing free variables, is not defined.

Note that $f \hat{=} \mathcal{X}$ means neither $f = \mathcal{X}$ nor $f \in \mathcal{X}$. The left-side argument of $\hat{=}$ is used intensionally. For example, to evaluate $f_1^2(\#1) \hat{=} D$ one

---
[30] That is, not containing any free variables.

first determines the denotation of $f_1^2(\#1)$, which is itself, and then checks whether r of that is D.

5.2.5 Model sets and entailment

Lexical-domain semantics is the main consideration, and uses the following definition of the model set. The *lexical model set* $[\![\Gamma]\!]_\mathcal{O}$ is the set of interpretations I for ν such that $\langle \mathcal{O}, I \rangle$ satisfies Γ, that is,

$$[\![\Gamma]\!]_\mathcal{O} = \{I \mid val[\alpha, \langle \mathcal{O}, I \rangle] = \mathrm{T} \text{ for all } \alpha \text{ in } \Gamma\}.$$

The *free-domain model set* $[\![\Gamma]\!]_f$ for a set of formulae Γ is defined quite conventionally by

$$[\![\Gamma]\!]_f = \{s \mid val[\alpha, s] = \mathrm{T} \text{ for all } \alpha \text{ in } \Gamma\},$$

where s ranges over all structures for σ of the form $\langle \mathcal{O}, I \rangle$ with arbitrary $\mathcal{O}$. The subscript f stands for 'free' and is therefore a fixed symbol, not a variable. $\mathcal{O}$ will often be omitted when it is clear from the context. Notice that $[\![\Gamma]\!]_\mathcal{O}$ and $[\![\Gamma]\!]_f$ are two different ways of adapting the definition of $[\![\Gamma]\!]$ from Chapter 4, subsection 4.3.1, when an object domain is present.

The axioms in subsection 4.2.4 continue to apply, as well as the standard equality axioms.

5.3 An example

This section contains an example of the LFL syntax and the use of the lexical domain for reasoning about a specific scenario. The reasoning exercise will involve object identification, which was defined in Chapter 2, subsection 2.4.3, but with the following modifications. Omitting the temporal aspect, object identification can be phrased as follows in LFL terms. For a given set Γ of LFL logic formulae, a given $\mathcal{O}$, and a given formula $P(o)$ characterizing a property for objects o ranging over $\mathcal{O}$, the task is to identify a member #i of $\mathcal{O}$ such that

$$\Gamma \models P(\#\mathrm{i}),$$

using lexical-domain semantics. This problem is of course equivalent to the problem of identifying an object #i for which

$$\Gamma, (o = \#\mathrm{i}) \models P(o),$$

where o is a constant symbol that does not occur elsewhere in Γ or P. If $\Gamma \models \exists o[P(o)]$, then that problem can be replaced by the problem of identifying a member #i of $\mathcal{O}$, such that

$$\Gamma, P(o) \models (o = \#\mathrm{i}).$$

It is seen at once that any solution to the latter problem is a solution to the former one, if $P(o)$ for some o, and that the two problems are equivalent iff $P(o)$ for exactly one o when Γ holds.

5.3.1 The assignment scenario

Consider the following instance of a classical kind of quiz:

Problem: There is a doctor, a lawyer, and a psychiatrist (one of each). They live in a red house, a yellow house, and a brown house (in some order, one of each). Likewise, they drive a Saab, a Volvo, and a BMW. The doctor lives in the red house. The man in the yellow house drives the Volvo. The psychiatrist drives the BMW. What is the occupation of the Volvo driver?

Vocabulary: The features are called Occ, Hou, and Car, respectively, and the feature-values will be chosen as the respective first letters (R for red house, S for Saab, etc). The similarity type is, therefore,
$$\sigma = \{Occ : \{D, L, P\}, Hou : \{R, Y, B\}, Car : \{S, V, B\}\}.$$
The object domain $\mathcal{O}$ is chosen as $\{\#1, \#2, \#3\}$.

Laws: The following premise schema may be used in each scenario for this type of quiz, and instantiates for every feature symbol f and for every $\xi \in \sigma[f]$:

    1: $\exists o[f(o) \hat{=} \xi] \land \forall o \forall o'[f(o) = f(o') \rightarrow o = o']$,

that is, there is exactly one house of each color, exactly one person of each profession, etc.

Scenario description: The objects are assigned arbitrarily so that the following premises apply:

    2: $Occ(\#1) \hat{=} D$
    3: $Occ(\#2) \hat{=} L$
    4: $Occ(\#3) \hat{=} P$

Also, the object constants o_1 through o_6 are introduced for 'the man in the red house', ..., 'the man who drives the BMW', giving the following premises:

    5: $Hou(o_1) \hat{=} R$
    6: $Hou(o_2) \hat{=} Y$
    7: $Hou(o_3) \hat{=} B$
    8: $Car(o_4) \hat{=} S$
    9: $Car(o_5) \hat{=} V$
    10: $Car(o_6) \hat{=} B$

The given facts can then be encoded as follows:

    11: $Hou(\#1) \hat{=} R$
    12: $Car(o_2) \hat{=} V$
    13: $Car(\#3) \hat{=} B$

Table 5.1. Successive conclusions for quiz

Line	Conclusion	Support
14	$o_5 = o_2$	1,9,12
15	$o_6 =$ #3	1,10,13
16	$o_5 \neq o_6$	9,10
17	$o_1 =$ #1	1,5,11
18	$o_2 \neq o_1$	5,6
19	$\neg(o_5 \hat{=}$#1#3$)$	14–18
20	$o_5 \hat{=}$#2	Domain

Note that B for brown and B for BMW are different items, as discussed before.

Problem statement: The problem is to find some $\xi \in \sigma[Occ]$ such that
$\Gamma \models \exists o[Car(o) \hat{=} \text{V} \land Occ(o) \hat{=} \xi]$.

Preliminary transformation of the problem: The natural approach is to first solve the object identification problem
$\Gamma \models Car(\text{\#i}) \hat{=} \text{V}$,
and, having obtained the solution #i, to identify the ξ that satisfies
$\Gamma \models Occ(\text{\#i}) \hat{=} \xi$.
In view of premise schema 1, there is exactly one #i and one ξ for the given Γ. Therefore, the first problem is replaced by the object identification problem
$\Gamma, (Car(o) \hat{=} \text{V}) \models (o = \text{\#i})$.
Since Γ contains $Car(o_5) \hat{=} \text{V}$ it follows using premise 1 that
$\Gamma, (Car(o) \hat{=} \text{V}) \models (o = o_5)$,
and the problem can again be equivalently rewritten as
$\Gamma \models (o_5 = \text{\#i})$,
and it is now on a form that is suitable for direct deduction.

Solution: The problem is solved by object identification for $o_5 =$ #i, after which $Occ(o_5)$ can be determined immediately. One obtains the conclusions shown in Table 5.1. The solution is obtained as $Occ(\text{\#2})$.

Answer: The lawyer drives the Volvo.

Notice the crucial role of axiom (1) and of the knowledge of the exact object domain in this deduction. Although this information may be expressed using unique-names and closed-world premises, it is more reasonable to let it be a built-in property of the logic and of the reasoning algorithm.

5.3.2 The elementary codesignation algorithm

The type of deduction that was performed in the last stage of the example arises in several ways in AI, for example in the Waltz algorithm for constraint propagation [8], and in the 'binding' of action arguments in the TWEAK algorithm for knowledge based planning [13]. This illustrates the importance of lexical object domains. Although a problem statement in predicate logic with finite lexical domains can always be translated into propositional logic, such a transformation does not necessarily make representational or computational sense. Reasoning operations and algorithms can make effective use of the structure that is inherent in lexical domains. This is significant if one considers the choice between the use of lexical domains or explicit premises for domain closure and unique names. A possible argument for the latter approach is that the problem is reduced to standard logic, but that argument is weakened if the direct computational use of the additional premises (for example in a theorem prover) is computationally senseless.

The computational topic is strictly speaking outside the scope of this book. It is also a very large one, in particular considering the present interest in finite-domain systems in logical programming and elsewhere. I shall anyway illustrate the computational usefulness of lexical domains by outlining a simple, obvious, and well-known algorithm for object identification that makes direct use of the known lexical domain and of the special role played by the feature-value relation $\doteq$ in LFL. It can be understood as a simplified Waltz algorithm. The algorithm, which I refer to as a *codesignation algorithm*, makes systematic use of equality and inequality relationships between object expressions, in particular object names and object constants as in the example.

The strategy in the codesignation algorithm is to partition the set of all object expressions into equivalence classes for expressions denoting the same object. Inferred equalities and inequalities allow one to merge equivalence classes, and to keep track of the classes that cannot be merged. This method depends essentially on the existence of a fixed lexical domain. The algorithm uses only the types of formulae allowed in LFL, and proceeds as follows.

Start with one separate line for each object expression. In the course of the algorithm, each line will contain a set of several object expressions. There is an operation of merging lines, meaning that a new line is formed containing the union set /f the expressions on two previous lines, and the old lines are discarded.

Whenever, for two object constants o_j and o_k which appear on different lines, it can be deduced that $o_j = o_k$, then merge the lines where they are located.

Fig. 5.1. Initial table

		#1	#2	#3
Doctor	#1		×	×
Lawyer	#2	×		×
Psychiatrist	#3	×	×	
Red	o1			
Yellow	o2			
Brown	o3			
Saab	o4			
Volvo	o5			
BMW	o6			

Also, have one column for each object name. Whenever it can be deduced that $o_k \neq \#i$, then cross out the column for #i in the line where o_k occurs.

When two lines are merged, then the column crossings of both the input lines are retained. Figures 5.1 through 5.4 illustrate the method for the quiz example.

Clearly, this codesignation analysis requires that there is some process which is able to use the current codesignation table and to generate additional equalities and inequalities between object expressions. In the example above, the one-to-one mapping between objects and feature-values provided the basis for that process.

If all columns but one have been crossed out in a given line, then the object of the one remaining column can be merged with the present line. It is then the only possible value for the present set of expressions. This is how knowledge of the fixed domain is used in the algorithm.

If the process stops, which occurs when no additional line merges are possible, then the reasoning process has to branch into several alternatives. This is the same kind of branching as in tableaux. Take a line which has a minimal number of uncrossed columns, and generate one branch for each uncrossed column.

Fig. 5.2. After three steps of deduction

			#1	#2	#3
Doctor	#1			×	×
Lawyer	#2		×		×
Psychiatrist, BMW	#3	o6	×	×	
Red	o1				
Yellow, Volvo	o2	o5			×
Brown	o3				
Saab	o4				

Fig. 5.3. After five steps of deduction

			#1	#2	#3
Doctor, Red	#1	o1		×	×
Lawyer	#2		×		×
Psychiatrist, BMW	#3	o6	×	×	
Yellow, Volvo	o2	o5	×		×
Brown	o3				
Saab	o4				

Fig. 5.4. After all six steps of deduction

			#1	#2	#3
Doctor, Red	#1	o1		×	×
Lawyer, Yellow, Volvo	#2	o2,o5	×		×
Psychiatrist, BMW	#3	o6	×	×	
Brown		o3			
Saab		o4			

5.4 Set-level approach

Terms in the lexical object-feature logic LFL characterize individual objects in the object domain. There is also an alternative approach where terms refer to sets of objects, and which is actually a variant of terminological languages. The special case with only unary, item-valued features is reviewed here for comparison. The more general case, also using 'roles', would be possible but more complicated, and is too remote from the present context. I will call this a **set-level** approach, as opposed to LFL which is a ground-level approach. (The term 'object level' would have been misleading.)

5.4.1 Syntax and semantics

Let $\nu = \langle \sigma, \mathcal{O} \rangle$ be a vocabulary where all features are item-valued and unary. A *set expression* is either of the following:

- $f \div \mathcal{X}$ where f is a feature symbol and $\mathcal{X}$ is a subset of the feature domain for f. It denotes the set of all o, members of $\mathcal{O}$, such that $f(o)$ is a member of $\mathcal{X}$.

- $\emptyset$ representing the empty set (of objects).

- $a \cup b$, $a \cap b$, $\bar{a}$, and $a - b$, where a and b are set expressions. These expressions have their usual meaning in set theory. The complement is taken relative to $\mathcal{O}$.

Thus if $I = \langle M, r \rangle$ is an interpretation for ν, then the value of $f \div \mathcal{X}$ in I is $\{o \mid r[f(o)] \in \mathcal{X}\}$.

A *set relationship* is either of the following expressions, where a and b are set expressions: $a = b$, $a \subset b$, $a \subseteq b$, $a \Updownarrow b$. These expressions have their usual meanings; $a \Updownarrow b$ says that a and b are disjoint sets. (The last symbol is chosen to suggest 'not overlap' or 'empty intersection').

A *logic formula* is either a set relationship, or is formed from set relationships by composition using the propositional connectives.

Note that the value of a set-level formula is defined for one particular object domain $\mathcal{O}$ and one interpretation. The semantics is extensional since the value of an expression formed using $\div$ is simply the set of all objects in $\mathcal{O}$ having the property. The literature on terminological languages generally uses an extensional semantics, although it is often also suggested that an intensional semantics for terminological expressions would be appropriate and useful.

5.4.2 Axioms

The following axiom schemata are evident, besides the axioms of set theory:

$f^1 \div \mathcal{Y} = (f^1 \div \mathcal{X}) \cup (f^1 \div \mathcal{X}')$, where $\mathcal{Y} = \mathcal{X} \cup \mathcal{X}'$,

$f^1 \div \mathcal{Y} = (f^1 \div \mathcal{X}) \cap (f^1 \div \mathcal{X}')$, where $\mathcal{Y} = \mathcal{X} \cap \mathcal{X}'$,

$\overline{f^1 \div \mathcal{X}} = f^1 \div \mathcal{Y}$, where $\mathcal{Y} = \overline{\mathcal{X}}$,

$f^1 \div \mathcal{X} \subseteq f^1 \div \mathcal{X}'$, provided that $\mathcal{X} \subseteq \mathcal{X}'$,

$f^1 \div \mathcal{X} \Updownarrow f^1 \div \mathcal{X}'$, provided that $\mathcal{X} \Updownarrow \mathcal{X}'$,

$f^1 \div \emptyset = \emptyset$.

Notice that as usual $\mathcal{X}$ must be explicitly specified in each formula: no variables or constants are admitted for feature-values. The present ontology assumes a concept hierarchy without exceptions, and a monotonic inheritance regime.

5.4.3 The assignment scenario revisited

The following is the same example as above:

Problem repeated: There is a doctor, a lawyer, and a psychiatrist (one of each). They live in a red house, a yellow house, and a brown house (in some order, one of each). Likewise, they drive a Saab, a Volvo, and a BMW. The doctor lives in the red house. The man in the yellow house drives the Volvo. The psychiatrist drives the BMW. What is the occupation of the Volvo driver?

Scenario description: The same vocabulary as before is used. The given facts are then formulated as follows:
1: $Occ \div D = Hou \div R$
2: $Hou \div Y = Car \div V$
3: $Occ \div P = Car \div B$

The given question is formulated as
Q: $Car \div V \subseteq Occ \div ?$

Solution: The following derivation is easily obtained using the axioms above, and ordinary set theory:
4: $Occ \div LP = Hou \div YB$
5: $Car \div V = Hou \div Y \subseteq Hou \div YB = Occ \div LP =$
$Occ \div L \cup Occ \div P = Occ \div L \cup Car \div B$
6: $Car \div V \pitchfork Car \div B$
7: $Car \div V \subseteq Occ \div L$

Line number 4 is obtained from premise number 1 by complementation. Line number 5 is the combination of several equalities and $\subseteq$ relationships in the obvious fashion.

Answer: The lawyer drives the Volvo.

Clearly, the set-level approach is more concise and convenient for this example. However, in other cases the situation may be reversed, as the following example shows.

5.4.4 Example 'mother of son'

Problem: There are three persons: a grandparent, a parent, and a child. The grandparent is a woman; the child is a boy. None of the persons has any additional child. Prove that one of the persons is the mother of a son.

Scenario description: We use the similarity type
$\{f_1 : \{M, F\}, p_2 : \{T, F\}, f_3 : 1\}$,
where f_1 indicates the sex of the person occurring as argument, p_2 is true iff the argument is the mother of a son, and f_3 maps a person to his/her child (or to himself if he does not have a child). The object domain is $\{\#1, \#2, \#3\}$.

Ground-level formulation: The given facts are formulated as follows:
1: $f_1(\#1) \triangleq F$

2: $f_1(\#3)\hat{=}M$
3: $f_3(\#1)\hat{=}\#2$
4: $f_3(\#2)\hat{=}\#3$
5: $\forall o[p_2(o) \leftrightarrow f_1(o)\hat{=}F \wedge f_1(f_3(o))\hat{=}M]$

or, correctly in LFL syntax,

5: $\forall o[p_2(o) \leftrightarrow f_1(o)\hat{=}F \wedge \exists o'[f_3(o) = o' \wedge f_1(o')\hat{=}M]]$

and the following solution is obtained:

6: $f_1(\#2)\hat{=}FM$
7: $f_1(\#2)\hat{=}F \vee f_1(\#2)\hat{=}M$

followed by a proof by cases leading to

x: $p_2(\#1) \vee p_2(\#2)$.

Set-level formulation: It does not appear to be possible to solve this problem in a reasonable fashion using the set-level formulation.

This example suggests that the set-level formulation has a narrower range of application, and that it may be more convenient in those cases where it does apply.

5.5 Summary

The elementary logic of the preceding chapter has been generalized to lexical-domain object-feature logic, or LFL, by allowing an object domain $\mathcal{O}$ to be associated with a similarity type σ, forming a vocabulary $\nu = \langle \sigma, \mathcal{O} \rangle$. The definitions of syntax and semantics have been extended accordingly. Members of the lexical object domain may appear both in logical formulae and in interpretations. This makes it possible to introduce denotations, that is, expressions which are formed by a feature symbol and its arguments, and where all arguments are fully evaluated. Denotations are used as an auxiliary construct in the semantics, and play a role from the IDS point of view as 'state variables' in IDS states. We have outlined how the use of a lexical-domain semantics is the rationale behind the codesignation algorithm, which is classical in AI. Finally, if the same semantics is retained but the language of logical formulae is replaced by a language that refers to sets of objects, rather than individual objects, then one obtains a simple terminological language.

6
Temporal feature logic for discrete time domains

Earlier chapters have defined IDS developments whose structure is the five-tuple $\langle \mathcal{B}, M, R, \mathcal{A}, \mathcal{C} \rangle$. They have also defined the operation of chronicle completion where a scenario description is given, and where it is sufficient to first identify the set of developments that satisfy the given description, then to extract the set of weakened models of the form $\langle M, R \rangle$ while throwing away the rest of the information, and then to use the set of weakened models as the set of intended conclusions from the given scenario description. In this way, one will only obtain conclusions about feature-values and changes of feature-values at various points in time, but not conclusions about, for example, the starting times and ending times of actions.

This chapter addresses the same topic from the perspective of logic, and in particular it defines the syntax and the semantics of a logic, called DFL-1, where interpretations take the form of $\langle M, R \rangle$, which has previously been introduced for weakened chronicle models. However, here the interpretations are seen as generalizations of the interpretations which were introduced for LFL in Chapter 5 and which had the form $\langle M, r \rangle$, where r is a state. This gives us a base logic whose interpretations have the structure that has already been identified for the intended models in $Mod(\Upsilon, \text{RM})$.

First of all, however, we discuss the choice of ontology for time in a logic for common-sense reasoning, and a framework allowing several different ontologies for time will be introduced. This may be considered as a digression from our main topic, since we are already committed to using the non-negative integers or reals as the time domain in IDS developments, but the framework will be used for introducing branching time in Section 6.6, and as a background for the analysis of different logical approaches in Volume II. Also, of course, we will define the syntax of logic formulae in DFL-1, that is, the syntax of formulae that can be used in the various partitions Γ_i of a scenario description as defined in Chapter 2, subsection 2.3.1.

Volume II will address temporal logics with richer interpretations, in particular logics that also use the $\mathcal{A}$ component of full developments, representing the actions actually performed. This is in line with our general strategy of starting with more limited cases and obtaining an understanding of them first, and only then proceed to the more complex cases.

6.1 Ontology for standard time

6.1.1 Choice and use of time domain

Several different choices of time domain are in common use, namely:

- continuous time, in particular the real numbers as time domain;
- discrete, linear, and metric time, in particular the integers as time domain;
- discrete, linear time without a metric;
- discrete, forward-branching time, where each timepoint may have several successors, but only one predecessor.

For convenience I will refer to the first alternative as 'real' time, and to the second alternative as 'integer' time. From an engineering point of view it may seem natural to distinguish the first two choices from the last two, since the concept of time as linear and metric is fundamental in physics. For the present purposes, however, it is more convenient to first consider the three discrete-time alternatives, and to postpone the case of continuous time to a later occasion. This will also require generalizing the underlying IDS to continuous time. For discrete time, each timepoint t has a predecessor timepoint θt, regardless of whether time is linear or forward-branching. For continuous time, the concept of a predecessor timepoint is of course meaningless, and other concepts will be used instead.

Let $\nu = \langle \sigma, \mathcal{O} \rangle$ be a vocabulary for LFL, as defined in Chapter 5, section 5.2.1. In Chapter 5, an interpretation for ν was defined as a tuple $\langle M, r \rangle$, where M assigns values in $\mathcal{O}$ to object constant symbols, and the state r assigns values to features (= rigid feature expressions) in ν. From that perspective, a DFL-1 interpretation $\langle M, R \rangle$ can be understood as the following straightforward generalization. M is be similar as before, but generalized to allow temporal constant symbols as well as object constant symbols. The other component, previously called r, will now have time as a second argument, and will be a mapping $R(f, t)$ from features times timepoints, to corresponding feature-values. Objects themselves or the value of object constant symbols will not depend on time. Feature domains will still be assumed to be discrete and finite.

For a given timepoint t, I will write $R(t)$ for the state r defined by $r(f) = R(f,t)$. Similarly for a given feature f, the function g defined by $g(t) = R(f,t)$ will be written as $R(f)$. Such a function g from timepoints to a feature-value domain will be called a fluent, as introduced in subsection 1.2.2. The term 'discrete fluent' will refer to the discreteness of the value domain of the fluent (and therefore, the value domains of the function R), not to the time domain.

The main choice of temporal logic in this book is characterized by the use of such a function R as the most significant component in interpretations, and will be referred to as *discrete fluent logic*. Several variants of the logic will be introduced and studied, in order to form a picture of the available options. This chapter will introduce a basic form of discrete fluent logic, DFL-1. As reasons for amendments are identified, modifed versions DFL-2 (in Chapter 11), DFL-3 (in Volume II), etc., will be introduced.

6.1.2 Rationale for standard time

A common characteristic of each of the choices of time domain is that the same time domain is used by all interpretations. It has already been established that a vocabulary contains the value domains for the features or fluents, so that all interpretations for a given vocabulary use the same items in the value domain. A similar policy will be used for time, so that the same domain of timepoints is also used by different interpretations in a given use of the logic. I use the term *standard time* for this choice. The alternative to standard time is to use axiomatically specified time or *axiomatic time*, where each interpretation has its own timedomain, and axioms are used to constrain functions and relations on that domain.

The use of standard time is fairly obvious when temporal logic is used for characterizing IDS, and in any situation where one specific time domain is intended. One does not have to worry about whether the axiomatization of the time domain determines it completely, or about the behavior of a preference relation on models having different time domains, isomorphic or not. Finally, any reasonable software program which operates on the temporal information is going to use special purpose mechanisms that are tailored to the characteristic properties of the time domain, for example the algorithm of Dechter et al [18]. Normally, such algorithms and software are based on a standardized time domain, and it is quite implausible that they would use an axiomatic characterization of time directly.

6.1.3 Timepoints

I shall use the term 'timepoint' in a broad sense for any temporal item at which the value of a fluent is specified. It is therefore possible, but not necessary, to choose the timepoints as points along a time axis in the sense of physics. However, the following assumptions are made, besides the fact that standard time is to be used:

- Each timepoint has a unique predecessor (this assumption will of course be dropped in the generalization to continuous time).

- Time is not cyclic. The time domain is allowed but not required to contain one distinguished timepoint Θ, called the *origo*, which has itself as its predecessor, and which is the direct or indirect predecessor of all other timepoints. No timepoint other than origo may have itself as direct or indirect predecessor. Time structures with an origo will be the major case under consideration, but time structures without an origo (for example, the domain of all positive and negative integers) may also be considered briefly.

- Any two timepoints have a common indirect predecessor. This is trivial for the case of time with an origo and for linear time, but the assumption is made for all cases.

The primary intuition behind the notation and the arguments here is to choose the integers or the non-negative integers as the time domain in IDS. However, there are other possibilities as well. Suppose we are describing a 'frequent flyer' who spends most of his time on airline flights, and that we are only interested in describing him or her during the stops between the flights: the same properties in flight may be irrelevant, or even undefined, in the present vocabulary. Suppose, furthermore, that these properties are constant during each stopover, so there is no need to characterize change within the time period between two flights.

A straightforward choice of time domain is then to use the timepoints 'initially', 'after the first flight', 'after the first two flights', and so on. All interpretations may use this time domain; each interpretation will specify the destination of the first flight, the destination of the second flight, etc. Different route choices will therefore be captured by separate interpretations. In this way one obtains linear, non-metric time (except to the extent that one may wish to measure time by the number of airplane trips).

Another possible choice of time domain is to let each finite sequence of destinations be a timepoint. Thus the empty sequence $\langle \rangle$ is used as origo; the sequence $\langle \text{copenhagen} \rangle$ represents the 'situation' that is obtained from origo by making a flight to Copenhagen, $\langle \text{copenhagen}, \text{oslo} \rangle$ represents the 'situation' that is obtained from origo by first making a flight to Copenhagen, then one from there to Oslo; ⟨frankfurt⟩ represents the 'situation' that is obtained from *origo* by making a flight to Frankfurt a.M., and so on. Both ⟨copenhagen⟩ and ⟨frankfurt⟩ have origo as their predecessor. With this choice of time domain, one single interpretation may contain all possible choices of itinerary for the frequent flier (for a given initial position).

More generally, if a domain of 'actions' is given, then one may either use linear time and define a way of associating current action(s) with each timepoint, or one may view each action as a function from a timepoint to one of its immediate successors. The latter choice will be referred to as *Herbrand time*, and is the usual choice in the situation calculus.

The assumption that every timepoint has a unique predecessor rules out the choice of intervals (for example, over the integer axis) as the time domain. Intervals will be represented as members of $\mathcal{T} \times \mathcal{T}$ where $\mathcal{T}$ is the time domain, but not as $\mathcal{T}$ itself.

6.1.4 States

I will make a strict distinction between timepoints and states. States have already been defined in Chapter 5, subsection 5.2.4, as the same as a feature assignment, that is, a mapping from features to corresponding values. Therefore, a state specifies a possible momentary configuration of the world, with respect to what are currently the values of all the fluents. A timepoint, on the other hand, is merely a 'handle' to which an interpretation may attach feature-value combinations. It is possible for an interpretation to map two different timepoints to the same state, that is, for each feature its value at one timepoint is the same as its value at the other timepoint. It can still be two different timepoints. However, if two states are such that the value of a feature in one equals the value of the same feature in the other, for all features, then the states are identical.

In a careful analysis of expressivity, there seems to be an interreducibility between timepoints and states, since timepoints can be designed to have a certain structure (for example, chronometric content in the case of integer time, or an action-sequence content in the case of Herbrand time). One can always define the vocabulary so that the same information is contained in some of the features, and thereby in the states. This does not contraindicate the strict formal distinction between timepoints and states.

6.1.5 Situations

The term 'situation' is usually used in the AI literature for what is here called a 'timepoint' in Herbrand time, but it is sometimes used instead for a family of states or a specialization of a state. In the former case there are several states to a situation; in the other case several situations correspond to the same state. Many authors have considered the case where actions generate successor situations, often using the notation

$result(E, s)$

for the successor situation obtained from situation s by applying the action E. This notation is consistent both with viewing s as a timepoint with Herbrand ontology, and with viewing it as a state. In the latter case one has to assume that all actions are deterministic. In other research areas the word 'situation' is used with yet other meanings, in particular for a partial state. This topic is further discussed in the overview article [75].

Some of the intuitions that are often associated with the term 'situation' seem to correspond to a third formal entity in our presentation namely to a finite development.

6.1.6 Arguments in favor of non-metric time

Given the importance of clocks in our civilization, as well as the familiar and central role of time in physics, one may well question the appropriateness of non-metric time which was used in the frequent flyer example. Why not use chronometric time, and for each flight identify the interval during which the flight takes place? This makes it much easier to characterize the velocity of flight, and to reason about the concurrent travel of several persons who might not arrive and depart at the same times, as well as to accommodate feature changes that occur during one of the stopovers.

Basically I agree with these objections, but in all fairness there are some arguments to the contrary. Our main concern here is with discrete-valued fluents. On a continuous and chronometric time axis, such fluents must necessarily be piecewise constant, unless they are extremely 'badly behaved'. If it is now possible to divide the time axis into segments such that all fluents are constant within each segment, then that would certainly facilitate the subsequent treatment. The use of 'integer' time does not achieve this precisely, and can only be seen as an approximation.

Also, a frequent problem with discrete-valued features is that they are sometimes indeterminate. Consider the feature of 'sitting'. For a given person, there are many moments where she is obviously sitting, and many other moments where she is obviously not sitting. However, there are also times where it is difficult to say whether she is sitting or not. One possible way out of this problem is to exclude such intermediate moments from the logical time domain. An interval of chronometric time during which all features are well defined and constant will be called a *stable period*, and segments between stable periods will be referred to as *temporal gaps*. Notice, however, that this method only works well when actions can be assumed to occur in strict sequence and without concurrency.

A possible ontology for time is, then, to let each stable period be a 'time point', and to exclude temporal gaps from consideration in the logic. Such an approach may be appreciated either because one feels it is the ontology that is the most 'natural' one to use, or if it turns out to have computational advantages. It may also offer a convenient way of dealing with causal chains. Consider the action of throwing a snowball, which may cause a foe's nose to start bleeding, a window-pane to break, or a teacher to come running. For reasoning about such scenarios in overview, one perhaps does not wish to identify all details of how the ball moves through the air, how the teacher reacts to being hit, etc. It is then convenient to allow a 'temporal gap' for the period of these successive detail changes. The brief period during which the snowball is thrown is seen as one timepoint, and the next period that one wishes to describe and reason about is seen as the resulting next timepoint.

Naturally, if one adopts a stable period ontology for time, it is relatively complicated to characterize all the changes that may take place from one timepoint to the next. However, describing indirect effects of actions is complicated whichever way one does it, and at least one may hope to be able to constrain the properties of the next timepoint, knowing the properties of the previous timepoint and what action was performed. The use of non-metric timepoints and allowing for temporal gaps is then a way of reasoning in overview about a scenario. It does not require complete information, and naturally it does not produce complete information about the scenario, but it is arguably a promising approach to dealing with incomplete information.

6.1.7   Intra-situation change and chronometric fluents

Let us return to the frequent flyer example, where we used a time domain consisting of the stopovers 'initially', 'after the first flight', 'after the first two flights', and so on. This time domain can be used by all the interpretations. In each time domain, there will be an arrival time and a departure time for each stopover. More generally, if timepoints are understood as stable periods of chronometric time, it may be useful to map those timepoints to points in chronometric time, which I shall call *timestamps*. The *chronometric fluent(s)* which are mappings from timepoint to timestamp cannot be made part of the time structure, since such mappings must be specific to each interpretation, and the time structure is to be common for all interpretations.

A chronometric fluent is therefore appropriately represented as one which has chronometric time as the value domain. One may consider using a single chronometric fluent for the time-stamp of the stable period, but one may also choose to work with two chronometric fluents, one for the beginning time and one for the ending time of the stable period.

From a formal point of view, chronometric fluents are not much different from any other fluent. The only differences are that one presumably needs an infinite domain of timestamps, and that one wants to have relations and functions on the value domain. However, both of these generalizations will be needed at some point for various other fluents as well.

A chronometric fluent is an obvious example of non-inert fluents, since its value changes continuously, the change is intrinsic, and it is not caused by any action.

6.1.8 Time structure

On the basis of these considerations, we can now proceed to the formal definitions. A *discrete time structure* is an algebraic structure
$$\mathbf{T} = \langle \mathcal{T}, \theta, <, +, -, ...\rangle,$$
consisting of a *time domain* $\mathcal{T}$ whose members are called *timepoints*, and a finite number of functions and relations on the timepoints. I assume that the algebraic structure defines both the symbols and the meaning for those functions and relations. The time structure is therefore used both for the syntax definition of DFL and for the semantics. It specifies elements from which composite expressions involving timepoints can be formed, and it also contributes to the specification of how such composite expressions are to be evaluated[31].

The function θ and the relation $<$ must always be present in a discrete time structure, and a function Θ of no arguments is optional there. They must satisfy the following conditions:

- θ is a function $\mathcal{T} \to \mathcal{T}$ called the *immediate predecessor function*.

- $<$ is a relation on $\mathcal{T} \times \mathcal{T}$ called the *precedes relation*. It is defined so that $t < t'$ iff for some positive integer n, $\theta^n t' = t$.

- If the origo Θ is present, it must satisfy $\Theta < t$ for all t and $\theta(\Theta) = \Theta$.

- No t, except Θ if present, may satisfy $t < t$.

- For every t and t' there exists a t'' such that $t'' < t$ and $t'' < t'$.

The different alternatives for the time domain, which were described at the beginning of this chapter, can now be rendered formally as follows.

---

[31] To be completely precise, the time structure should have the form
$$\{\ulcorner\mathcal{T}\urcorner \mapsto \mathcal{T}, \ulcorner\theta\urcorner \mapsto \theta, ...\},$$
containing both the symbols for use in formulae, and the corresponding values to be used in the evaluation of formulae. These details will be taken for granted.

Integer time

The special case of two ways infinite, integer time domain is represented as the discrete time structure without origo,
$$\langle T, \theta, <, +, - \rangle,$$
where T is the set of integers, $\theta(t) = t - 1$, and $<$, $+$, and $-$ are as usual for integers.

Finite integer time

For finite developments as used in previous chapters, it is appropriate to use the time structure
$$\langle T, \theta, <, \Theta, n \rangle,$$
where n is a non-negative integer, T is the set of integers between 0 and n, $\theta(t) = max(t - 1, 0)$, and $\Theta = 0$.

Herbrand time

The special case of a forward-branching time domain, with origo, which is generated by actions, is represented as the discrete time structure,
$$\langle T, \theta, <, \Theta, N_1, N_2, ..., N_n \rangle,$$
where T is essentially the set of sequences of the successor-forming symbols N_i, $\theta(\Theta) = \Theta$, and $\theta(N_i(t)) = t$ for all i and t.

Linear, non-metric time

The case of a linear, non-metric time domain with origo is formally a special case of Herbrand time with $n = 1$, that is,
$$\langle T, \theta, <, \Theta, N \rangle,$$
where N is the successor function and $<$ is a total order.

Additional cases are also possible, for example forward-branching time where actions are relations over timepoints so that each 'situation' may have several successors for a given action. Note, however, that within this framework it is not possible to use pairs of integers (representing intervals) for the timepoint domain.

The major part of this volume will use linear chronometric time. A variety of branching time will be introduced in Section 6.6. It turns out that the results in later chapters generalize quite easily to the case of branching time.

6.1.9 Non-determinism in Herbrand time

A major characteristic of the situation calculus is that its expressivity is restricted. In particular, it is difficult and inconvenient to characterize concurrent actions with duration in the situation calculus. This restrictiveness may be a disadvantage or an asset depending on the character of the application at hand.

However, there is one particular expressivity restriction that the situation calculus does not have, provided that it is interpreted with a semantics based on Herbrand time, but where it is easy to be misled by the standard notation: the use of the operator name *result* should not lead one to infer that the situation calculus requires all actions to be deterministic. For example, suppose that the action N_2 has the effect that fluent f_4 obtains either of the values C or E, but the actual effect may be different at different times when N_2 is applied. One would then write
$$\forall t [[N_2(t)] f_4 \hateq \text{CE}]$$
or, equivalently,
$$\forall t [[N_2(t)] f_4 \hateq \text{C} \vee f_4 \hateq \text{E}].$$
This premise certainly allows, for example, an interpretation where
$$[N_2(\Theta)] f_4 \hateq \text{C}$$
and, anyway,
$$[N_2(N_1(\Theta))] f_4 \hateq \text{E}.$$
This illustrates how timepoints are merely *points at which* interpretations assign values to features. Although several ontological variants for time are admitted, none of them allows the domain of states[32] to be used as the time domain. 'The situation $N_2(\Theta)$' is understood as a precisely defined time-*point*, but the values assigned to features at that timepoint need not be defined precisely, and can differ between different models.

However, one can also make another semantical interpretation of the situation calculus. If situations are identified with states then only deterministic actions can be considered; if a situation is considered as a state together with some additional information (for example the history leading up to this state) then nondeterministic actions can again be represented.

6.2 Reification of temporal formulae

Before proceeding to the syntax and semantics of the temporal logic, I want to discuss briefly the available options with respect to both the syntax and the semantics. Consider a simple treatment of time in a many-sorted first-order logic with time as one of the sorts, allowing logic formulae such as
$$R(f(a,t), b, t) \vee S(g(a), t),$$
where the variable t has temporal type, and other constants and functions have other types. Consider the obvious possibility of re-expressing this formula as
$$[t] R(f(a), b) \vee S(g(a)),$$

---

[32] See Baker [6] for an approach that uses a state-oriented semantics for situations.

which is intended to mean 'the atemporal statement $R(f(a),b) \vee S(g(a))$ holds at time t'. Consider also the similar rewriting of the formula
$$R(f(a,t),b,t) \vee S(g(a),\theta t)$$
as
$$[t]R(f(a),b) \vee \Lambda S(g(a)),$$
where θ is the predecessor operator in the time-domain as defined above, and Λ is a modal operator on logic formulae which shifts the reference point for time one step back.

This transformation is commonly characterized as a *reification* because atemporal statements, such as $R(f(a),b) \vee S(g(a))$, can then be seen as objects in themselves. This becomes particularly evident if one uses a syntax that suggests a binary relation between the statement and the time where it is valid, such as
$$Holds(t, R(f(a),b) \vee S(g(a))),$$
as used for example by [33].

However, regardless of whether or not one chooses to use this transformation, one uses *explicit time* (sometimes also called reified time), that is, the timepoints are one of the sorts. The alternative to explicit time is to use a *tense syntax* allowing us to write, for example,
$$\Diamond R(f(a),b) \vee S(g(a)),$$
saying that 'at some point in time, the statement $R(f(a),b) \vee S(g(a))$ holds'.

Consider now the possible reasons why one may wish to use reified statements. These are:

- Notational convenience. If the same temporal argument occurs repeatedly in large expressions, it is convenient to factor it out and write it once at the beginning of the formula.

- Other uses of reified statements. Once one has introduced the separate syntax for atemporal statements, one may wish to use them for other purposes as well, such as
  $$Believes(john, R(f(a),b) \vee S(g(a)))$$
  or
  $$Causes(R(f(a),b), S(g(a)))$$
  One may also wish to quantify over reified statements, for example
  $$\forall t \forall p \forall q [Holds(t,p) \wedge Causes(p,q) \rightarrow \exists t'[t < t' \wedge Holds(t',q)]]$$

- Avoiding axiomatization of time. One may argue that the use of explicit time requires the use of axioms for characterizing the time domain explicitly. With tense logic that is not necessary, but then of course one must use reified statements. Notice, however, that the use of standard time is another way of avoiding axiomatization of time.

In turn, the reasons for not axiomatizing time may be either that one is interested in a time domain (such as the integers) which cannot be axiomatized in a satisfactory fashion, or because one wishes to use general theorem-proving techniques and assumes that a tense logic will have performance advantages.

Choice for this work

Here and in the following chapters, I will use reified statements, but only for the first-mentioned reason of notational convenience. I shall also, of course, use explicit time. The following is an example of a formula in the logic used here.
$$[t]p(o_1, o_2) \vee \Lambda q(\#4),$$
Since there are no other uses of reified statements, they are only a syntactic sugar. The semantics of the logic defined here could equally well have been used for a first-order logic with time as one of the sorts. Formally, a formula such as
$$[t]p(o_1, o_2) \vee \Lambda q(\#4)$$
will be considered simply as an abbreviation for
$$[t]p(o_1, o_2) \vee [t]\Lambda q(\#4).$$
Therefore, the core syntax is only defined for the case where the expression following the time expression is an elementary fluent formula. Composite atemporal statements are introduced in the subsection for abbreviations (6.3.2).

Previous discussion of these issues

The topic of how to approach and represent time in a temporal logic is virtually inexhaustible. Some relevant articles in the AI literature are those by Allen [2], McDermott [50], Shoham [77], and Reichgelt [59], and the handbook article by Sandewall and Shoham [75].

6.3 Main syntax

We now proceed to the syntax of formulae in DFL-1 logic. The logic is defined on the assumption that the object domain $\mathcal{O}$ is kept fixed over time, that is, there is no **T** characteristic. Formulae for characterizing occurrences are considered as a 'side language' which is introduced in Section 6.5, and are outside the main syntax.

There will be no allowance for functions on or between the value domains of features. Adding such functions is a trivial extension, and it is required for many applications, but from the present semantic point of view it would not add any essential aspect to the logic.

6.3.1 Core of the main syntax for DFL-1

The main syntax will be introduced in two steps: first a *core* for which the semantics is directly defined as well, and then the *abbreviations* which are only defined by rules for how to expand them into the core.

A *vocabulary* for DFL [33] has the same structure as for LFL in Chapter 5. DFL formulae for a given vocabulary $\nu = \langle \sigma, \mathcal{O} \rangle$ and a given time structure are formed using the following syntactic elements:

- object names in $\mathcal{O}$, object constant symbols and object variables as defined in Chapter 5,
- feature symbols defined in σ,
- items defined in σ,
- timepoints (members of $\mathcal{T}$) and the symbols defined in the time structure,
- timepoint constants, denoted t_i where i is a positive integer,
- timepoint variables, denoted t_i where i is a positive integer,
- the special operator Λ,
- the relation symbols $\hat{=}$ and $=$,
- the propositional connectives T, F, $\neg$, $\wedge$, $\vee$, $\rightarrow$, and $\leftrightarrow$ and the quantifiers $\forall$ and $\exists$,
- parentheses and brackets.

This suffices for DFL-1; each later version will make a few minor extensions.

An *object expression* is either an object name (member of $\mathcal{O}$), an object constant symbol, or an object variable. An *elementary feature expression* is either of the following:

- a feature symbol f_k^0,
- $f_k^i(\omega_1, ..., \omega_i)$ where each argument is an object expression.

The *domain* of such an elementary feature expression is defined as $Dom_k(\nu)$ as in Chapter 5.

A *feature expression* is either an elementary feature expression, or an expression Λf where f is a feature expression. In the latter case the domain of f is the domain of Λf as well.

---

[33] This applies to DFL-i for all i.

It is actually possible to manage without the symbol Λ in the case of discrete time, but in the forthcoming generalization to continuous time its counterpart[34] is essential. The use of the symbol Λ also offers a certain notational convenience even for discrete time.

A *rigid feature expression* is a special case of an elementary feature expression and is defined in the same way as in Chapter 5, that is, all arguments must be object names.

An *elementary fluent formula* in DFL has the form
$$f \hateq \mathcal{X},$$
where f is a feature expression, and $\mathcal{X}$ is a subset of its domain. An elementary formula in LFL formed using $\hateq$ is also an elementary fluent formula in DFL, therefore. It is intended to mean that at the present time the value of f is a member of $\mathcal{X}$.

A *fluent formula* in DFL is simply an elementary fluent formula. The syntax for fluent formulae will be extended by abbreviations in the next subsection.

A *timepoint expression* is one of the following:

- a member of $\mathcal{T}$,
- a timepoint constant symbol,
- a timepoint variable,
- an expression formed from timepoint expressions using any of the functions in the time structure (including θ).

An *elementary fixed formula* has one of the forms
$$[\tau]\phi,$$
$$\omega = \omega',$$
$$\tau = \tau',$$
or it is formed from timepoint expressions using any of the relations defined in the time structure, including $<$. Here τ and τ' are timepoint expressions, ϕ is a fluent formula, and ω and ω' are object expressions. These formulae are intended to mean, in turn: 'ϕ holds at time τ'; 'the object ω equals the object ω'', and 'the timepoint τ equals the timepoint τ''.

A *logic formula* in DFL is obtained by composition of elementary fixed formulae using the propositional connectives in the usual fashion, or as
$$\forall t[\alpha],$$
$$\forall o[\alpha],$$
or the corresponding existentially quantified expressions. Here t is a timepoint variable, o is an object variable, and α is a logic formula. Parentheses

---

[34] Namely, the left limit value in a timepoint.

are used as needed to avoid ambiguity. Quantifiers range over only the succeeding bracketed expression, so, for example, in $\forall o[\alpha] \vee \beta$ the quantification does not range over β.

For DFL-1 these are the only formulae there are. Additional syntactic constructs will be introduced for DFL-2 in Chapter 11.

6.3.2 Convenience abbreviations

The main syntax is defined to consist of the core as defined in subsection 6.3.1, plus the following abbreviations.

Abbreviations for fluent formulae

The class of *fluent formulae* in DFL-1 is extended by composition of elementary fluent formulae using the propositional connectives and quantification over objects in the usual fashion. These abbreviations are defined by distribution of the time as follows.

$[\tau]\mathrm{T} \equiv \mathrm{T},$

$[\tau]\mathrm{F} \equiv \mathrm{F},$

$[\tau]\neg\phi \equiv \neg[\tau]\phi,$

$[\tau](\phi \wedge \phi') \equiv [\tau]\phi \wedge [\tau]\phi',$

$[\tau](\phi \vee \phi') \equiv [\tau]\phi \vee [\tau]\phi',$

$[\tau](\phi \to \phi') \equiv [\tau]\phi \to [\tau]\phi',$

$[\tau](\phi \leftrightarrow \phi') \equiv [\tau]\phi \leftrightarrow [\tau]\phi',$

$[\tau]\forall o[\phi] \equiv \forall o[[\tau]\phi],$

$[\tau]\exists o[\phi] \equiv \exists o[[\tau]\phi].$

This definition is sufficient since the only way to use a fluent formula in the DFL-1 syntax is for forming a fixed formula of the form $[\tau]\phi$. Therefore, if the abbreviation $\phi \wedge \phi'$ is formed, it can only be used for forming larger abbreviation constructs, and then ultimately it must be preceded by $[\tau]$ for some timepoint expression τ. The resulting formula can then be reduced to an abbreviation-free formula using the definition above.

Note that if f and f' are features, $f \wedge f'$ is not a feature, just a syntactic substructure in an abbreviation. This approach is not affected by the problem discussed by Lifschitz in [42], therefore.

For convenience, $[\tau]\phi, \phi'$ is defined to be synonymous with $[\tau](\phi \wedge \phi')$. It will only be used when ϕ and ϕ' are elementary or negated fluent formulae. Finally, we define $f = f'$ where f and f' are feature expressions with the same value domain, as an abbreviation for

$$\bigvee [f \doteq d_i \wedge f' \doteq d_i \mid d_i \in \sigma[f]].$$

Abbreviations for logic formulae

The class of *logic formulae* in DFL-1 is extended by the following constructs:

- $\tau \leq \tau'$, where τ and τ' are timepoint expressions. This is an abbreviation for
  $$\tau < \tau' \vee \tau = \tau'.$$

- $\tau < \tau' < \tau''$ abbreviating $\tau < \tau' \wedge \tau' < \tau''$, and similarly for $\tau \leq \tau' \leq \tau''$, etc.

- using the symbols ∞ and $-\infty$ in the argument positions of $<$ and $=$, the resulting abbreviations are defined as follows, where τ is a timepoint expression:
  $$-\infty < \tau \equiv T,$$
  $$\tau < -\infty \equiv F,$$
  $$\infty < \tau \equiv F,$$
  $$\tau < \infty \equiv T,$$
  $$\tau = -\infty \equiv F,$$
  $$\tau = \infty \equiv F,$$
  $$-\infty = \tau \equiv F,$$
  $$\infty = \tau \equiv F.$$

- $(\tau, \tau') \phi$, where τ is a timepoint expression or the symbol $-\infty$, and τ' is a timepoint expression or the symbol ∞. This is an abbreviation for
  $$\forall t[\tau < t < \tau' \rightarrow [t]\phi],$$
  where t is a variable that does not occur free in ϕ.

- $\Box \phi$, as an abbreviation for $(-\infty, \infty) \phi$, that is, 'always ϕ'

- $[\tau, \tau']\phi$, where τ and τ' are timepoint expressions or $\pm\infty$, which is an abbreviation for
  $$\forall t[\tau < t < \tau' \rightarrow [t]\phi],$$
  that is, the fluent formula ϕ holds throughout the closed time interval from τ to τ'.

- The notation for left half-open intervals $(\tau, \tau']$ and right half-open intervals $[\tau, \tau')$ is defined similarly.

Notice that this notation is equally applicable for branching time. Note, furthermore, that $-\infty$ and ∞ are *not* members of the timepoint domain; they are just symbols defined as parts of abbreviations. This is not only to satisfy those who claim that infinity does not exist, but also because in branching time there would be one infinity along each path. It follows also that an expression such as $-\infty < \infty$ is not a legal abbreviation.

The expressions $[-\infty, \tau]$ and $(-\infty, \tau]$ are equivalent by these definitions, since nothing is equal to $-\infty$ anyway, and similarly for $[\tau, \infty]$. Finally, notice that by these rules, $[\tau, \tau'](\phi \vee \phi')$ can only expand to
$$\forall t[\tau \leq t \leq \tau' \to [t]\phi \vee [t]\phi'],$$
and not to $[\tau, \tau']\phi \vee [\tau, \tau']\phi'$.

Priority rules

The potential ambiguity in an expression of the form $[t]\phi \otimes \psi$ for a connective $\otimes$ is resolved by the convention that it is parsed as $[t](\phi \otimes \psi)$ if $\otimes$ is the comma, and as $([t]\phi) \otimes \psi$ for any other connective. For example, $[t]\phi_1, \phi_2 \wedge \phi_3$ is parsed as $([t]\phi_1, \phi_2) \wedge \phi_3$.

6.4 Semantics and axioms for the base logic

6.4.1 Semantics

The lexical-domain semantics for DFL-1 will now be defined; free domain semantics is analogous. Let $\nu = \langle \sigma, \mathcal{O} \rangle$ be a vocabulary as usual. A *history* for ν is a function from rigid feature expressions times $\mathcal{T}$, to values in the domain for each feature expression. Thus a history is a generalization of a feature assignment (= state) which also has a second argument for time.

A *valuation* for ν in DFL is a mapping which assigns to each temporal constant a member of $\mathcal{T}$, and to each object constant a member of $\mathcal{O}$. An *interpretation* for ν in DFL-1 is a tuple
$$\langle M, R \rangle,$$
where M is a valuation and R is a history.

A history R will be viewed alternatively as a curried function mapping timepoints to states in the obvious fashion. This means that for each timepoint t, $R(t)$ is a state, that is, a mapping from features to corresponding values which are supposed to represent the values of the fluents at time t. Conversely, $R(f)$ is a fluent, that is, a mapping from timepoints to the value domain for f.

A DFL *structure* for σ is a pair $\langle \mathcal{O}, I \rangle$ as before, where $\mathcal{O}$ is an object domain, and I is a DFL interpretation for $\langle \sigma, \mathcal{O} \rangle$.

Let $\nu = \langle \sigma, \mathcal{O} \rangle$ be a vocabulary, let α be a closed DFL-1 formula for ν, let f be a feature expression, let I be an interpretation for ν, and let $s = \langle \mathcal{O}, I \rangle$ be the resulting structure. The functions $val[\alpha, s]$ and $den[f, s]$ are used like in previous chapters, and in addition there is an auxiliary function $mval[\phi, s, t]$ representing the momentary value of a fluent formula ϕ or a feature expression f in s at time t.

Semantics and axioms for the base logic

The value of the closed logic formula α in s, $val[\alpha, s]$, where $s = \langle \mathcal{O}, \langle M, R \rangle \rangle$ is defined recursively as follows. The first definitions are as before:

If ω is an object name, then $val[\omega, s] = \omega$.
If ω is an object constant symbol, then $val[\omega, s] = M[\omega]$.
The denotation of a feature expression is defined by
$den[f_k^0, s] = f_k^0$,
$den[f_k^i(\omega_1, \omega_2, ..., \omega_i), s] = \ulcorner f_k^i(\underline{val}[\omega_1, s], \underline{val}[\omega_2, s], ..., \underline{val}[\omega_i, s])\urcorner$.

The underlining of val indicates that the evaluation is to be done inside the Quine quotes, as defined in Chapter 4.

The following amendments are made for this chapter:

If τ is a timepoint then $val[\tau, s] = \tau$.
If τ is a timepoint constant symbol then $val[\tau, s] = M[\tau]$.
If τ is a composite timepoint expression formed using function symbols in the time structure, then its value is obtained by applying the function definitions in the time structure to the values of the subexpressions of τ.

Values of elementary fixed formulae are defined as follows:
$val[\omega = \omega', s] = \text{T}$ iff $val[\omega, s] = val[\omega', s]$,
$val[\tau = \tau', s] = \text{T}$ iff $val[\tau, s] = val[\tau', s]$,
$val[[\tau]\phi, s] = mval[\phi, s, val[\tau, s]]$,

where $mval$ is defined by
$mval[f \hat{=} \mathcal{X}, s, t] = \text{T}$ iff $mval[f, s, t] \in \mathcal{X}$,
$mval[f, s, t] = R(den[f, s], t)$, iff f is an elementary feature expression,
$mval[\Lambda f, s, t] = mval[f, s, \theta t]$.

The value of an elementary fixed formula τ op τ' formed using a relation op which is defined in the time structure, is obtained in the obvious fashion by applying the relation to $(val[\tau, s], val[\tau', s])$, and similarly for time-domain relations of other arity.

The value of a logic formula constructed using propositional connectives is defined as in Chapter 4, section 4.2.3, and for quantified formulae as in subsection 5.2.4, as follows:

$val[\forall o[\alpha], s] = \min_{n \in \mathcal{O}} val[\alpha_o^n, s]$,

$val[\exists o[\alpha], s] = \max_{n \in \mathcal{O}} val[\alpha_o^n, s]$,

$val[\forall t[\alpha], s] = \min_{\mathbf{t} \in \mathcal{T}} val[\alpha_t^{\mathbf{t}}, s]$,

$val[\exists t[\alpha], s] = \max_{\mathbf{t} \in \mathcal{T}} val[\alpha_t^{\mathbf{t}}, s]$,

where $\alpha_t^{\mathbf{t}}$ is the result of substituting $\mathbf{t}$ for t throughout α.

If α is a logic formula then
$$[\![\alpha]\!]_{\mathcal{O}} = \{\langle M, R\rangle \mid val[\alpha, \langle \mathcal{O}, \langle M, R\rangle\rangle] = \mathrm{T}\},$$
as usual. We write
$$[\![\alpha]\!] = \bigcup_{\mathcal{O}} [\![\alpha]\!]_{\mathcal{O}},$$
where $\mathcal{O}$ ranges over all domains that at least contain the objects mentioned in α. Also, if ϕ is a fluent formula we write $[\![\phi]\!]_{\mathcal{O}}$ for the set of states where ϕ is true, that is, retaining the LFL meaning. Abbreviations and constructs in the side language naturally do not need any definition of val.

6.4.2 Some axioms

The following rules are obtained for the abbreviations:
$$[\tau, \tau'](\phi \wedge \phi') \leftrightarrow [\tau, \tau']\phi \wedge [\tau, \tau']\phi',$$
$$\Box(\phi \wedge \phi') \leftrightarrow \Box\phi \wedge \Box\phi'.$$
Similar rules are also obtained for $\forall o$ instead of $\wedge$, but not for $\vee$, $\neg$, etc.

6.5 The side language for occurrences

6.5.1 Approach

Chronicle completion as defined in Chapter 2, subsection 2.4.2, does not treat statements about feature-values and statements about occurrences symmetrically. The given scenario description can contain both kinds of statements, but the possible conclusions are restricted so that statements about occurrences can not be included. Correspondingly, it is sufficient to use intended models of the form $\langle M, R\rangle$.

This asymmetry will be exploited in the treatment of formulae for occurrences in the logic. Initially, we have defined logic formulae in DFL-1 so that they only allow statements about feature-values and/at timepoints. This language has already been referred to as the *main language*. Formulae for occurrences will be defined as a separate *side language*, that is, a separate syntax. The side language is, however, parasitical on the main language, in the sense that main language expressions can be included as subexpressions in formulae of the side language.

Several alternative ways of dealing with the side language will be used. The first and simplest approach is to consider the action laws (that is, the laws describing the effects of actions) as a translation function from the side language into the main language. In this way the actions and the action laws can be dealt with entirely syntactically, which allows a particularly simple treatment.

In order to generalize the range of applicability and the expressiveness of the first approach, we then proceed to extend the main language (obtaining

DFL-2) with an additional and more powerful construct while retaining the view of action laws as a translation. In a further generalization the main language and the side language are merged, obtaining DFL-3 logic, and the status of action laws is then changed so that they become formulae in the unified language rather than a translation between languages. This allows the treatment of a broader class of reasoning problems. In particular, translation into a main syntax does not seem to be sufficient when causal chains are used for characterizing delayed effects of actions.

For example, in the Yale shooting scenario[35], the side-language formula for the occurrence "firing the gun from time 14 to time 16" can be translated into the formula for "the gun is unloaded at time 16, and if the gun is loaded at time 14 then the turkey is dead at time 16, and the changes for loadedness and liveness happen arbitrarily within the time interval [14, 16]". In this way the logic only needs to deal with ways of characterizing feature change, and everything that directly has to do with the actions or other occurrences is handled by the translation function. This technique also avoids the problem that was identified by Morgenstern and Stein [51], namely, that when inertia is assumed one must both minimize the set of occurrences and the set of changes which are brought about by each occurrence.

6.5.2 The elementary occurrence language

We now proceed to the precise specification of the side language for occurrences, or the occurrence language for short. It allows formulae such as $[4,6]A_{12}$, saying that the action A_{12} (an action symbol without arguments) takes place during the two timesteps from time 4 to time 6. Formulae in the occurrence language are not DFL-1 main formulae, although they may use constructs in DFL-1 main language as subexpressions, for example $[t_1, t_1 + 2]A_{12}$.

In this section we define the elementary occurrence language. Chapter 12 will extend it to a composite occurrence language which allows composite actions such as $[10, 16]A_1; A_2$ representing the fact that the sequence of the two actions named A_1 and A_2 is performed during the interval [10, 16].

Occurrences will be described using *occurrence symbols* $A_1, A_2, ..., A_k$. Let a vocabulary $\nu = \langle \sigma, \mathcal{O} \rangle$ be given. An *occurrence vocabulary* π for ν is a mapping which to each occurrence symbol A_i assigns $\pi[A_i]$ as a finite sequence of type descriptors. Each type descriptor is either a feature-value domain occurring in σ, or the number 1 representing $\mathcal{O}$ as usual.

An *elementary occurrence expression* for ν and π is an expression of the form
$$A(\omega_1, \omega_2, ...),$$

---

[35] See Chapter 7, subsection 7.2.1

where A is an occurrence symbol in π, each ω_i is either an object expression or an item, and the number and type of the arguments agree with $\pi[A]$ in the same way as for feature expressions.

For example, if $\sigma = \{f : \{\text{R, Y, G, B}\}\}$, $f(o)$ indicates the color of o as Red, Yellow, Green, or Blue, $\mathcal{O} = \{\#1, ..., \#5\}$, and $\pi = \{A_1 : \langle 1, \{\text{R, Y, G, B}\}\rangle\}$, then $A_1(\#4, \text{G})$ is an occurrence expression for $\langle \sigma, \mathcal{O}\rangle$ and π. It may, for example, represent the action of painting the object #4 with green color.

For the elementary occurrence language, an *occurrence expression* is the same as an elementary occurrence expression. The composite occurrence language will add more types of occurrence expressions.

An *occurrence designator* or rigid occurrence expression is an occurrence expression where all arguments are either objects or items. For example, $A_1(\#4, \text{G})$ is an occurrence designator, but $A_1(o_2, \text{G})$ is not. We have thereby provided the formal definition of the domain $\mathcal{E}$ of occurrence designators which was introduced in Chapter 1, subsection 1.3.2.

An *occurrence formula* is a formula of the form
$$[\varsigma, \tau]\varepsilon,$$
where ε is an occurrence expression, and ς and τ are timepoint expressions. A *rigid occurrence formula* is an occurrence formula where ς and τ are timepoints and ε is a rigid occurrence expression. An *occurrence* is a rigid occurrence formula where in addition $\varsigma < \tau$. As anticipated in Chapter 1, subsection 1.3.2, we identify the triple $\langle s, E, t\rangle$ in $\mathcal{T} \times \mathcal{E} \times \mathcal{T}$ and the rigid occurrence formula $[s, t]E$.

An occurrence formula $[\tau, \tau']\varepsilon$ is intended to say that the corresponding rigid occurrence formula (obtained by evaluating arguments) is present in the development or interpretation at hand.

An *occurrence definition* Π is a mapping from occurrence formulae to their *expansions* which are logic formulae in DFL-i for some i. Occurrence definitions will be written in a straightforward fashion, where $\Rightarrow$ indicates the translation, and s and t are often used as meta-variables for timepoint expressions, for example:
$$[s, t]\,Load \Rightarrow [t]\,l,$$
whereby it is defined, for example, that $\Pi[[4, 6]\,Load]$ is translated into $[6]\,l$, saying that at time 6 the propositional feature l is true.

Note that the bold square brackets enclosing two timepoints [-,-] have different meanings when they precede a fluent formula or an occurrence expression. Consequently, $[\varsigma, \tau](\phi \wedge \varepsilon)$ or $[\varsigma, \tau](\phi \to \varepsilon)$ are not syntactically well-formed if ϕ is a fluent formula and ε is an occurrence expression. Since we are only going to analyze chronicles, but not the more general types of scenarios, there will be no opportunity to use formulae with conditional actions or disjunctions between actions.

Note also that, although ε is used as a meta-variable for occurrence expressions, in the special case of rigid elementary occurrence expressions the meta-variable E is usually used as in Chapter 1.

6.5.3 Reassignment formulae as abbreviations

The translation of occurrence formulae into main syntax formulae will be realized as a two-step process via the intermediate abbreviation := that represents value change of a fluent at a timepoint or within a temporal interval. For example, the obvious rule for the *Fire* action in the Yale Shooting Scenario (see Chapter 7, subsection 7.2.1) may be defined by

$$[t,t']\,Fire \Rrightarrow ([t]\,loaded \rightarrow [t,t']\,alive := \mathrm{F})$$

so that

$$[10,12]\,Fire$$

translates into

$$[10]\,loaded \rightarrow [10,12]\,alive := \mathrm{F},$$

saying that if the gun is loaded at time 10 when the action starts, the turkey dies somewhere in the interval between 10 and 12.

The operator := is in turn defined as an abbreviation, with different meanings in DFL-1 and DFL-2. In DFL-1 it is defined so that $[\tau,\tau']\,f := \mathcal{X}$ is expanded into $[\tau']\,f \hat{=} \mathcal{X}$. DFL-2, which will be introduced in Chapter 11, uses a more powerful expansion. When composite expressions contain := in a subexpression, then the subexpression is expanded locally in the usual fashion.

Formally, a *reassignment expression* is an expression of the form $f_i := \mathcal{X}_i$, where f_i is an elementary feature expression, and $\mathcal{X}_i$ is a non-empty subset of the domain for f_i. A *reassignment formula* is a formula of the form

$$[\varsigma,\tau]\,f_1 := \mathcal{X}_1 \wedge ... f_k := \mathcal{X}_k,$$

where ς and τ are timepoint expressions and each $f_i := \mathcal{X}_i$ is a reassignment expression. A reassignment formula

$$[\varsigma,\tau]\,f_1 := \mathcal{X}_1 \wedge ... f_k := \mathcal{X}_k$$

is an abbreviation for

$$([\varsigma,\tau]\,f_1 := \mathcal{X}_1) \wedge ... \wedge ([\varsigma,\tau]\,f_k := \mathcal{X}_k).$$

Commas may be used instead of $\wedge$ in the same way as for fluent formulae. A *conditional reassignment formula* over the timepoint expressions ς and τ is a formula of the form

$$[\varsigma]\,\phi \rightarrow [\varsigma,\tau]\,\alpha,$$

where ϕ is a fluent formula (not necessarily an elementary fluent formula), but not using $\wedge$, and $[\varsigma,\tau]\,\alpha$ is a reassignment formula.

In simple standard cases, action laws will map action statements to conjunctions of conditional reassignment formulae with mutually exclusive ϕ. The range of syntax that action statements can expand into depends on the ontological family being considered.

To summarize, an occurrence expression in the sidelanguage for occurrences is translated to the core language of DFL-1 or DFL-2 through a two-step process. The first step translates it to a composite expression containing reassignment expressions, namely into a conjunction of conditional reassignment formulae. This translation is specific for each occurrence symbol, and actually it is the way in which the effects of the occurrence are characterized. However, the first step is identical for DFL-1, DFL-2, and higher DFL-i, and generates formulae in the non-core part of the main language. The second step is the same across all occurrence symbols and occurrence types, but it differs between different DFL-i. It expands every reassignment expression into a formula in the core syntax.

6.6 Branching time

The temporal logic defined in this chapter can be used for all the types of time structures that were defined in subsection 6.1.8. When the logic is used for characterizing IDS developments it is necessary to choose a time domain that is compatible with the IDS structure. The definition of IDS in Chapters 1-3 used integer time, so we must either restrict ourselves to the integer time domain, or generalize the definition of IDS to allow branching time. It turns out that the generalization is quite straightforward, but it does make the proofs a bit more difficult to follow. The formal material in the following chapters will therefore be presented in terms of integer time, but the present section describes how branching time can be performed.

6.6.1 The linear and branching time domains

The time structure for integer time will be called $\mathbf{T}_\emptyset$. The time structure for branching time is similar in spirit to Herbrand time, but it also contains embedded timestamps, which gives it a metric component. This time structure will be called $\mathbf{T}_\mathcal{E}$. For uniformity, both time structures will be defined to have the structure
$$\mathbf{T} = \langle \mathcal{T}, \theta, <, \Theta, \asymp, \smile, N_0, N_1, ..., N_k \rangle,$$
where for each action designator $E_i \in \mathcal{E}$ there is a corresponding operator N_i. (This means in particular that $\mathbf{T}_\mathcal{E}$ is different for different choices of $\mathcal{O}$). The latter operators in the sequence are motivated by their use in $\mathbf{T}_\mathcal{E}$. For $\mathbf{T}_\emptyset$ they are defined as follows.
$$\mathbf{T}_\emptyset = \langle \mathcal{T}, \theta, <, \Theta, \asymp, \smile, N_0, N_1, ... N_k \rangle,$$
where $\mathcal{T}$ is the domain of non-negative integers, and if $t \in \mathcal{T}$, then $\theta t = max(t-1, 0)$, $t < t'$ has its ordinary meaning, $\Theta = 0$, $\asymp$ represents equality, $\smile$ is always true, and $N_i(t) = t + 1$ for all i.

If $\mathcal{E}$ is the domain of action designators, then
$$\mathbf{T}_\mathcal{E} = \langle \mathcal{T}_\mathcal{E}, \theta, <, \Theta, \asymp, \smile, N_0, N_1, ... N_k \rangle,$$

where $\mathcal{T}_\mathcal{E}$ is the set of all pairs $\langle m, \{\langle m_i, E_i\rangle\}_i\rangle$, where m and all m_i are non-negative integers, all m_i are less than m and different, and all $E_i \in \mathcal{E}$. The operators are defined as follows.

$\theta\langle 0, C\rangle = \langle 0, C\rangle$,
$\theta\langle m+1, C\rangle = \langle m, \{\langle m_i, E_i\rangle \in C \mid m_i < m\}\rangle$,
$\langle m, C\rangle < \langle m', C'\rangle$ iff $m < m' \wedge C \subseteq C'$,
$\Theta = \langle 0, \emptyset\rangle$,
$\langle m, C\rangle \asymp \langle m', C'\rangle$ iff $m = m'$,
$\langle m, C\rangle \smile \langle m', C'\rangle$ iff $C = C'$,
$N_0(\langle m, C\rangle) = \langle m+1, C\rangle$,
$N_i(\langle m, C\rangle) = \langle m+1, C \cup \{\langle m, E_i\rangle\}\rangle$ when $E_i \in \mathcal{E}$.

Clearly, this is a correctly formed time structure. The intuition is that, for example,

$\langle 10{:}40, \{\langle 8{:}15, \ulcorner Load\urcorner\rangle, \langle 8{:}40, \ulcorner Drive(\text{Sth})\urcorner\rangle\}\rangle$

is the timepoint at 10:40 hours in a branch of time where the ego initiated a *Load* action at 8:15 hours and a *Drive*(Sth) action at 8:40 hours. In this way we obtain branching time, but in a fashion that preserves the metric aspect in a well-defined way. The operator N_i for $i > 0$ represents an invocation of the action A_i, and N_0 represents waiting a single timestep.

In order to have an intuitive notion for the branching time domain, it is best to imagine a tree where each node t has an immediate or 'vertical' successor $N_0(t)$ representing that one timestep has passed, and that no action has been invoked. In addition each node t may also have 'sideways' successors $N_i(t)$ for $i > 0$. The sideways successors have the same timestamp as $N_0(t)$, that is, one greater than the timestamp of t, but they are differentiated by the fact that an E_i action was invoked at time t. The function θ obtains the predecessor of a node in the tree so formed. The relation $<$ is defined over $\mathcal{T}_\mathcal{E}$, in such a way that $s < t$ iff t is a direct or indirect successor of s, by an arbitrary combination of vertical and sideways steps. The definition of $s \leq t$ as an abbreviation for $s < t \vee s = t$ applies of course for the present definition of $<$ as well.

The relation $\asymp$ holds between two nodes that have the same timestamp, which means that they are 'horizontally equal' and are located at an equal number of steps from the root of the tree. The relation $\smile$ holds between two nodes if they are on the same vertical path, that is, one of them is a direct or indirect vertical successor of the other. For example, the fact that there is a wait period of arbitrary length, but no action invocation, between s and t can be expressed as $s \leq t \wedge s \smile t$.

For convenience, I shall use the following abbreviations, where t is a timepoint and i and k are integers: $t + k = N_0^k(t) = N_0(N_0(...N_0(t)...))$ with k applications, $N_i^k(t) = N_i(t) + (k-1)$, and $t - k = \theta^k(t)$. This means that if $i \neq 0$, then $N_i^k(\langle m, C\rangle) = \langle m + k, C \cup \{\langle m, E_i\rangle\}\rangle$, that is, it is obtained by first taking one 'sideways' successor step and then $k - 1$ 'vertical' steps.

6.6.2 Intervals and trees in branching time

The following example illustrates how branching time is intended to be used. Suppose a history starts without any action. At time 4 there is a fork between action E_1, action E_2, and no action. In the branch resulting from action E_2, there is an E_3 action starting at time 8. In the branch where no action was taken at time 4, there is an E_4 action at time 9. All actions take three timesteps. This action tree can be described by the following formulae, using temporal relations that are defined in the time structure.

scd1 $[s_1, t_1] E_1$
scd2 $[s_1, t_2] E_2$
scd3 $[s_3, t_3] E_3$
scd4 $[s_4, t_4] E_4$
scd5 $\Theta \smile s_1$
scd6 $t_2 \smile s_3$
scd7 $\Theta \smile s_4$
scd8 $s_1 \asymp \langle 4, \emptyset\rangle$
scd9 $s_3 \asymp \langle 8, \emptyset\rangle$
scd10 $s_4 \asymp \langle 9, \emptyset\rangle$
scd11 $t_1 = N_1^3(s_1)$
scd12 $t_2 = N_2^3(s_1)$
scd13 $t_3 = N_3^3(s_3)$
scd14 $t_4 = N_4^3(s_4)$

More compact representations may also be obtained by introducing a few additional, often-useful abbreviations, for example, for writing $s_1 \asymp 4$ to mean $s_1 \asymp \langle 4, \emptyset\rangle$. The following are the timepoints which are necessary in a model for this schedule. The branch without any actions in it consists of

$$\langle 0, \emptyset\rangle,$$
$$\langle 1, \emptyset\rangle,$$
$$\ldots$$
$$\langle 9, \emptyset\rangle,$$

which is where the last action was invoked in that branch. The branch starting as a result of the invocation of E_1 at time 4 consists of

$$\langle 5, \{\langle 4, E_1\rangle\}\rangle,$$
$$\langle 6, \{\langle 4, E_1\rangle\}\rangle,$$
$$\langle 7, \{\langle 4, E_1\rangle\}\rangle.$$

The parallel branch, starting as a result of the invocation of the action E_2 at time 4, is

$$\langle 5, \{\langle 4, E_2\rangle\}\rangle,$$
$$\langle 6, \{\langle 4, E_2\rangle\}\rangle,$$
$$\langle 7, \{\langle 4, E_2\rangle\}\rangle,$$
$$\langle 8, \{\langle 4, E_2\rangle\}\rangle.$$

It must be extended to time 8, where an E_3 action is invoked from it. That branch consists of

$$\langle 9, \{\langle 4, E_2\rangle, \langle 8, E_3\rangle\}\rangle,$$
$$\langle 10, \{\langle 4, E_2\rangle, \langle 8, E_3\rangle\}\rangle,$$
$$\langle 11, \{\langle 4, E_2\rangle, \langle 8, E_3\rangle\}\rangle,$$

and then nothing more happens in it. Finally the branch which resulted from the invocation of E_4 at time 9, contains

$$\langle 10, \{\langle 9, E_4\rangle\}\rangle,$$
$$\langle 11, \{\langle 9, E_4\rangle\}\rangle,$$
$$\langle 12, \{\langle 9, E_4\rangle\}\rangle.$$

Intervals are defined in a natural way in such a structure. For example, the interval $[s_3, t_3]$ during which the event E_3 takes place, is a set with the following four members:

$$\langle 8, \{\langle 4, E_2\rangle\}\rangle$$
$$\langle 9, \{\langle 4, E_2\rangle, \langle 8, E_3\rangle\}\rangle,$$
$$\langle 10, \{\langle 4, E_2\rangle, \langle 8, E_3\rangle\}\rangle,$$
$$\langle 11, \{\langle 4, E_2\rangle, \langle 8, E_3\rangle\}\rangle.$$

It should be clear how a set of timepoints which are formed in this fashion represents a tree structure with $\Theta = \langle 0, \emptyset\rangle$ as its root. The following definitions capture the intended concepts. All the definitions suppose that a domain $\mathcal{E}$ of event designators is previously given.

Definition. An **interval** in $\mathcal{T}_\mathcal{E}$ is a pair $\langle\langle m, C\rangle, \langle m', C'\rangle\rangle \in \mathcal{T}_\mathcal{E} \times \mathcal{T}_\mathcal{E}$, such that $m < m'$, and either $C' = C \cup \{\langle m, E\rangle\}$ for some $E \in \mathcal{E}$, or $C = C'$ and $m' = m + 1$. ⋈

Definition. A **tree** in $\mathcal{T}_\mathcal{E}$ is a set B of intervals in $\mathcal{T}_\mathcal{E}$, such that

- $\langle \Theta, t\rangle \in B$ for some $t \in \mathcal{T}_\mathcal{E}$.

- If $\langle s, t\rangle \in B$ and $s \neq \Theta$, then $\langle u, s\rangle \in B$ for some $u \in \mathcal{T}_\mathcal{E}$. ⋈

These definitions guarantee that B has the intended structure. The following notations will also be used. If $\langle s, t\rangle = \langle\langle m, C\rangle, \langle m', C'\rangle\rangle$ is an interval in $\mathcal{T}_\mathcal{E}$, then the *interval sets* $[s, t]$ and $(s, t]$ are sets of timepoints defined by $(s, t] = \{\langle k, C'\rangle \mid m < k \leq m'\}$ and $[s, t] = \{s\} \cup (s, t]$.

If B is a tree in $\mathcal{T}_\mathcal{E}$, then $\eta(B)$ is defined as the set of timepoints consisting of any n or n' which occur in an interval $\langle n, n'\rangle$ of B, that is, as

$$\{\Theta\} \cup \{n' \mid \langle n, n'\rangle \in B\}.$$

Furthermore, $[B]$ is the union of all $[s, t]$ for all intervals $\langle s, t\rangle \in B$. It is therefore the set of all timepoints that occur in the tree, so the R

component of a development will now be defined for all members of $[B]$ for a certain tree B. It follows from the definition that $\eta(B) \subseteq [B]$ and that $[B] = [B']$ implies $B = B'$. B is therefore completely determined by $[B]$.

A succession of nodes through $[B]$ will be called a path, with the following definition.

Definition. *If B is a tree and $t \in [B]$, then the **path** in B through t is defined as*
$$\{s \in [B] \mid s \leq t \vee \exists k[s = t + k]\}.$$
⋈

The path contains, therefore, all the successive invocations of actions that lead up to the point t, and from there it continues with successor nodes not involving any additional action until the end of the tree.

We need a counterpart of the $\triangleright$ operator that is used for extending a finite history with a trajectory in linear time. The counterpart must also mention the timepoint s in R which is to be extended by the trajectory, since the history tree may grow in arbitrary directions. Furthermore, it must mention the event designator that is used for extending the tree. The counterpart of $R \triangleright v$ will be written $R \triangleleft s, E_i \triangleright v$, and is defined as
$$R \cup \{N_i^j(s) \mapsto R(s) \oplus r_j' \mid 1 \leq j \leq k\},$$
where $v = \langle r_1', ... r_k' \rangle$. It is required that R is defined for s and not for $N_i(s)$. It follows that if R is defined over $[B]$, where B is a tree, and if $R \triangleleft s, E_i \triangleright v$ is defined then its domain is $[B \cup \{\langle s, N_i^k(s)\rangle\}]$. The particular case of extending R without action from a timepoint $s \in [B]$ (corresponding to $R \triangleright \langle \emptyset \rangle$) will be written $R \diamond s$, and is defined by $R \diamond s = R \cup \{s+1 \mapsto R(s)\}$. Notice that here it does not matter whether $R(s+1)$ is already defined, since its value is completely determined by the assumption of strict inertia.

Example. Let $s = \langle 12, \{\langle 4, E_1\rangle, \langle 6, E_2\rangle\}\rangle$, that is, it is the timepoint with the timestamp 12 and where an E_1 action was invoked at time 4 and an E_2 action at time 6. Let $R(s) = r$, and let $v = \langle r_1', r_2'\rangle$ be a trajectory. Then, $R \triangleleft s, E_4 \triangleright v$ is obtained by adding the following two maplets to R:
$$\langle 13, \{\langle 4, E_1\rangle, \langle 6, E_2\rangle, \langle 12, E_4\rangle\}\rangle \mapsto r \oplus r_1'$$
and
$$\langle 14, \{\langle 4, E_1\rangle, \langle 6, E_2\rangle, \langle 12, E_4\rangle\}\rangle \mapsto r \oplus r_2'.$$
⋈

6.6.3 Inhabited dynamical systems with branching time

We proceed to generalize the definition of IDS to the case of branching time, using the possibility of maintaining several concurrent time-threads within one development. The threads are distinguished by the difference in the C components of the timepoints. This means that the ego is allowed to invoke an action at any timepoint where R is defined, except one where an action with the same designator has already been invoked.

First of all, we modify the definition of a development $\langle \mathcal{B}, M, R, \mathcal{A}, \mathcal{C} \rangle$, so that $\mathcal{B}$ is now a tree and not a set of timepoints. If the new definition is applied to a development where the actions occur sequentially, it is intended that if $\mathcal{B}$ is the tree according to the new definition, then $\eta(\mathcal{B})$ shall equal the set $\mathcal{B}$ of timepoints according to the old definition.

The distinction beetween finite and infinite developments (Chapter 1, subsection 1.3.2) is now inconvenient since one tree-formed history may contain both finite and infinite branches. It is replaced by the following unified definition, which is modelled on the old definition for finite developments.

Definition. *A* **development** *(over images in branching time) is a tuple* $\langle \mathcal{B}, M, R, \mathcal{A}, \mathcal{C} \rangle$,
where

- $\mathcal{B}$ *is a tree in* $\mathcal{T}_{\mathcal{E}}$,
- M *is a mapping which assigns values in* $\mathcal{T}_{\mathcal{E}}$ *to some or all of the temporal constant symbols* t_i, *and values in* $\mathcal{O}$ *to all the object constant symbols* o_i,
- R, *the* **history**, *is a mapping from* $[\mathcal{B}]$ *to* $\mathcal{R}$,
- $\mathcal{A}$, *the* **past action set**, *is a set of tuples* $\langle s, E, t \rangle$ *where* s *and* t *($s < t$) are members of* $\eta(\mathcal{B})$,
- $\mathcal{C}$, *the* **current action set**, *is a set of tuples* $\langle s, E \rangle$ *where* s *is a member of* $\eta(\mathcal{B})$. ⋈

We use $\mathcal{J}^\circ$ to denote the domain of branching-time developments. Notice that this definition works even when $\mathcal{B}$ contains an infinite progression of timepoints $\langle n_i, C \rangle$ with increasing n_i but fixed C, which will be the case in each path after all actions in that path have been performed, and onwards to infinity. Accordingly:

Definition. *A tree B is* **complete** *iff $\langle m, C \rangle \in [B]$ and $m < m'$ implies $\langle m', C \rangle \in [B]$. A branching-time development $\langle \mathcal{B}, M, R, \mathcal{A}, \mathcal{C} \rangle$ is complete iff $\mathcal{B}$ is complete.* ⋈

The definition of correct revision is modified as follows.

Definition. *Let $J = \langle \mathcal{B}, M, R, \mathcal{A}, \mathcal{C} \rangle$ and $J' = \langle \mathcal{B}', M', R', \mathcal{A}', \mathcal{C}' \rangle$ be two developments for branching time. J' is said to be a* **correct revision** *of J iff the following conditions hold. $\mathcal{B}' = \mathcal{B}$ or $\mathcal{B}' = \mathcal{B} \cup \{b\}$, where $b = \langle n, n' \rangle$*

is an interval for some $n \in \eta(\mathcal{B})$, $M \subseteq M'$, $\mathcal{A} \subseteq \mathcal{A}'$, $R \subseteq R'$, and the restriction of R' to $[\mathcal{B}]$ shall equal R. For every $\langle s, E, t \rangle \in (\mathcal{A}' - \mathcal{A})$, it must be the case that $t = n'$, and that $\langle s, E \rangle$ is a member of $\mathcal{C}$ but not of $\mathcal{C}'$. For every $\langle s, E \rangle \in (\mathcal{C} - \mathcal{C}')$, it must be the case that $s = n$ or $\langle s, E, n' \rangle \in \mathcal{A}'$. Finally, for every $\langle s, E_i \rangle \in (\mathcal{C}' - \mathcal{C})$, it must be the case that $s = n$ and $N_i(n) \notin [\mathcal{B}]$. ⋈

Comparing this definition to the one for linear time, we notice that the present one is more concise, and that it actually works as well for linear time as for branching time. However, it is a bit more difficult to grasp due to the greater complexity of timepoints in the branching time domain.

The definition of ego remains unchanged, except for the change of domain:

Definition. *A* **ego K** *in branching time is a mapping* $\mathcal{J}° \to \mathcal{J}°$, *where* $\mathbf{K}(J)$ *is a correct revision of* J, *and if*
$$\mathbf{K}(\langle \mathcal{B}, M, R, \mathcal{A}, \mathcal{C} \rangle) = \langle \mathcal{B}', M', R', \mathcal{A}', \mathcal{C}' \rangle,$$
then it must be the case that $\mathcal{B} = \mathcal{B}'$. ⋈

The definition of world is modified to the following.

Definition. *A* **world W** *in branching time is a transformation from* $\mathcal{J}°$ *to* $\mathcal{J}°$, *that is, a subset of* $\mathcal{J}° \times \mathcal{J}°$, *such that if* $J \in \mathcal{J}$, *then* $\mathbf{W}(J, J')$ *for some* J', *and if*
$$\mathbf{W}(\langle \mathcal{B}, M, R, \mathcal{A}, \mathcal{C} \rangle, \langle \mathcal{B}', M', R', \mathcal{A}', \mathcal{C}' \rangle),$$
then the second argument is a correct revision of the first one, and the set $\mathcal{C}' \subseteq \mathcal{C}$. ⋈

This definition is also simpler than the one for linear time, partly because some of the essence of the definition has now been moved to the definition of correct revision, and partly because this definition does not allow the case where the world extends the history to infinity in one stroke. There has to be an infinite sequence of finite history extensions.

With these definitions, each history is defined over a tree that may contain finite and/or infinite branches. The moves of the ego consist of choosing a suitable timepoint and direction where the tree can be extended, but without actually making the extension. The ego invokes an action by adding a member to $\mathcal{C}$, as before. Technically, the ego may invoke several actions, but then only one of them will be executed. The world identifies the action (or an action) that has been invoked by the ego, and extends the tree in an appropriate manner from the point chosen by the ego.

On this basis we specify the adaptation of the definition of trajectory-semantics worlds from Chapter 3, subsection 3.2.4.

Definition. *An IDS branching-time world* **W** *is the* **corresponding world** *of a world description* $\langle \text{Infl}, \text{Trajs} \rangle$ *iff it satisfies the following* conditions. If $J = \langle \mathcal{B}, M, R, \mathcal{A}, \mathcal{C} \rangle$ and R is defined over $[\mathcal{B}]$, then $\mathbf{W}(J, J')$ iff $J' = \langle \mathcal{B} \cup \{\langle n, n' \rangle\}, M, R', \mathcal{A}', \emptyset \rangle$, where either of the following holds:

- $\mathcal{C} = \emptyset$, $n' = n + 1$, $R' = R \diamond n$, and $\mathcal{A}' = \mathcal{A}$;
- For some timepoint n and some action designator E_i it holds that $\langle n, E_i \rangle \in \mathcal{C}$, and for some trajectory $v = \langle r'_1, r'_2, ... r'_k \rangle$ which is a member of $\mathrm{Trajs}(E, R(n))$, it holds that $n' = N_i^k(n)$, $R' = R \triangleleft n, E_i \triangleright v$, and $\mathcal{A}' = \mathcal{A} \cup \{\langle n, E_i, n' \rangle\}$.

A branching-time **trajectory-semantics world** *is a branching-time IDS world which corresponds to some world description.* ⋈

The definition of trajectory-semantics ego is unchanged:

Definition. *A* **trajectory-semantics ego** *is an IDS ego* $\mathbf{K}$ *such that, if* $\mathbf{K}(\langle \mathcal{B}, M, R, \mathcal{A}, \mathcal{C} \rangle) = \langle \mathcal{B}', M', R', \mathcal{A}', \mathcal{C}' \rangle$, *then* $\mathcal{A} = \mathcal{A}'$, $\mathcal{C} \subseteq \mathcal{C}'$, *and* $\mathcal{C}' - \mathcal{C}$ *has at most one member.* ⋈

We are now only interested in infinite games since, with the present definitions, one can not obtain a finished development after a finite game. The definition of infinite game in Chapter 1, subsection 1.3.4, is revised as follows.

Definition. *An* **IDS game** *for a branching time ego* $\mathbf{K}$, *a branching time world* $\mathbf{W}$, *an initial state* r_0, *and an initial valuation* M_0, *is an infinite sequence of developments,*
$$J_0, J'_0, J_1, J'_1, ...,$$
where
$$J_0 = \langle \{\Theta\}, M_0, \{\Theta \mapsto r_0\}, \emptyset, \emptyset \rangle,$$
$$J'_i = \mathbf{K}(J_i) \text{ for } i \geq 0,$$
$$\mathbf{W}(J'_i, J_{i+1}) \text{ for } i \geq 0.$$
If $J_i = \langle \mathcal{B}_i, M_i, R_i, \mathcal{A}_i, \mathcal{C}_i \rangle$ *for every* i, *then a* **resulting development** *from the game is any complete development of the form*
$$\langle \lim_{i \to \infty} \mathcal{B}_i, M, \lim_{i \to \infty} R_i, \lim_{i \to \infty} \mathcal{A}_i, \lim_{i \to \infty} \mathcal{C}_i \rangle,$$
where M *is a complete valuation and* $\lim_{i \to \infty} M_i \subseteq M$. ⋈

The set of developments $Mod(\Upsilon)$ corresponding to a chronicle Υ was defined in Chapter 2, subsection 2.3.2, and the same definition holds without modification:

Definition. *For a given scenario description* Υ *and time domain* $\mathcal{T}$, $Mod(\Upsilon)$ *shall denote the set of complete developments which are correctly described by* Υ. ⋈

The informal description of 'correctly described by' that was found there, is equally applicable for linear and for branching time-domains $\mathcal{T}$. At this point, we have caught up with the account for linear time in the previous chapters. A comparison of the definitions for linear time and for branching time, item by item, shows that they are very similar, and in fact the definitions for branching time are often more concise. It would be possible to write all definitions for branching time right from the start, and to treat linear time as a special case, but the greater complexity of the branching-time domain would probably make such a presentation more difficult to follow.

The same observation holds for the chapters that come next. Rather than developing the whole text for branching as well as linear time, I have written the main parts of the presentation for linear time. At certain points there are special subsections, analogous to the present one, which describe how to adapt definitions and theorems in the immediately preceding chapter or section to the case of branching time. I hope that this will give maximum clarity both for the reader who wishes to focus on the simpler case of linear time, and for the reader who wants to verify that the generalization works properly.

6.7 Summary

A general definition of discrete time structure has been introduced. The base logic DFL-1 has been defined, with a syntax and a semantics that use timepoints in a discrete time structure as one of its sorts. A separate side language for occurrences has also been introduced, but only in terms of its syntax. Finally, the discrete time structure for branching metric time has been introduced, and those concepts and definitions in previous chapters which had previously only been stated for integer time, have now been extended to the case of branching metric time.

7
Chronicle completion in $\mathcal{K}$-IA

As defined in Chapter 2, subsection 2.3.1, a chronicle in $\mathcal{K}$-**IA** is a tuple of the form

$$\langle \mathcal{K}, \mathcal{O}, \langle \mathbf{IA}, \mathrm{LAW} \rangle, \mathrm{SCD}, \mathrm{OBS} \rangle.$$

The DFL logic from Chapter 6 provides a formal language for expressing the action laws, the schedule, and the observations. The present chapter uses the trajectory semantics to define the relationship between a $\mathcal{K}$-**IA** chronicle consisting of DFL formulae on one hand, and a set of IDS developments on the other. It applies equally for linear and for branching time.

7.1 Chronicle completion

Let us begin with an example. The Stockholm Delivery Scenario was introduced in Chapter 2, subsection 2.1.1. To rewrite its statements in DFL-1 syntax, we use the similarity type $\{a : \{\mathbf{T}, \mathbf{F}\}, b : \{\mathbf{Sth}, \mathbf{Lkp}\}, c : \{\mathbf{Sth}, \mathbf{Lkp}\}\}$, where a represents whether the box is currently loaded into the car or not, b represents the current location of the box, and c represents the current location of the car. The locations are represented as **Sth** for Stockholm and **Lkp** for Linköping; it is straightforward to introduce additional locations. The premises can then be written as:

    obs1 $[0] \neg a$
    obs2 $[0] b \hat{=} \mathbf{Lkp}$
    obs3 $[0] c \hat{=} \mathbf{Lkp}$
    law1 $[s,t] \mathit{Load} \Rrightarrow [s,t] a := \mathbf{T}$
    law2 $[s,t] \mathit{Drive}(\mathbf{Sth}) \Rrightarrow [s,t] c := \mathbf{Sth} \wedge$
                     $([s]a \to [s,t]b := \mathbf{Sth})$
    scd1 $[8{:}15, 8{:}20] \mathit{Load}$
    scd2 $[8{:}40, 11{:}15] \mathit{Drive}(\mathbf{Sth})$

This is a direct translation of the natural-language formulation in Chapter 2, which was as follows:

obs1 The box is not in the car at time 0
obs2 The box is in Linköping at time 0
obs3 The car is in Linköping at time 0
law1 If the box is loaded into the car from time s to time t, then the box is in the car at time t
law2 If the car is driven to L during the interval from time s to time t, then the car is in L at time t.
If the box was in the car at time s, then the box will also be in L at time t.
scd1 The box is loaded from time 8:15 to time 8:20
scd2 The car is driven to Stockholm from time 8:40 to time 11:15

The notation with one tagged formula per line is just a convenient way of writing out the chronicle description. The above defines the set OBS of formulae to be the set

$$\{\ulcorner[0]\neg a\urcorner, \ulcorner[0]b \hat{=} \mathrm{Lkp}\urcorner, \ulcorner[0]c \hat{=} \mathrm{Lkp}\urcorner\},$$

and similarly for the other formula-set components of the chronicle.

Since DFL-1 does not allow quantification over items (feature-values), it was necessary to specialize the action law for *Drive* to the case of one particular location argument. In general, one would of course have an action-law schema for that purpose.

Let Υ be this chronicle. The set $Mod(\Upsilon, \mathrm{RM})$ of weakened models for Υ which was defined in Chapter 2 will be used as the set of intended models. The main problem under study is how it can be expressed using LAW, SCD, OBS, and the customary operations on sets of formulae and of models.

According to definitions, $Mod(\Upsilon, \mathrm{RM})$ can be constructed as follows in this particular case. First, determine that IDS world **W** which is defined by the set {law1, law2} of action laws under the ontological assumption **IA**. Using the trajectory semantics from Chapter 3, construct the world description $\langle \mathtt{Infl}, \mathtt{Trajs} \rangle$ that corresponds to these two laws, and convert the world description to a corresponding IDS world **W**.

The step of going from the two action laws to the corresponding world description $\langle \mathtt{Infl}, \mathtt{Trajs} \rangle$ has not yet been defined; it will appear in Chapter 8, section 8.1. Here I will just write it out for the specific SDS example, so that the general idea is clear. The intended $\mathtt{Infl}$ component satisfies $\mathtt{Infl}[Load, r] = \{\ulcorner a \urcorner\}$ for all r, and $\mathtt{Infl}[Drive(d), r] = \{\ulcorner b \urcorner, \ulcorner c \urcorner\}$ in case $r[a] = \mathtt{T}$, that is, if the box is loaded into the car in state r. For other r, $\mathtt{Infl}[Drive(d), r] = \{\ulcorner c \urcorner\}$ since only the car changes its position as the result of the driving action in such a case.

Furthermore, the intended $\mathtt{Trajs}$ component is chosen so that $\mathtt{Trajs}[Load, r]$ is the set of trajectories of arbitrary finite length ≥ 1 which are defined only over $\ulcorner a \urcorner$, and where the last element of the sequence is $\{a : \mathtt{T}\}$, that is, the box is inside the car at the end of the trajectory. The value of a in earlier elements of the sequence is unconstrained.

Similarly, Trajs[$Drive(d), r$] for any d (for example Sth) is a set of trajectories assigning values to b and c or only to c, as determined by Infl[$Drive(d), r$]. The last element of the sequence must be $\{b : \text{Sth}, c : \text{Sth}\}$ or just $\{c : \text{Sth}\}$, representing that the car and (when loaded) the box is in Stockholm at the end of the action. As for Load, the values of b and c in earlier states during the trajectory are unconstrained.

After ⟨Infl, Trajs⟩ has been constructed, one obtains the corresponding IDS world **W** according to the definition in Chapter 3, subsection 3.2.4. Next, consider the set of all games that can be played between **W** and an arbitrary ego. According to the observation in Chapter 3, it is sufficient to consider games with trajectory-semantics egos.

Let W be the set of developments that result from those games. Now obtain a subset W' of W consisting of those developments ⟨$\mathcal{B}, M, R, \mathcal{A}, \mathcal{C}$⟩ which satisfy two conditions. $\mathcal{A}$ must be

$$\{⟨8{:}15, \ulcorner Load \urcorner, 8{:}20⟩, ⟨8{:}40, \ulcorner Drive(\text{Sth})\urcorner, 11{:}15⟩\},$$

expressing that the set of actions that were performed during the development exactly equals the set of the two actions mentioned in SCD. Also, the three observations for time 0 must be true in R, that is,

$$R(0) = \{a : \text{F}, b : \text{Lkp}, c : \text{Lkp}\}.$$

The fact that the development has been obtained from a game played by the world **W** constructed above, means that the constraint on $\mathcal{A}$ is a constraint indirectly as well R during the time periods $[8{:}15, 8{:}20]$ and $[8{:}40, 11{:}15]$. Furthermore, because of the construction of **IA** worlds, there cannot be any changes of value in R except those which are defined by the trajectories of the actions.

In the Stockholm Delivery Scenario there is no use of constant symbols for timepoints or objects, so the choice of M is unconstrained. However, if the first action were replaced by, for example,

scd1 $[t_1, 8{:}20]$ *Load*,

then the requirement on $\mathcal{A}$ would be that it must equal

$$\{⟨M[t_1], \ulcorner Load \urcorner, 8{:}20⟩, ⟨8{:}40, \ulcorner Drive(\text{Sth})\urcorner, 11{:}15⟩\},$$

so the same timepoint must appear as the value of t_1 according to M, and as the starting point of the *Load* action according to $\mathcal{A}$.

Finally, from W' one constructs $Mod(\Upsilon, \text{RM})$ as

$$\{⟨M, R⟩ \mid ⟨\mathcal{B}, M, R, \mathcal{A}, \mathcal{C}⟩ \in W' \text{ for some } \mathcal{B}, \mathcal{A}, \mathcal{C}\}.$$

The semantics for DFL-1 has been defined in such a way that this is also a set of DFL-1 models, and in particular it is the set of intended models for the given scenario description Υ.

This example illustrated the general idea of how to obtain the set of intended models for a given chronicle. It remains for this chapter to formulate the same ideas in general form, and in particular to specify how to get from a set of action laws to the corresponding IDS world. However, first we shall also discuss chronicle completion from the point of view of common-sense reasoning.

The actions in the Stockholm Delivery Scenario are only described in terms of their preconditions and postconditions. Notice, however, that the DFL logic is expressive enough to allow constraints on feature values within the action trajectory, or on the duration of the action. For example, the action law for $Drive(\mathtt{Sth})$ might be refined to

$$\mathtt{law2} \quad [s,t]\,Drive(\mathtt{Sth}) \Rrightarrow t = s + 2{:}35 \land [s,t]c := \mathtt{Sth} \land$$
$$\exists t_1 \exists t_2 [s < t_1 < t_2 \leq t \land [s,t_1)\,c \hat{=} \mathtt{Lkp} \land$$
$$[t_1,t_2)\,c \hat{=} \mathtt{Enroute} \land [t_2,t]\,c \hat{=} \mathtt{Sth}] \land$$
$$([s]a \to [s,t]b := \mathtt{Sth} \land [s,t]b = c),$$

which says that the trip takes $2{:}35 = 155$ minutes, the successive locations of the car are $\mathtt{Lkp}$, $\mathtt{Enroute}$, and $\mathtt{Sth}$, and if the box was loaded when the action started, it will have the same location as the car at each point in time during the action. The more specific action law corresponds to more restricted trajectory sets $\mathtt{Trajs}[Drive(d),r]$.

In practice, of course, it is unlikely that one will choose this way of characterizing the detailed execution of an action. It will probably be more convenient to characterize actions as aggregates of simpler actions, thus obtaining the ontological family **IAH** where hierarchical actions are allowed. For example, one may consider the use of modified chronicle descriptions in order to specify complex actions. Such modified extensions are best addressed after the present basic case has been analyzed and understood, however.

7.2 Examples of chronicle completion in common-sense domains

The usual methodology in knowledge representation research is to assume that the set of intended models is given by intuition, and to let the search for an adequate logic be guided mostly by representative examples and counterexamples. Although the definition of intended models as $Mod(\Upsilon,\mathrm{RM})$ has the advantage of being much more precise, it is useful to double-check the work against the kinds of common-sense scenarios that are used in the conventional methodology. The following is a set of representative examples of chronicle completion for $\mathcal{K}$-**IA**, several of which have been published previously and are being extensively used in the AI literature. Although these examples of common-sense logical reasoning may appear to be very simple, it turns out that few of the approaches to non-monotonic temporal reasoning that have been published to date can obtain the intended conclusions even in all of these examples.

Actions will generally be written on the form $[s_i,t_i]E_i$, where the timing of actions is given by temporal constant symbols. Such examples are

Examples of chronicle completion in common-sense domains 151

therefore not specific to single-timestep actions, but may be specialized to that case by adding premises of the form $s_i = \theta t_i$. In some other examples I write out explicit (numerical) timepoints, in order to emphasize that the example regards reasoning about the duration of actions, or simply for concreteness. In several of the examples I do not write out actions (such as *Load*, *Wait*, or *Fire*) explicitly; those examples are expressed directly using reassignment formulae for the relevant fluents. It is understood that such formulae in the schedule have been obtained as translations of action statements.

Since it would be inconvenient to write out the entire set of models for a given chronicle, and equally inconvenient to read it, I will characterize the selected sets using a solution of the chronicle completion problem for the respective chronicle Υ. In other words, I specify one or more extensions α_i, each of which is a set of simple formulae, where $\bigcup_i [\![\alpha_i]\!] = Mod(\Upsilon, \text{RM})$. By the nature of things we must leave to the reader to judge whether, or to what extent, the set of these extensions corresponds to the intended common-sense conclusions from the chronicle.

7.2.1 Yale shooting scenario (YSS)

This example is due to Hanks and McDermott [32]. The Stockholm Delivery Scenario is essentially a variant of the Yale shooting scenario.

World

The world is characterized by two truth-valued features, specifying whether a certain gun is loaded, and whether a certain bird is alive. More specifically, the bird is often described as a turkey, and referred to as Fred. There are two actions, namely loading the gun and firing it. Depending on the choice of base logic, a third action for 'wait' may also be needed technically. Loading the gun has the effect that the gun is loaded, regardless of whether it was loaded before or not. If the gun is loaded at the beginning of firing, then the bird is not alive and the gun is unloaded at the end of firing; otherwise firing has no effect.

Schedule and observations

Initially, the bird is alive and the gun is not loaded. Then the events of gun-loading, waiting, and firing take place in succession.

Intended conclusion

At the end of firing, the bird is not alive.

Significance and pitfalls

The YSS illustrates temporal prediction, that is, reasoning from a known fact at an earlier point in time to a concluded fact at a later point in time. The example is significant in spite of its apparent simplicity because straightforward application of some early non-monotonic logics obtains unintended models [32, 33].

DFL formalization

Let a for 'alive' and l for 'loaded' be truth-valued feature symbols, and let $Load$ and $Fire$ be action symbols. The scenario is

    obs1 $[0]\,a, \neg l$
    scd1 $[s_1, t_1]\,Load$
    scd2 $[s_2, t_2]\,Fire$
    scd3 $t_1 < s_2$

The appropriate action laws are:

    law1 $[s, t]\,Load \Rightarrow [s, t]\,l := \text{T}$
    law2 $[s, t]\,Fire \Rightarrow [s]\,l \rightarrow [s, t]\,a := \text{F}, l := \text{F}$

Expansion of the abbreviations for actions gives the premises as follows:

    obs1 $[0]\,a, \neg l$
    scd1 $[s_1, t_1]\,l := \text{T}$
    scd2 $[s_2]\,l \rightarrow [s_2, t_2]\,a := \text{F}, l := \text{F}$
    scd3 $t_1 < s_2$

The following single extension is intended:

    obs1 $[0, s_1]\,a, \neg l$
    obs2 $(s_1, t_1)\,a$
    obs3 $[t_1, s_2]\,a, l$
    obs4 $[t_2, \infty)\,\neg a, \neg l$
    obs5 $s_1 < t_1 < s_2 < t_2$

Please recall that a comma between literal fluent formulae or reassignment expressions indicates conjunction, for example in $[t_1, s_2]\,a, l$.

7.2.2 Hiding turkey scenario (HTS)

This example was first published in a review version of the manuscript for the present book, [73]. An analogous example had previously been described by Schubert in [76]: a brittle object breaks if it falls; we do not know whether a particular object is brittle.

World

The world is as for the YSS, with the following additions. There are two additional truth-valued features, representing whether the turkey is deaf or

Examples of chronicle completion in common-sense domains 153

not, and whether it is hiding or not. If the turkey is not deaf, then when the gun is loaded it will go into hiding (because it hears the sound which is made when loading the gun). Firing only kills the turkey if it is not hiding.

Schedule and observations

Initially the turkey is alive and not hiding, and the gun is not loaded. It is unknown whether the turkey is deaf. Then the events of gun-loading, waiting, and firing take place in succession, as in YSS.

Intended conclusion

At the end of firing, either the turkey is deaf and dead, or non-deaf and alive.

Significance

An example where the initial state is incompletely specified, in a way that matters for the continued development.

DFL formalization

Let d for 'deaf' and h for 'hiding' be additional, truth-valued feature symbols, otherwise as before. After expansion, the premises are as follows:

obs1 $[0] a, \neg l, \neg h$
scd1 $[s_1, t_1] l := \text{T}$
scd2 $[s_1] \neg d \rightarrow [s_1, t_1] h := \text{T}$
scd3 $[s_2] l \rightarrow [s_2, t_2] l := \text{F}$
scd4 $[s_2] l, \neg h \rightarrow [s_2, t_2] a := \text{F}$
scd5 $t_1 < s_2$

The following two extensions are intended:

obs1 $[0, \infty) d, \neg h$
obs2 $[0, s_1] a, \neg l$
obs3 $(s_1, t_1) a$
obs4 $[t_1, s_2] a, l$
obs5 $[t_2, \infty) \neg a, \neg l$
obs6 $s_1 < t_1 < s_2 < t_2$

and

obs1 $[0, \infty) \neg d, a$
obs2 $[0, s_1] \neg l, \neg h$
obs3 $(s_1, t_1) l, h$
obs4 $[t_1, s_2] h$
obs5 $[t_2, \infty) \neg l, h$
obs6 $s_1 < t_1 < s_2 < t_2$

7.2.3 Stanford murder mystery (SMM)

This example is due to Ginsberg and Baker [5].

World

Same world as for the Yale shooting scenario.

Schedule and observations

The turkey is alive in the initial situation. The actions of firing and waiting are performed in succession, and then the turkey is not alive.

Intended conclusions

The gun was initially loaded, and the turkey was not alive at the end of the firing action.

Significance and pitfalls

A simple example of temporal postdiction: given some facts at a later point in time, obtain conclusions about an earlier point in time. Some proposed approaches obtain unintended models.

DFL formalization

Same feature symbols as for YSS. Premises:
- obs1 $[0]a$
- scd1 $[s_1]l \rightarrow [s_1, t_1]a := F, l := F$
- obs2 $[t_2]\neg a$
- obs3 $t_1 < t_2$

Intended single extension:
- obs1 $[0, s_1]a, l$
- obs2 $[t_2, \infty) \neg a, \neg l$
- obs3 $s_1 < t_1 < t_2$

7.2.4 Ferryboat connection scenario (FCS)

This example has previously only been published in the first review version of this manuscript [73]. It was inspired by a short story by the Danish author and Nobel prize laureate, Johannes V. Jensen.

World

Motorcycle having position 'On Island Fyen', 'At ferryboat landing', 'On ferryboat', or 'In Jutland'.

Examples of chronicle completion in common-sense domains 155

Schedule and observations

A motorcycle is initially driving along a road on island Fyen, in the direction of a ferryboat landing. The ferryboat departs at time 23:01, so the last time for arrival is 23:00. If the motorcycle is on board at time 23:01 then it will arrive to Jutland between the time 23:45 and 23:50, otherwise it stays. The motorcycle is known to reach the landing some time between time 22:55 and 23:05.

Intended conclusions

The motorcycle is either at the ferryboat landing or in Jutland at time 23:55.

Significance

An example of reasoning about the duration of actions.

DFL formalization

The motorcycle position m is either F (on Fyen and not at landing), L (at landing and not on board), B (on board), or J (in Jutland). The premises are as follows (if the motorcycle boards as late as possible):

obs1 $[22:00]\, m \hat{=} F$
scd1 $[\theta t_1, t_1]\, m := L$
scd2 $22{:}55 \leq t_1 \leq 23{:}05$
scd3 $[23{:}00]\, m \hat{=} L \to [23{:}00, 23{:}01]\, m := B$
scd4 $[23{:}01]\, m \hat{=} B \to [t_2]\, m := J$
scd5 $23{:}45 \leq t_2 \leq 23{:}50$

7.2.5 Russian turkey scenario (RTS)

This is a modification of problems used by Haugh [34] and by Goodwin and Trudel [30]. It was first published in an early review version of this book [73].

World

Same as for the Yale shooting scenario, but with the addition of one more action for spinning the gun's chamber. The effect of this action is that, randomly, the gun may or may not be loaded after the action, regardless of whether it was loaded before or not.

Schedule and observations

Initially the gun is unloaded and the turkey is alive. The actions of loading, spinning, and firing take place in succession.

Intended conclusion

Two distinct completions are intended: one where the gun becomes unloaded as the result of spinning, and firing leads to no change; and one where spinning leads to no change, and the effect of firing is as in the original YSS.

Significance

An example with a nondeterministic action.

DFL formalization

Assume that the loading takes place during the time interval $[s_1, t_1]$, the spinning during the time interval $[s_2, t_2]$, and the firing takes place during the time interval $[s_3, t_3]$:

    obs1 $[0] a, \neg l$
    scd1 $[s_1, t_1] l := \text{T}$
    scd2 $[s_2, t_2] l := \text{TF}$
    scd3 $[s_3] l \rightarrow [s_3, t_3] a := \text{F}, l := \text{F}$
    scd4 $s_1 < t_1 \leq s_2 < t_2 \leq s_3 < t_3$

The following two extensions are intended:

    obs1 $[0, s_1] a, \neg l$
    obs2 $(s_1, t_1) a$
    obs3 $[t_1, s_2] a, l$
    obs4 $(s_2, t_2) a$
    obs5 $[t_2, \infty) a, \neg l$

and

    obs1 $[0, s_1] a, \neg l$
    obs2 $(s_1, t_1) a$
    obs3 $[t_1, s_2] a, l$
    obs4 $(s_2, t_2) a$
    obs5 $[t_2, s_3] a, l$
    obs6 $[t_3, \infty) \neg a, \neg l$

7.2.6 Stolen car scenario (SCS)

This example was first described without solution by Kautz in [38]. It was also used by Baker [5], who proposed a solution that obtains the intended

Examples of chronicle completion in common-sense domains 157

conclusions for this example.

World

One propositional fluent, for 'the car is in my possession'. One action, for 'leaving the car overnight in the garage'. This action normally has no effect at all, but in unusual cases the car gets stolen, that is, the proposition that the car is in my possession, goes from true to false. Once stolen, the car will not become unstolen by leaving it in the garage again. Cars are absolutely never stolen in the daytime.

Schedule and observations

Initially the car is in my possession. I leave it in the garage for two successive nights. The next evening it is not in my possession.

Intended conclusions

In the morning after the second night, the car was not in my possession. No conclusion is intended with respect to whether the car was still in my possession after the first night in the garage.

Significance

An example of an action with surprises, and with postdiction. This example belongs to the ontological family $\mathcal{K}$-**IS** and is therefore outside the topic of this chapter.

7.2.7 Ticketed car scenario (TCS)

This is a new example.

World

The world is similar to the stolen car scenario, but 'car gets stolen' is replaced by 'car gets a parking ticket'. It is assumed that the car is left in a place where daytime parking is allowed, but night parking is not allowed. The key difference from the previous case is that getting a ticket is not an exception: in any given night the car will, randomly, be ticketed or not be ticketed. Once ticketed it does not become unticketed by leaving it one more night.

Schedule and observations

The car is unticketed on January 25 at noon. It is left in the street during two successive nights. On January 27 at 17:00 hours it is ticketed.

Intended conclusions

In the morning of January 27, the car was already ticketed. No conclusion is intended with respect to whether the car was still unticketed on January 26.

Significance

An example of an action with random effects, and with postdiction.

DFL formalization

Let the feature symbol p represent 'the car is not ticketed'. The notation 25:12 is used, ad hoc, to represent the day and the hour. The premises are as follows.

    obs1 $[25{:}12]\,p$
    scd1 $[25{:}23]\,p \to (25{:}23, 26{:}06]\,p := \mathsf{TF}$
    scd2 $[26{:}23]\,p \to (26{:}23, 27{:}06]\,p := \mathsf{TF}$
    obs2 $[27{:}17]\,\neg p$

7.2.8 Furniture assembly scenario (FAS)

This is a new example.

World

The world consists of a furniture kit which is to be assembled by the customer. There are two features: either the kit is assembled or not, and either the kit contains the assembly instructions or it does not. There is an action for assembling the kit. Provided that the kit was previously unassembled, this action takes 20 minutes if the instructions were included, and 60 minutes otherwise.

Schedule and observations

Initially the kit is unassembled, and it is not known whether the instructions are included or not.

Intended conclusions

Two cases: either the instructions were included and the kit is assembled within 20 minutes, or the instructions were not included and the kit is assembled within 60 minutes.

Significance

An example of an action with conditional duration.

DFL formalization

The propositional fluents a and i are used for 'the furniture is assembled' and 'the instructions were included', respectively. The action symbol *Asmbl* is used for the action of assembling the furniture. The premises are:

    obs1 $[0] \neg a$
    scd1 $[100, t]\, Asmbl$

The appropriate action law is:

    law1 $[s,t]A \Rrightarrow ([s]\neg a, i \rightarrow t = s + 20 \land [s,t]a := \mathrm{T}) \land$
                  $([s]\neg a, \neg i \rightarrow t = s + 60 \land [s,t]a := \mathrm{T}) \land$
                  $([s]a \rightarrow t = s + 1)$

7.3 The **IA** family of IDS worlds

We return now to the systematic methodology. The characteristics and subcharacteristics that were introduced in the first two chapters will serve as the classification whereby the range of applicability of each entailment method can be identified. Since the trajectory semantics has been introduced in Chapter 3, it is possible to give an entirely precise definition of the nodes in the ontological taxonomy. The present section will define the family of strictly inert IDS worlds without dependencies, which is represented as **IA**, and its subcharacteristics. This ontological family will be the topic of the following chapters.

7.3.1 Definition of the IA ontological family

IA was defined in general terms as the set of IDS worlds where there is strict inertia, non-determinism, actions with extended duration, but none of the complications mentioned in Table 1.1 in Chapter 1, such as concurrent actions, surprises, dependencies (ramification), etc. We now proceed to the precise definition of **IA** using trajectory semantics, which was defined in Section 3.2. The use of this semantics makes a number of general assumptions. It assumes that the image-level description of the IDS world is exhaustive, that is, either the material level equals the image level, or at least the material level does not provide any additional information for the purpose of deduction. Time is assumed to be discrete, and chosen as the non-negative integers.

In addition there is an inherent assumption which becomes significant when the set $\texttt{Trajs}(E, r)$ of possible trajectories is infinite, namely that the set must be *describable* using the present logic. Since the members of the set are sequences of states over $\texttt{Infl}(E, r)$, and the number of such states is finite, the set $\texttt{Trajs}(E, r)$ can be considered as a formal language over a finite alphabet. For example, if the words in that language are unrestricted except for their last letter, which is the case in the subset of trajectory semantics that corresponds to encapsulated semantics, then there is no problem. However, one is free to consider any formal language there, and naturally if the logic is not able to characterize the set or language as such, there is no hope that it will be able to characterize precisely the possible histories that are formed from these actions.

In practical cases this does not seem to be problematic at all, so it is probably harmless to assume that every $\texttt{Trajs}(E, r)$ is a describable set.

An additional assumption is that the identities of objects and timepoints are unimportant for world descriptions $\langle \texttt{Infl}, \texttt{Trajs} \rangle$. For example, it does not matter if one permutes the names of two objects #1 and #2. (Formally it would be appropriate to define the world description as a quotient set with respect to $\mathcal{O}$). Furthermore, an action cannot have different sets of possible results depending on the time at which it is started.

Constraints due to the $\mathcal{K}$ epistemological assumption

An IDS world described as $\langle \texttt{Infl}, \texttt{Trajs} \rangle$ is appropriate for the $\mathcal{K}$ epistemological assumption iff $\texttt{Trajs}(E, r) \neq \emptyset$ for every E and r. The purpose of this restriction is to assert that for every combination of action designator and current state, there is a definition (although possibly a nondeterministic one) of how to perform the action from that state. In other words, actions are always feasible. Without this restriction, one obtains the type of problems which are customarily studied in knowledge-based planning, where actions are only feasible if certain preconditions are satisfied.

Notice the use of the expression 'is appropriate for' rather than 'satisfies'. The epistemological assumption is made by us when writing out and interpreting a chronicle; it is not meaningful to ask whether an IDS world satisfies $\mathcal{K}$ or not. However, if $\texttt{Trajs}(E, r) = \emptyset$ for some E and r then it is not possible to make the $\mathcal{K}$ assumption in the context of that world, within the underlying trajectory semantics as defined here.

*Definition of the **IAD** family*

The ontological family **IAD** is defined as the set of all trajectory-semantics worlds where $\text{Trajs}(E, r) \neq \emptyset$ for every E and r. This restriction is made in order that **IAD** worlds will be appropriate for use with the $\mathcal{K}$ epistemological assumption.

*Definition of the **IA** family*

The more restricted ontological family **IA** is obtained from **IAD** by imposing a restriction of strict locality on the world, as follows.

If O is a set of objects then let $\mathcal{F}/O \subseteq \mathcal{F}$ be the set of those features where all arguments are members of O. If r is a state then let $r/O \subseteq r$ be the partial state which is obtained from r by restricting its arguments to $\mathcal{F}/O$.

Let $E = \ulcorner A(o_1, ..., o_k)\urcorner$ be an action designator, that is, the arguments are either objects or items. I write $args(E)$ for the set of objects which occur as arguments in E. A world description $\langle \text{Infl}, \text{Trajs} \rangle$ is said to have *strict locality* iff for all E, r, and r'

$r/args(E) = r'/args(E) \rightarrow$

$\text{Infl}(E, r) = \text{Infl}(E, r') \subseteq \mathcal{F}/args(E) \wedge$

$\text{Trajs}(E, r) = \text{Trajs}(E, r')$.

This means that the set of possible outcomes of the action E cannot depend on anything other than the features of the arguments (including global features, that is, features having no argument), and only these features can be affected by the action. Furthermore an IDS world **W** is said to have strict locality iff its widening-minimal world description (compare Chapter 3, subsection 3.2.4) has strict locality.

The family **IA** is defined as the set of worlds in **IAD** which have strict locality[36].

*Definition of the **I** family*

The designator **I** is primarily intended to indicate integer time, so that the **RA** family is analogous to **IA** but for real-valued time. However, it is also very useful to assign an independent meaning to **I** as an ontological family contained in **IA**. I define it as the very restricted family which allows inertia but where actions lack alternatives, and which of course also requires strict locality. It is defined from **IA** using an additional restriction on the world, namely that the result of an action is unconditional and deterministic with

---

[36] To be precise: as the set of worlds in **IAD** for which the widening-minimal world description has strict locality.

respect to the final result of the action. If the action of turning on the light is defined so that if the light is off it goes on, and if it is on it stays on, then it is an example of an action which lacks alternatives. This restricted family will not be analyzed in the present volume. It is of interest as the natural range of applicability for the original event calculus, and since classical knowledge-based planning of the STRIPS type operates in Q-**I**.

The restriction from **IA** to **I** corresponds roughly to what Lin and Reiter call context-free actions in [44].

Formally, a world **W** is in **I** iff it is in **IA**, all members of $\mathtt{Trajs}(E, r)$ have the same last element, and $\mathtt{Trajs}(E, r) = \mathtt{Trajs}(E, r')$ for arbitrary E, r, and r', in any uniform world description $\langle \mathtt{Infl}, \mathtt{Trajs} \rangle$ of **W**. The restriction to uniform world descriptions is made in order to ensure that trajectories in $\mathtt{Trajs}(E, r)$ and in $\mathtt{Trajs}(E, r')$ are defined over the same set of features.

7.3.2 Subcharacteristics for I

Although there are some entailment criteria that obtain the correct results throughout the family $\mathcal{K}$-**IA**, a number of proposed methods work only for more restricted sets of problems. They may be of interest anyway, in particular if they allow more efficient implementation than the fully general methods. In order to identify precisely where a method is applicable and where it is not, we introduce a number of **subcharacteristics** under **I** and **A**. Table 7.1 lists the ontological subcharacteristics under **I**.

Subcharacteristics are additional constraints within characteristics. In this chapter we shall introduce a number of subcharacteristics whose definitions are simple and easy to grasp, such as 'actions are deterministic' or 'actions take a single timestep'. We shall also show how these subcharacteristics can be formally defined as properties of the world description $\langle \mathtt{Infl}, \mathtt{Trajs} \rangle$. The intended usage of subcharacteristics is for characterizing the range of applicability of various entailment methods. In Chapter 9 we shall show that more precise estimates of the range of applicability can be obtained by using a second generation of subcharacteristics, which are more closely related to the entailment methods themselves, although still defined in terms of the world description.

The simple subcharacteristics that are introduced here will be represented by single, lower-case letters. One particular subcharacteristic is when actions are assumed always to take place during a single timestep; this will be considered as a special case of **I**, and will be written as **Is**. In terms of the trajectory semantics, all members of $\mathtt{Trajs}(E, r)$ must have length 1 in this subcharacteristic.

The ontological family **Is** is obtained, therefore, by taking those IDS worlds in **I** where all actions take a single timestep. The ontological family

Table 7.1. Subcharacteristics of **I**

Code	Subcharacteristic	Absence
Is	Single-step actions. *All actions are performed from one timepoint to the next.* This implies that both **Id** and **If** apply.	Extended actions
Ie	Encapsulated actions. *The actions are only characterized in terms of their preconditions and postconditions: there are no constraints on the duration of the action, or the timing of the changes within the action.* This implies that **If** applies.	Non-encapsulated actions
If	Duration-specific encapsulated actions. *The actions are only characterized in terms of their possible durations, and for each choice of duration, the preconditions and postconditions. For given duration, precondition, and postcondition, there are no constraints on the timing of the changes within the action.*	Actions without duration-specific encapsulation
Id	Direct change in all actions. *When an action changes a feature from an old to a new value, then the feature only changes value once within the duration of the action.*	Indirect change
Ib	Binary features. *All features take the values T or F only.*	Multi-valued features
Ip	Post-uniqueness. *There are not two different action types which change the same feature to the same value.*	Lacking post-uniqueness
Iu	Unary actions. *Each action affects exactly one feature.*	Non-unary actions

IsA is obtained by taking those IDS worlds in **IA** where all actions take a single timestep. When we mention imposing the subcharacteristic **Is** on the ontological family **IA** then the **I** in **Is** is only there in order to differentiate between several possible subcharacteristics called s.

Equivalent world descriptions can not differ with respect to the **Is** subcharacteristic. However, this does not hold for all subcharacteristics, and in principle a subcharacteristic characterizes a world description and not necessarily a world.

A more general subcharacteristic, which will be written as **Ie**, is when actions may take several timesteps, but each action is only characterized in terms of preconditions and postconditions. What happens within the interval of the action is unrestricted. In trajectory semantics, this restriction says that if $\langle r'_1, r'_2, ..., r'_{k'} \rangle$ is a member of $\texttt{Trajs}(E,r)$, then any other trajectory $\langle r''_1, r''_2, ..., r''_{k''} \rangle$ where $r'_{k'} = r''_{k''}$ is also a member of $\texttt{Trajs}(E,r)$, under the restriction, of course, that the r''_i are also restricted to the members in $\texttt{Infl}(E,r)$.

Note the difference between **I**, **Ie**, **IA**, and **IeA**. In **I** if two trajectories are members of $\texttt{Trajs}(E,r)$ for the same E, they must have the same last element. **IA** does not demand the same. In **Ie**, compared with **I**, there is the additional converse requirement that if two trajectories using the same set of features have the same last element, and one is a member of $\texttt{Trajs}(E,r)$, then the other one is so as well. Finally, **IeA** is obtained from **IA** by imposing the converse requirement.

The **If** subcharacteristic is similar to **Ie** except that the requirement only applies when $k' = k''$: if the action has a trajectory of length k', then any other trajectory with the same feature set, the same length and the same last element is also a member of $\texttt{Trajs}(E,r)$. For **Ie**, $\texttt{Trajs}(E,r)$ must also contain all trajectories of any other length if only the last element is the same as in the given one.

It follows from these definitions that both **Is** and **Ie** are special cases of **If**, and that **Ie** consists exactly of those developments which can be obtained in encapsulated semantics. Notice that adding subcharacteristics makes the ontological family more specific, whereas adding characteristics makes it less specific. For example, **Is** and **Ie** are more specific than **I**, which in turn is more specific than **IA**.

Similarly, the subcharacteristic **Id** represents the cases where the change in a feature, caused by an action, always goes directly from the old value to the new value. Formally, if $\langle r'_1, r'_2, ..., r'_k \rangle$ is a member of $\texttt{Trajs}(E,r)$, then for each $f \in \texttt{Infl}(E,r)$ there is some non-negative $k' \leq k$ such that $r(f) = r'_1(f) = ... r'_{k'-1}(f)$ and $r'_{k'}(f) = ... r'_k(f)$. In particular, this excludes the case where features represent disconnected 'pockets' with non-stable areas between them; actions can only go from one pocket to another, adjacent pocket. It follows that **Is** is a special case of **Id** as well. Outside **Is**, **Id** and **Ie** are for the most part mutually exclusive. (The only additional overlap seems to be for actions taking exactly two timesteps, and having the necessary change in binary features).

The subcharacteristics **Id**, **Ie**, **If**, and **Is** will be used in this and the following chapters. The three remaining subcharacteristics **Ib**, **Ip**, and **Iu** are used for characterizing classes where planning algorithms have favorable complexity properties, according to the results of Bäckström et al. [3]. They are included here in order to collect all subcharacteristics of **I** in one place.

Most of the subcharacteristics apply to each action individually, and I shall sometimes use them in this way as well. They apply for an IDS as a whole iff they apply to each action in the IDS. The only exception is **Ip**.

7.3.3 Subcharacteristics for A

Table 7.2 lists the ontological subcharacteristics under **A**. Here we distinguish between deterministic change **Ad** and non-deterministic change. The difference is that in the deterministic case the values of features when an action terminates are completely determined by their values when the action starts. Of course, there may still be a deterministic underlying primary system, but there is simply no way to relate to it.

For example, in the goldfish hotel, suppose there is an action 'move to neighbor' where the fish manager moves a given goldfish to the neighboring bowl to the left or right (cyclically) depending on which direction the fish is swimming at the time the action is called. If there is no image-level feature indicating the direction of the fish, then this action will be nondeterministic from the point of view of the world, although it is perfectly deterministic in the IDS material system.

The Yale shooting scenario is an example of **Ad**; the Russian turkey scenario is an example outside that subfamily. Formally in **Ad**, for every E and r there is some r^* such that the final element in every member of $\text{Trajs}(E, r)$ equals r^*. However, unlike the case for the **I** family, this r^* may be different for different r.

The subcharacteristic **At** represents strongly deterministic change, where not only the final state, but also the intermediate states of the action are uniquely determined by E, r, and the duration of the action. In other words, for each choice of E and r, $\text{Trajs}(E, r)$ must not contain two members that have the same length. It follows that **IsAd** $\subset$ **IAt** $\subset$ **IAd**.

I also distinguish between "optional" and "necessary" change, the difference being that for necessary change all the alternative new values differ from the value of the same feature when the action started, whereas for optional change it may also be that the feature remains unchanged. Formally, for necessary change, if $\langle r'_1, r'_2, ..., r'_k \rangle$ is an arbitrary member of

Table 7.2. Subcharacteristics of **A**

Code	Subcharacteristic	Absence
Ad	Deterministic change. *The final effects of an action are completely determined by the state of the world at the time when the action starts.*	Non-deterministic change
An	Necessary change. *For non-deterministic actions: for a given state of the world when the action starts, there is never any choice between having or not having a change. Non-determinacy only applies between several different changes.* **IdAd** *implies* **An**; **IbAn** *implies* **Ad**.	Optional change
Au	Uniform change. *The set of fluents that are allowed to change as a result of the action is the same regardless of the state of the world when the action starts. Alternatives are only obtained by different choices of new values for these fluents.*	Non-uniform change
At	Strongly deterministic change. *Each action is deterministic with respect to its final effects, and the intermediate states while the action is performed are also completely determined by the starting state and the duration of the action.*	Weakly deterministic or nondeterministic change
Ae	Equidurational change. *The set of possible durations of an action is independent of the initial state.* **IsA** *implies* **IsAe**.	Polydurational change.

Trajs(E,r), then for every feature f in Infl(E,r), $r(f) \neq r'_k(f)$. Necessary change will be characterized using **An**.

The subcharacteristic **Au** is formally defined by the restriction that Infl(E,r) must be independent of r, that is, the world description is uniform. Notice that equivalent world descriptions may differ with respect to both **An** and **Au**. Notice, also, that a given world always has some world description in **Au**, which is of interest since there is one entailment method which is correct for chronicles in **Au**. It is not the case that every world has a world description in **An**.

For example, the Yale shooting scenario is in the category of **IdAd**, since the world is inert (nothing changes without an action), change is direct, and the outcome of each action depends deterministically on the state when the action starts (the turkey dies iff the gun is loaded)[37]. The widening-minimal description of the YSS world is not in the subfamily **IAu**, since the liveness of the turkey may or may not be affected depending on the state of the world when the firing starts. World descriptions of the Russian turkey scenario are neither in **IAd** nor possibly in **IAn**, since, nondeterministically, either the turkey dies or it does not. The case of **IAn** but not **IAd** would be obtained by a random-walk device which can be in either of a number of different positions, and in each step it moves randomly to a neighboring position but never stays in the same position.

Finally, I define the equidurational subcharacteristic **Ae** as follows. Let Pdur(E,r) (possible durations) be the set of the lengths of all members of Trajs(E,r). An action E is *equidurational* iff Pdur(E,r) is independent of r. In other words, if the action could take k time units when it starts in state r, it could also take k time units if started in any other state r'. The Furniture assembly scenario is an example that is not in $\mathcal{K}$-**IAe**.

7.4 The $\mathcal{K}$-**IA** family of chronicles

7.4.1 Epistemological properties

The following epistemological properties were introduced in Chapter 2, subsection 2.2.2, and can now be given their precise definitions:

$\mathcal{K}$p: All members of OBS have the form $[0]\phi$.

$\mathcal{K}$s: Those members of OBS having the form $[0]\phi_i$ determine the state of the world at time zero as a function only of the valuation M. In other words, for each choice of M the LFL model set for $\{\phi_i\}$ has at most one member.

---

[37]This classification requires, of course, that there are no intermediate states in the description of how the gun goes from unloaded to loaded, or how the turkey goes from alive to dead.

One more epistemological property will be defined in Chapter 9, subsection 9.2.4.

7.4.2 Classification of scenario examples

The following is a classification of the scenario examples in terms of the taxonomy that has now been defined. The **Is** restriction appears to be natural in those scenarios involving loading and firing a gun, but it requires the introduction of additional formulae that state each action to take a single timestep. Therefore, the second column below gives the classification of the chronicle as stated above, and the third column gives the classification after formulae of the form $s_i = \theta t_i$ have been added to the respective schedules.

SDS	$\mathcal{K}$sp-**IAd**	
YSS	$\mathcal{K}$sp-**IAd**	$\mathcal{K}$sp-**IsAd**
HTS	$\mathcal{K}$p-**IAd**	$\mathcal{K}$p-**IsAd**
SMM	$\mathcal{K}$-**IAd**	$\mathcal{K}$-**IsAd**
FCS	$\mathcal{K}$sp-**IAd**	
RTS	$\mathcal{K}$sp-**IA**	$\mathcal{K}$sp-**IsA**
TCS	$\mathcal{K}$s-**IA**	
FAS	$\mathcal{K}$p-**IAd**	

The Stolen car scenario is an example of $\mathcal{K}$s-**IAdS** with surprises, which will be addressed in Volume II.

7.5 Summary

The first part of the chapter exemplified chronicle completion using a number of simple chronicles in the $\mathcal{K}$-**IA** family. The second part contained formal definitions of the $\mathcal{K}$-**IA** family and its subcharacteristics.

8
Intended models for chronicles in $\mathcal{K}$-IA

By the definitions in chapter 2, subsection 2.3.1, a chronicle in $\mathcal{K}$-IA is a tuple $\langle \mathcal{K}, \mathcal{O}, \langle \mathbf{IA}, \text{LAW}\rangle, \text{SCD}, \text{OBS}\rangle$. Some of the previous chapters have defined the syntax of formulae which may occur in LAW, SCD, and OBS. Other chapters have defined an underlying semantics in terms of the ego-world game and the trajectory concepts. We are now ready to establish the connection between those two backgrounds, and to define the intended set of models for a given chronicle in terms of the underlying semantics. In this context we shall also introduce a number of auxiliary concepts, including syntactical constraints on the formulae in the premise partitions LAW, SCD, and OBS. Similarly we introduce some subsidiary model-theoretic concepts, such as the classical and intended models restricted to a finite time interval, which will be used in the subsequent proofs.

8.1 Full trajectory normal form for action laws

From the trajectory-level description of an **IA** world as a pair $\langle \texttt{Infl}, \texttt{Trajs}\rangle$, we shall construct a corresponding logical formula called the *full trajectory normal form* (FTNF) for the action laws. This normal form will be used in the proofs in the following chapters.

8.1.1 An example

Before proceeding to the general definition, let us consider a simple example to illustrate the principle of the normal form. We continue with the same example as was used in Chapter 3, namely a shooting world with three propositional features, which are a ('Fred is alive'), l ('the gun is loaded'), and p ('it is raining'). Table 8.1 specifies the effects of the action *Fire* by indicating, for each choice of r, the value of $\texttt{Infl}(\textit{Fire}, r)$ and the single member of $\texttt{Trajs}(\textit{Fire}, r)$. It is identical with Table 3.1 in Chapter 3.

Table 8.1. The world description $\langle \text{Infl}, \text{Trajs} \rangle$

Starting state r	$\text{Infl}(Fire, r)$	Member of $\text{Trajs}(Fire, r)$
$\{a:T, l:T, p:T\}$	$\{a, l\}$	$\langle \{a:T, l:F\}, \{a:F, l:F\} \rangle$
$\{a:T, l:T, p:F\}$	$\{a, l\}$	$\langle \{a:T, l:F\}, \{a:F, l:F\} \rangle$
$\{a:T, l:F, p:T\}$	$\emptyset$	$\langle \emptyset \rangle$
$\{a:T, l:F, p:F\}$	$\emptyset$	$\langle \emptyset \rangle$
$\{a:F, l:T, p:T\}$	$\{l\}$	$\langle \{l:F\} \rangle$
$\{a:F, l:T, p:F\}$	$\{l\}$	$\langle \{l:F\} \rangle$
$\{a:F, l:F, p:T\}$	$\emptyset$	$\langle \emptyset \rangle$
$\{a:F, l:F, p:F\}$	$\emptyset$	$\langle \emptyset \rangle$

The action law for *Fire* may be written for example as follows:
$[s, t] Fire \Rightarrow$
$([s]a, l \to t = s + 2 \wedge [s, t]a := F \wedge [s+1]a, \neg l) \wedge$
$([s](\neg a \vee \neg l) \to t = s + 1) \wedge$
$([s]l \to [s, t]l := F),$

or as follows, using an obvious syntax extension for conditional expressions:
$[s, t] Fire \Rightarrow$
$(t = \text{if } [s]a, l \text{ then } s + 2 \text{ else } s + 1) \wedge$
$([s]a, l \to [s, t]a := F \wedge [s+1]a, \neg l) \wedge$
$([s]l \to [s, t]l := F).$

The following is the same action law in full trajectory normal form:
$[s, t] Fire \Rightarrow$
$([s]a, l, p \to$
$\quad t = s + 2 \wedge [s, t]a := F, l := F \wedge [s+1]a, \neg l) \wedge$
$([s]a, l, \neg p \to$
$\quad t = s + 2 \wedge [s, t]a := F, l := F \wedge [s+1]a, \neg l) \wedge$
$([s]a, \neg l, p \to t = s + 1) \wedge$
$([s]a, \neg l, \neg p \to t = s + 1) \wedge$
$([s]\neg a, l, p \to t = s + 1 \wedge [s, t]l := F) \wedge$
$([s]\neg a, l, \neg p \to t = s + 1 \wedge [s, t]l := F) \wedge$
$([s]\neg a, \neg l, p \to t = s + 1) \wedge$
$([s]\neg a, \neg l, \neg p \to t = s + 1).$

We notice that these versions of the action law are classically equivalent, and that the FTNF version is a direct translation of the information in Table 8.1. This is analogous to how full disjunctive normal form in propositional logic corresponds to a truthtable. The structural similarity is a technical advantage in the formal proofs that are to follow, and therefore FTNF will be used throughout. However, an action law is only used for its set of models, when intended or selected models are obtained, so for practical usage there is no restriction against utilizing shorter versions like the ones that were mentioned first.

Clearly, one can go from the trajectory structure to the corresponding FTNF formula, and from there to other, logically equivalent versions of the action law. That is the approach that we shall take. One might also have considered a less syntax-oriented approach, where one first defines when a world description $\langle \text{Infl}, \text{Trajs} \rangle$ satisfies a given set of action laws. The reason why we avoid that approach is that it requires a number of syntactic constraints on the action laws in order to obtain a correct correspondence with the structure of the world description. Therefore, it would not actually represent any simplification.

The step from the FTNF version of an action law to a more compact version may be thought of as a transformation in the direction of the presumed common-sense formulation of the action law. Volume II is intended to introduce and analyze additional transformations, which bring the premises further in the direction of their presumed common-sense formulations. We now proceed to the general definition of the FTNF.

8.1.2 Action laws for actions without arguments

Proceeding from the example towards the general definition, we first consider the case of actions without arguments. The example in the previous subsection belongs to this class. Let a vocabulary ν, an occurrence vocabulary π, and a world description $\langle \text{Infl}, \text{Trajs} \rangle$ for ν and π be given. If $[\varsigma, \tau] E$ is an occurrence formula, where E is an occurrence symbol in π without arguments, and ς and τ are arbitrary timepoint expressions, then its translation into a DFL-2 formula will be written $\Pi[[\varsigma, \tau] E]$. Continuing the example of the previous subsection, and with a minimum of simplification,

$\Pi[[4,6] Fire] =$
$\ulcorner ([4] a, l, p \to$
$\qquad (6 = 4 + 2 \wedge [4,6] a := \mathbf{F}, l := \mathbf{F} \wedge [4+1] a, \neg l) \wedge$
$\qquad \ldots$
$([4] \neg a, \neg l, \neg p \to 6 = 4 + 1)\urcorner.$

Continued simplification is of course possible. The expressions to the left of the implication signs in each conjunct may appear to be full specifications of a state. However, in the general case of $\mathcal{K}$-**IA** chronicles, but still restricted to actions without arguments, those expressions shall only specify partial states which are defined exactly for those features that do not have any arguments. This is because the family **IA** was defined so that $\texttt{Trajs}(E,r)$ can only depend on the values that r assigns to features of objects which occur as arguments of E. I use the notation $\mathcal{R}_0$ for the set of partial states obtained from the members of $\mathcal{R}$ by restricting them to zeroary features. Consequently, we define $\texttt{Trajs}(E,r)$ even when $r \in \mathcal{R}_0$, if E is an action designator without arguments.

The construction of the function Π corresponding to a given world description depends crucially on a function ψ defined as follows. If ς and τ are timepoint expressions and V is a set of trajectories, then $\psi(\varsigma, \tau, V)$ shall be a logic formula which expresses the condition that some trajectory in V has length $\tau - \varsigma$ and is performed over the interval $[\varsigma+1, \tau]$, although only the state of the world in the interior of the interval needs to be specified. The exact construction of ψ is left open. In the example above the first branch was characterized by $V = \{\langle\{a : \text{T}, l : \text{F}\}, \{a : \text{F}, l : \text{F}\}\rangle\}$, and we used
$$\psi(\varsigma, \tau, V) = \ulcorner \tau = \varsigma + 2 \land [\varsigma+1]a, \neg l \urcorner.$$
Clearly, it is trivial to define the function ψ for small finite sets V, and it is possibly nontrivial if V is infinite or very large. In particular, $\psi(\varsigma, \tau, \emptyset) \equiv$ F. Then, the following simple auxiliary functions are defined. If v is a trajectory, then $last(v)$ represents the last element of v, which is a partial state. If V is a set of trajectories, defined over the same set of features $dom(V)$, then

$last(V) = \{last(v) \mid v \in V\}$,

$resl(V, r) = \{v \mid v \in V \land last(v) = r\}$,

$assign(r) \equiv \bigwedge [f_i \hat{=} x_i \mid (f_i : x_i) \in r]$,

$change[r', \varsigma, \tau, V] \equiv$

$([\varsigma, \tau] \bigwedge [f_i := x_i \mid (f_i : x_i) \in r']) \land \psi(\varsigma, \tau, resl(V, r'))$,

$pt[\varepsilon, r, \varsigma, \tau] \equiv \bigvee [change[r', \varsigma, \tau, \texttt{Trajs}(\varepsilon, r)] \mid r' \in last(\texttt{Trajs}(\varepsilon, r))]$,

$\Pi[[\varsigma, \tau]\varepsilon] \equiv \bigwedge [[\varsigma]assign(r) \to pt[\varepsilon, r, \varsigma, \tau] \mid r \in \mathcal{R}_0]$.

The definition of Π generates a conjunction over the state domain $\mathcal{R}_0$, as was illustrated by the example in the previous subsection. The definition of pt generates a similar disjunction over the set of possible ending states, when the action has r as its starting state. In particular, for deterministic actions, the disjunction generated by pt has only one branch.

For example, if $r = \{f_1 : \text{G}, f_2 : \text{B}\}$, then $assign(r)$ is the formula $f_1 \hat{=} \text{G} \land f_2 \hat{=} \text{B}$. If $r' = \{a : \text{T}, l : \text{T}\}$ and V is as above, then $change[r', s, t, V]$ is the formula
$$t = s + 2 \land [s, t]a := \text{F}, l := \text{F} \land [s+1]a, \neg l.$$
The rest of the definition is easily understood by reference to the example in the previous subsection.

8.1.3 Action laws for actions with arguments

We proceed to the more general definition of Π, which applies even for actions with arguments. Given $\nu = \langle \sigma, \mathcal{O} \rangle$ like before, let E be an occurrence designator formed only using objects from $\mathcal{O}$, and let r be a complete (non-partial) state formed using ν. Then $\mathtt{Trajs}(E,r)$ is a set of trajectories. Using the notation from subsection 7.3.1, r/O is the partial state obtained from the state r by only retaining those feature-value maplets where all arguments of the feature are members of the set O of objects. Also, $\mathcal{R}$ being the domain of all states, $\mathcal{R}/O$ is defined in the natural way as $\{r/O \mid r \in \mathcal{R}\}$. In particular, $\mathcal{R}_0 = \mathcal{R}/\emptyset$. The functions $\mathtt{Infl}(E,r)$ and $\mathtt{Trajs}(E,r)$ are generalized in the obvious way to the case where $r \in \mathcal{R}/args(E)$.

If E is an action symbol having one or more arguments which are items (feature-values which are not objects), one makes a separate construction according to the previous subsection for each possible choice of argument or argument combination. For example, if E is formed as $A(p)$, where p may be either T or F, one defines
$$\Pi[[\varsigma,\tau]A(p)] = \ulcorner(p \wedge \underline{\Pi}[[\varsigma,\tau]A(\mathtt{T})]) \vee (\neg p \wedge \underline{\Pi}[[\varsigma,\tau]A(\mathtt{F})])\urcorner,$$
where the two alternatives on the right-hand side can be obtained directly from the definition for actions without arguments. Alternatively, this can be written as an action law of the form

$[s,t]A(\mathtt{T}) \Rightarrow \alpha_\mathtt{T},$

$[s,t]A(\mathtt{F}) \Rightarrow \alpha_\mathtt{F},$

where the formulae $\alpha_\mathtt{T}$ and $\alpha_\mathtt{F}$ are constructed as before.

If E is formed using an action symbol A that takes one object argument and no item argument, one proceeds instead as follows. Let $\mathcal{O}$ be the object domain consisting only of the object #1. Π is defined similarly as before, but using $\mathcal{R}/\{\#1\}$ instead of $\mathcal{R}/\emptyset$. This is necessary, because the effects of the action may depend on the values of features $f(\#1)$, and because the action may influence some of these features since they can be included in $\mathtt{Infl}(E,r)$ for some r. Anyway, each step is finite, so it is possible to construct the normal form of the law as a finite formula. Construct the formula

$[s,t]A(\#1) \Rightarrow \alpha,$

where α is as before, but now it may contain features having #1 as argument. Replace #1 by the variable o_1 throughout this formula, obtaining the action law in FTNF form as

$[s,t]A(o_1) \Rightarrow \alpha^{o_1}_{\#1}.$

The corresponding direct definition of Π is

$\Pi[[\varsigma,\tau]A(\omega)] = \Pi[[\varsigma,\tau]A(\#1)]^\omega_{\#1}.$

This generalization from #1 to (the meta-variable) o_1 or ω is correct, because of the assumption of strict locality, and because of the assumption that the exact choice of object name is unimportant. In other words, it does not matter if the object is called #1, #2, or any other #n.

If E is constructed from an action symbol A which takes several object arguments, then this process has to be repeated several times because the arguments may or may not be identical objects. For example, if A has two object arguments, then the action law on FTNF is

$$[s,t]A(o_1,o_2) \Rightarrow (o_1 = o_2 \to \alpha_1) \land (o_1 \neq o_2 \to \alpha_2),$$

where α_1 is constructed using a domain with one object which is used for both argument positions, and α_2 is constructed using a domain with two objects, each of which is used for one argument position.

The generalization to n arguments is obvious, and since it would be difficult to express it formally without additional notation it is omitted here.

Naturally, the standard-form formulae that are obtained with these constructions are not suitable for practical use, and in all practical cases one can find a more concise formulation for the action law. However, for the continued analysis we need the present, general formulation of the rewriting scheme.

8.1.4 Expressiveness of FTNF for infinite trajectory sets

The subformula $\psi(V,\varsigma,\tau)$ in the FTNF construction is required in order to characterize the set of all trajectories in V. This immediately raises the question whether such a formula $\psi(V,\varsigma,\tau)$ exists in all the cases that may arise. The trajectory set $\mathtt{Trajs}(E,r)$ for given E and r can be considered as a formal language, where states over the feature set $\mathtt{Infl}(E,r)$ are used as the (finite) alphabet. It is clear that the use of ordinary logic imposes some restrictions on what class of formal languages can be characterized, but it is far from obvious what those restrictions are. For example, the following statement

$$\exists u[t = s + u + u \land$$
$$\forall v[0 < v \leq u \to [s+v]f = [s+u+v]f]]$$

expresses that the f fluent in the trajectory must consist of two successive and equal parts, but is otherwise unconstrained. (This is a correct formula in DFL-1 syntax if $+$ is included in the time structure). It therefore characterizes a language that is not even context-free. On the other hand, there is no easy way to characterize an arbitrary context free set of state trajectories in DFL-1 logic.

The expressiveness of FTNF can be viewed from two perspectives. One possibility is that the world description $\langle \mathtt{Infl}, \mathtt{Trajs} \rangle$ has been derived from some underlying representation, for example from a quantitative description of the physical world and the low-level robot. In this case, one is concerned that the FTNF shall be able to re-express whatever trajectory sets can be obtained from the underlying representation.

The other possibility, which is the relevant one from the point of view of common-sense reasoning, is that the action law has been expressed in logic but not in FTNF form. In this case it is not necessary for the FTNF to characterize arbitrary sets of trajectories; it is just desirable that it should be able to characterize any set of trajectories that can be represented in a logic which has been relieved from the constraints of the FTNF format. However, even in that case there must still be some constraints on the logic, in order to avoid, for example, that different trajectories in the same trajectory set are defined over different sets of features.

In order to address this question with precision, one must therefore first specify a weaker set of restrictions than FTNF on the logical language for expressing action laws and world descriptions, and one must specify what world description is designated by a particular construct in the broader language. Only then is it meaningful to compare the capabilities of FTNF and of the broader language with respect to the description of infinite sets of trajectories. No such analysis will be undertaken here, for the simple reason that I do not think this will be a problem in practice.

8.2 Chronicle structure in $\mathcal{K}$-IA

The exact structure of chronicles in $\mathcal{K}$-IA can now be defined. The DFL-1 syntax from Chapter 6 and the FTNF from the previous section provide the pieces, and we proceed to putting them together.

8.2.1 The schedule in a chronicle

The term *statement* will be used for a formula which can appear as a member of a premise partition in a scenario description. In DFL-1 and DFL-2 this means that potential members of SCD and of OBS are included, but not those of LAW. All statements are required to be closed formulae. A *timing statement* is a closed DFL-1 formula which is entirely formed using temporal constant and variable symbols, the operations in the time structure, propositional connectives, and quantifiers. In other words, it is a formula which can be formed using the empty similarity type and without any reference to objects.

An *action statement* is an occurrence formula. It therefore has the form

$$[\varsigma, \tau]\varepsilon,$$

without combination by propositional connectives or quantifiers. ς and τ must be variable-free timepoint expressions, and ε a variable-free occurrence expression. *Schedule statements* are either action statements or timing statements.

The premise partition SCD in a $\mathcal{K}$-**IA** chronicle is required to be a finite set of schedule statements as just defined. The following are examples of action statements:
$$[14, 16] A,$$
$$[14, t_1 + 1] A.$$
In the **IsA** family, it may be convenient to further restrict the action statements to the form $[\theta\tau, \tau]\varepsilon$, representing that each action takes a single timestep.

This syntax allows some flexibility in the specification of the action sequence. The starting time and termination time of an action may be incompletely specified, namely if they are specified using temporal constant symbols (for example, t_1) rather than timepoints (for example, 14). The timing statements may then be constructed so that even the order of occurrence between the actions is incompletely determined, as in the following schedule:
$$[s_1, t_1] A_1,$$
$$[s_2, t_2] A_2,$$
$$t_1 \leq s_2 \vee t_2 \leq s_1.$$
The following are constructions that the present syntax does not allow in chronicles.

- Conditional actions (if a certain condition occurs then a certain action will be performed).

- Cycles of repeating actions: infinite cycles, or cycles which recur until a certain condition has been achieved. (Such cycles are of interest, for example, in manufacturing cell applications, where the same action is performed in sequence on successive workpieces. See [69] for earlier work on this problem).

- Cycles where an action is performed for several objects in the object domain.

- 'Maybe' actions: it may or may not be the case that the action is performed.

8.2.2 The observation set in a chronicle

An *observation* or observation statement is an arbitrary closed formula in the DFL-1 main language[38]. The premise partition OBS is required to be a finite set of observation statements.

---

[38]When the more expressive DFL-2 logic is introduced in a Chapter 11, observations will still be restricted to DFL-1 formulae.

8.2.3 The action laws in a chronicle

Action laws are written using the main operator $\Rightarrow$. For DFL-1, action laws are used as rewrite rules. For example the action law from the SDS,
$$[s,t]\,Load \Rightarrow [s,t]\,a := \mathsf{T},$$
is used as a rewrite rule that transforms $[8{:}15, 8{:}20]\,Load$ to the formula $[8{:}15, 8{:}20]\,a := \mathsf{T}$. Formally, a rewrite rule is a mapping from action statements to DFL-1 formulae. Some of the ontological assumptions expressed by **IA** are implicit in that mapping.

I shall use the notation LAW to represent a set of action laws, expressed as formulae with $\Rightarrow$ as the main connective, and Π for the corresponding action rewrite function from occurrence formulae to DFL formulae. From an informal point of view, it may be natural to think of LAW as the primary structure and of Π as a derived construct, particularly since the description of an application will use such a set LAW of action laws. However, from the point of view of the underlying semantics, it is natural to define the transformation Π directly from the world description $\langle \mathtt{Infl}, \mathtt{Trajs} \rangle$, as we did above. Once Π has been obtained, it is straightforward to also construct the corresponding set LAW of action laws. For example, if E is an action symbol without arguments and Π is given, then the action law for E is constructed as
$$[s,t]\,E \Rightarrow \underline{\Pi}[[s,t]\,E].$$
For convenience, we extend the definition of Π so that $\Pi(\phi) = \phi$ for any formula ϕ in the DFL-1 main language, and define $\Pi(\mathrm{SCD}) = \{\Pi(\phi) \mid \phi \in \mathrm{SCD}\}$. This rewrite function can serve as a description of the IDS world in a chronicle description. (Another way of defining Π from LAW is to let $\Pi(\phi)$ be $\bigwedge \Gamma$ where Γ is a suitably restricted subset of the set of conclusions from LAW $\cup \{\phi\}$, after having extended the main-language syntax so that LAW is included).

If LAW has the corresponding action rewrite function Π, then the **IA** chronicle
$$\langle \mathcal{K}, \mathcal{O}, \langle \mathbf{IA}, \mathrm{LAW} \rangle, \mathrm{SCD}, \mathrm{OBS} \rangle$$
will equally be written
$$\langle \mathcal{K}, \mathcal{O}, \Pi, \mathrm{SCD}, \mathrm{OBS} \rangle.$$
Furthermore, in a context where the epistemological assumption $\mathcal{K}$ is used throughout, the first element will be omitted, so a $\mathcal{K}$-**IA** chronicle can be written
$$\langle \mathcal{O}, \Pi, \mathrm{SCD}, \mathrm{OBS} \rangle.$$
The definition of reassignment statements as abbreviations for logical formulae in DFL will differ for DFL-1 and DFL-2, but the FTNF representation of $\langle \mathtt{Infl}, \mathtt{Trajs} \rangle$ is the same. Thus if $\langle \mathcal{O}, \Pi, \mathrm{SCD}, \mathrm{OBS} \rangle$ is a $\mathcal{K}$-**IA** chronicle, then $\Pi(\mathrm{SCD})$ is a set of DFL formulae, but after expansion of := it contains different formulae in DFL-1 and DFL-2.

8.2.4 Definition of 𝒦-IA chronicles

The precise definition of 𝒦-IA chronicles can now be formulated.

Definition. *Let σ be a similarity type, and let π be an occurrence vocabulary. A 𝒦-IA chronicle for σ and π is a fivetuple $\langle \mathcal{K}, \mathcal{O}, \Pi, \text{SCD}, \text{OBS} \rangle$, often abbreviated as the fourtuple $\langle \mathcal{O}, \Pi, \text{SCD}, \text{OBS} \rangle$, where $\mathcal{O}$ is a finite set of objects, Π is an action rewrite function obtained from the description $\langle \texttt{Infl}, \texttt{Trajs} \rangle$ of some 𝒦-IA world for σ and π, SCD is a set of schedule statements for $\langle \mathcal{O}, \sigma \rangle$ and π, and OBS is a set of observation statements for $\langle \mathcal{O}, \sigma \rangle$.* ⋈

This definition indicates in particular that the objects in $\mathcal{O}$ may be used in the action statements and the observations, but not in the action laws. The same action laws are supposed to be usable in different chronicles using different object domains.

Notice, also, that according to the construction of the FTNF, there can be no occurrence of a constant symbol or an object name in LAW. Therefore, $\Pi(\text{SCD})$ cannot contain any additional constant symbols or object names besides those occurring in SCD. For a similar reason, $\Pi(\text{SCD})$ can not contain any free variables.

The different incarnations of an IDS world using trajectory semantics are illustrated in Figure 8.1. The world description as a pair $\langle \texttt{Infl}, \texttt{Trajs} \rangle$ is the starting point. Such a world description determines a corresponding IDS world **W** according to the definition in Chapter 3. It also determines [39] a corresponding set LAW of action-law formulae on FTNF form, according to the definition given earlier in this chapter. The set LAW in turn determines an action-formula transformation Π.

All three transformations are reversible. The steps from Π to LAW and from LAW to $\langle \texttt{Infl}, \texttt{Trajs} \rangle$ have already been discussed. Also, for each **IA** world **W**, there is a corresponding world description $\langle \texttt{Infl}, \texttt{Trajs} \rangle$, simply because that was the definition of the ontological family **IA**. Notice, by the way, that two different world descriptions $\langle \texttt{Infl}, \texttt{Trajs} \rangle$ cannot possibly have the same FTNF representation.

If $\Upsilon = \langle \mathcal{O}, \Pi, \text{SCD}, \text{OBS} \rangle \in$ 𝒦-**IA**, then the trajectory-semantics world characterized by Π will be written as $\mathbf{W}_\Upsilon$. $\mathbf{W}_\Upsilon$ is uniquely determined by Π; this is a consequence of the definitions of 𝒦 and **IA**.

8.2.5 The labelled-formula layout

Examples of 𝒦-**IA** chronicles will actually be written in the format that has already been used, with one formula per line and each formula prefixed by a tag that indicates which premise partition it belongs to.

---

[39] Except for trivial variations of the order of conjuncts in the formula.

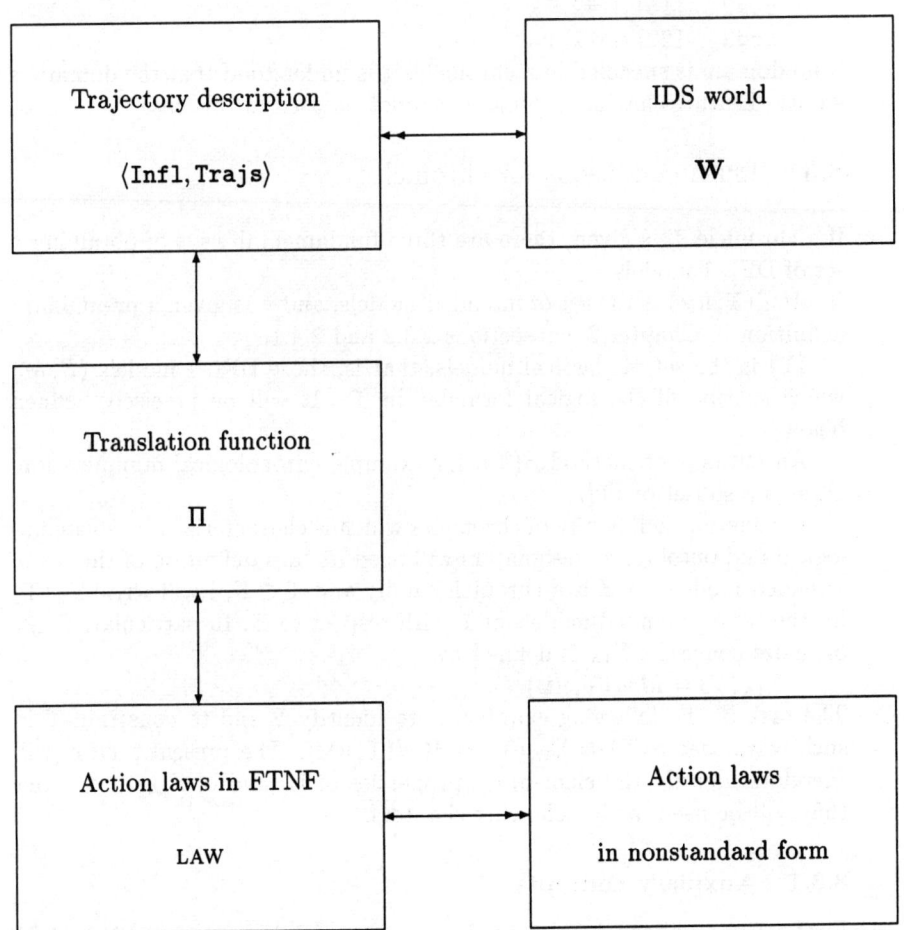

Fig. 8.1. Incarnations of IDS worlds using trajectory semantics

The choice of domain is part of the chronicle declaration. Few of our examples make use of an object domain, but when it is needed it will be written using a special group prefix dom, for example

    dom   #1, #2, #3
    obs1 $[0]\, f_1(\#2) \doteq G$
    scd1 $[10, 15]\, f_1(\#2) := R$
    obs2 $[15]\, f_1(\#2) \doteq R$
    scd2 $[22]\, f_1(\#2) := G$

If no domain is specified in a chronicle, it is understood that the domain is empty. Features having arguments cannot be used in this case.

8.3 Truth conditions for chronicles

If a chronicle Υ is given, there are three fundamental ways of obtaining a set of DFL-1 models:

$Mod(\Upsilon, \text{RM})$ is the set of intended models, and was given a preliminary definition in Chapter 2, subsections 2.3.2 and 2.4.1.

$[\![\Upsilon]\!]$ is the set of classical models, that is, those DFL-1 models $\langle R, M \rangle$ which satisfy all the logical formulae in Υ. It will be precisely defined below.

An entailment method $S(\Upsilon)$, for example chronological minimization, obtains a subset of $[\![\Upsilon]\!]$.

In general, each family of chronicles which is characterized by epistemological and ontological designators will need its own definition of the set of intended models. If $\mathcal{Z}$ is a chronicle family and $\Upsilon \in \mathcal{Z}$, I will write $\Sigma_{\mathcal{Z}}(\Upsilon)$ for the set of intended models of Υ with respect to $\mathcal{Z}$. In particular, $\Sigma_{\mathcal{K}\text{-IA}}$ or, more concisely, Σ_{IA} is defined by

$$\Sigma_{\text{IA}}(\Upsilon) = Mod(\Upsilon, \text{RM}).$$

The task of the following chapters is to identify S and to constrain Υ in such ways that $S(\Upsilon) = \Sigma_{\text{IA}}(\Upsilon) = Mod(\Upsilon, \text{RM})$. The present section will therefore identify the elementary properties of $[\![.]\!]$ and of Σ_{IA}; properties that will be used in the following chapters.

8.3.1 Auxiliary concepts

First a few auxiliary definitions. If ϕ is an expression or a formula and M is a valuation, I write $M[\phi]$ for the modified formula obtained by replacing each constant symbol in ϕ by its value according to M, and simplifying all subexpressions that are formed using the functions in the time structure. For example, if $M[t_3] = 4$ then $M[t_3 + 3] = 7$, and $M[[0, t_3 + 1]A] = \ulcorner[0, 5]A\urcorner$. Also, if Γ is a set of formulae, then I write $M[\Gamma]$ for the set of all those $M[\phi]$, where $\phi \in \Gamma$ and $M[\phi] \not\equiv \text{T}$. In other words, M is applied to all members of Γ, and those formulae which simplify to 'true' are removed.

Truth conditions for chronicles

Because of its syntax, the value of a timepoint expression in a DFL-1 interpretation $\langle M, R \rangle$ depends only on M and is independent of R. Since timing statements are obtained by composition of timepoint expressions, $M[\phi]$ for any timing statement ϕ can be simplified completely and must be either T or F. Therefore, and since the members of the set SCD are schedule statements, that is, either timing statements or action statements, $M[\text{SCD}]$ must consist of actions (= rigid action statements), plus possibly the formula F.

Example. Suppose SCD has the following members

$[10, t_1] A_1$,

$[s_2, 20] A_2$,

$t_1 < s_2$.

If $M = \{t_1{:}14, s_2{:}18\}$ then $M[\text{SCD}] = \{\ulcorner [10, 14] A_1 \urcorner, \ulcorner [18, 20] A_2 \urcorner\}$. If $M = \{t_1{:}14, s_2{:}12\}$ then $M[\text{SCD}] = \{\ulcorner [10, 14] A_1 \urcorner, \ulcorner [12, 20] A_2 \urcorner, \text{F}\}$. The occurrence of F in $M[\text{SCD}]$ is therefore used to indicate that at least one of the time restrictions is not satisfied. ⋈

For $\Upsilon = \langle \mathcal{O}, \Pi, \text{SCD}, \text{OBS} \rangle$ we define $M[\Upsilon] = \langle \mathcal{O}, \Pi, M[\text{SCD}], M[\text{OBS}] \rangle$. It is observed that $M[\Pi(\text{SCD})]$ and $\Pi(M[\text{SCD}])$ are equivalent and essentially equal. The only difference is that some simplifications may be possible in $M[\Pi(\text{SCD})]$ which are not performed in $\Pi(M[\text{SCD}])$.

A valuation M is said to be *good* for a schedule SCD iff all timing statements in SCD satisfy M. Therefore, M is good for SCD iff $M[\text{SCD}]$ consists entirely of actions; in the opposite case $M[\text{SCD}]$ also has F as a member.

If SCD is a schedule not containing any constant symbols (for example formed as $M[\text{SCD}']$ for an arbitrary schedule SCD'), then it is said to be *sequential* iff $s_i < t_i$ for every action $[s_i, t_i] E_i$ in SCD, and $(t_i \leq s_j) \vee (t_j \leq s_i)$ for every pair $[s_i, t_i] E_i$ and $[s_j, t_j] E_j$ of distinct actions in SCD.

If SCD is a schedule, M is a good valuation for SCD, and $M[\text{SCD}]$ is sequential, then the set $\text{At}(M, \text{SCD})$ of *accessible timepoints* is defined as

$\{n \mid n \in \mathcal{T} \wedge \neg \exists i [s_i < n < t_i \wedge \ulcorner [s_i, t_i] E_i \urcorner \in M[\text{SCD}]]\}$,

that is, the set of all timepoints except those that are members of the time interval (s_i, t_i) for some action in $M[\text{SCD}]$.

An observation is called *initial* iff it is a DFL-1 formula of the form $[0]\phi$. A chronicle is an *initial-observation chronicle* iff all members of its observation set are initial. Initial-observation chronicles are characterized by the epistemological property $\mathcal{K}$p. I will write the set of observations in initial-observation chronicles mnemonically as IOBS rather than as OBS.

As before, $R_{0:t}$ denotes the restriction of R to $[0, t]$. $\mathcal{A}_{0:t}$ (where $\mathcal{A}$ is any set of actions, obtained, for example, as $M[\text{SCD}]$) is defined as the subset of $\mathcal{A}$ containing those members $[s_i, t_i] E_i$ where $t_i \leq t$. $\langle M, R \rangle_{0:t}$ is defined as $\langle M, R_{0:t} \rangle$, and $\langle \mathcal{B}, M, R, \mathcal{A}, \emptyset \rangle_{0:t}$ as $\langle \mathcal{B} \cap [0, t], M, R_{0:t} \mathcal{A}_{0:t}, \emptyset \rangle$. If

W is a set of interpretations, then $W_{0:t}$ is defined by applying the restriction to each member of the set.

8.3.2 Model set for an IA chronicle

The model set $Mod(\Upsilon)$ for a scenario description Υ was broadly defined in Chapter 2, subsection 2.3.2, as the set of complete developments that are correctly described by Υ. It was also illustrated for the concrete SDS example in Section 7.1. We are now in a position to give the precise definition for the case of $\mathcal{K}$-**IA** scenario descriptions.

Definition. *Let a similarity type σ and an occurrence vocabulary π be given, and let $\Upsilon = \langle \mathcal{O}, \Pi, \text{SCD}, \text{OBS} \rangle$ be a chronicle in $\mathcal{K}$-**IA** over σ and π. The **model set** $Mod(\Upsilon)$ is defined as the set of all complete developments over $\mathcal{O}$, σ, and π,*
$$\langle \mathcal{B}, M, R, \mathcal{A}, \mathcal{C} \rangle,$$
which result from a game between $\mathbf{W}_\Upsilon$ and some trajectory-semantics ego, with arbitrary r_0 and M_0, where $M[\text{SCD}] = \mathcal{A}$ and $\langle M, R \rangle \in [\![\text{OBS}]\!]$. ⋈

The restriction to trajectory-semantics egos is harmless according to Propositon 3.1 in Chapter 3. The condition $M[\text{SCD}] = \mathcal{A}$ means that the actions that occurred during the game exactly equals the actions that are mentioned in the schedule SCD, and that all the timing conditions are satisfied. Recall that the members of $\mathcal{A}$ are written $\langle s, E, t \rangle$ and the members of $M[\text{SCD}]$ are rigid action statements of the form $[s, t] E$, but recall, also, that these two configurations of the same information have already been declared as identical.

Notice, furthermore, that even if the schedule SCD should allow some actions to occur concurrently, such a development will not be obtained in $Mod(\Upsilon)$ since a trajectory-semantics world will not generate such a development. This will be a consideration for the entailment methods so that they do not allow models with concurrent actions.

Note that if, for example,
$$\text{SCD} = \{ \ulcorner [s_1, t_1] A \urcorner, \ulcorner [s_2, t_2] A \urcorner \},$$
then some members of $Mod(\Upsilon)$ may have a singleton $\mathcal{A}$ component and satisfy $M[s_1] = M[s_2]$ and $M[t_1] = M[t_2]$. Such unintended models will be eliminated by the sequentiality requirement in Chapter 9, section 9.1.

The weakened model set $Mod(\Upsilon, \text{RM})$ was defined from $Mod(\Upsilon)$ in Chapter 2, subsection 2.4.1, so the precise definition of Mod for $\mathcal{K}$-**IA** is also a precise definition of $Mod(\Upsilon, \text{RM})$ which is going to be used as the set of intended models.

When the definition of Mod is used, it is convenient to have restricted the range of possible egos as much as possible. One step in that direction was taken by introducing the restriction to trajectory-semantics egos, which did not reduce the model set at all. The following is another similar step.

Definition. *An ego* **K** *is said to be* **predeterminate** *iff it satisfies the condition that if*
$$\mathbf{K}(\langle \mathcal{B}, M, R, \mathcal{A}, \mathcal{C}\rangle) = \langle \mathcal{B}', M', R', \mathcal{A}', \mathcal{C}'\rangle),$$
then $M = M'$. ⋈

In other words, the moves of a predeterminate ego do not change the valuation component of a development. Combining the definitions of ego, trajectory-semantics ego, and predeterminate, a predeterminate trajectory-semantics ego satisfies the condition that if
$$\mathbf{K}(\langle \mathcal{B}, M, R, \mathcal{A}, \mathcal{C}\rangle) = \langle \mathcal{B}', M', R', \mathcal{A}', \mathcal{C}'\rangle,$$
then $\mathcal{B} = \mathcal{B}'$, $M = M'$, $R = R'$, $\mathcal{A} = \mathcal{A}'$, and either $\mathcal{C}' = \mathcal{C}$ or $\mathcal{C}' = \mathcal{C} \cup \{\langle n_B, E\rangle\}$ for some action designator E. In a game played by a predeterminate ego, the initial valuation for the game is therefore also the final one:

Definition. *Let a similarity type σ and an occurrence vocabulary π be given, and let $\Upsilon = \langle \mathcal{O}, \Pi, \text{SCD}, \text{OBS}\rangle$ be a chronicle in $\mathcal{K}$-**IA** over σ and π. The **predetermined model set** of Υ is defined as the set of all those complete developments, of the form*
$$\langle \mathcal{B}, M, R, \mathcal{A}, \mathcal{C}\rangle$$
over $\mathcal{O}$, σ, and π, which result from a game between $\mathbf{W}_\Upsilon$ and some predeterminate trajectory-semantics ego, with arbitrary initial state r_0 and an arbitrary complete (all constants defined) initial valuation M_0, where $M[\text{SCD}] = \mathcal{A}$ and $\langle M, R\rangle \in [\text{OBS}]$. ⋈

It follows immediately that the predetermined model set of Υ equals $Mod(\Upsilon)$ if $\Upsilon \in \mathcal{K}$-**IA**, and I will often use this in place of the definition of Mod. In particular, as we also define intended models over finite time intervals we restrict the definition to games against predeterminate egos.

If M is good for the schedule in Υ, we write $Mod(M, \Upsilon)$ for the subset of $Mod(\Upsilon)$ consisting of those members having M as their valuation element. Therefore, of course,
$$Mod(\Upsilon) = \bigcup_M Mod(M, \Upsilon).$$

Definition. *If $\Upsilon = \langle \mathcal{O}, \Pi, \text{SCD}, \text{IOBS}\rangle \in \mathcal{K}$p-**IA**, M is a good valuation for SCD, and $t \in \text{At}(M, \text{SCD})$, we write $Mod^t(M, \Upsilon)$ for the set of finite developments over $\mathcal{O}$, σ, and π, of the form $\langle \mathcal{B}, M, R, \mathcal{A}, \mathcal{C}\rangle$ which result from a finite game between $\mathbf{W}_\Upsilon$ and some predeterminate trajectory-semantics ego with arbitrary r_0 and complete M_0, $\mathcal{A} = (M[\text{SCD}])_{0:t}$, IOBS is satisfied in $\langle \mathcal{O}, \langle M, R\rangle\rangle$, and $\mathcal{B} = [0, t] \cap \text{At}(M, \text{SCD})$.* ⋈

It follows that the R component must be defined over the time interval $[0, t]$. Notice that this notation is only defined for initial-observation schedules. Some such restriction is necessary since in general one observation may refer to several timepoints, so allowing any set of observations with reasonable definitions would require the use of a threevalued logic. Similarly, it would not be worth while to define $Mod^t(\Upsilon)$ without specifying M, since some of the candidate models would then have t within the interval of an action.

It follows

Proposition 8.1. *If* $\Upsilon = \langle \mathcal{O}, \Pi, \text{SCD}, \text{IOBS} \rangle \in \mathcal{K}\text{p-}\mathbf{IA}$, *then*
$Mod(M,\Upsilon) = \{J \mid J_{0:t} \in Mod^t(M,\Upsilon) \text{ for all } t \in \text{At}(M, \text{SCD})\}$.

Proof According to the definition of Mod above and the definitions in Chapter 1, subsection 1.3.4, the members of $Mod(M,\Upsilon)$ and only them can be obtained as
$$\lim_{i \to \infty} J_i = \langle \mathcal{B}_i, M, R_i, \mathcal{A}_i, \mathcal{C}_i \rangle,$$
where J_i with varying i are the resulting developments from successively longer, finite games between $\mathbf{W}_\Upsilon$ and a predeterminate trajectory-semantics ego, with M as the initial valuation and the initial state r_0 selected so that IOBS is satisfied. But the set of such Υ_i for a given i is, by definition, the set $Mod^i(M,\Upsilon)$. The result follows by the definition of lim for developments (Chapter 1). □

It also follows that

Lemma 8.2. *If* $\Upsilon = \langle \mathcal{O}, \Pi, \text{SCD}, \text{IOBS} \rangle \in \mathcal{K}\text{p-}\mathbf{IA}$, *and* $t \in \text{At}(M, \text{SCD})$ *then*
$Mod(M,\Upsilon)_{0:t} \subseteq Mod^t(M,\Upsilon)$.

Proof Suppose $J \in Mod(M,\Upsilon)_{0:t}$. There is then some $J' \in Mod(M,\Upsilon)$ such that $J'_{0:t} = J$. J' is the resulting development from a game played between $\mathbf{W}_\Upsilon$ and some predeterminate trajectory-semantics ego, and clearly J is the resulting development from the initial part of that game up to time t. Therefore, $J \in Mod^t(M,\Upsilon)$. □

This lemma says that if we identify all complete models and then truncate them at t, we obtain a subset ($\subseteq$) of $Mod^t(M,\Upsilon)$.

Example. The following example illustrates why equality is not obtained in Lemma 8.2. Let SCD consist of
$[s_1, t_1]E, s_1 < t_1,$
$[s_2, t_2]E, s_2 < t_2,$
$t_1 \leq s_2,$
and assume action laws which require E to always have the duration 4. Choose the valuation M as $\{s_1 : 4, t_1 : 8, s_2 : 10, t_2 : 15, ...\}$, where the other components of M do not matter for the example. Then $Mod(M,\Upsilon)_{0:10} = \emptyset$ because $Mod(M,\Upsilon) = \emptyset$, but $Mod^{10}(M,\Upsilon)$ consists of all developments over $[0, 10]$ where E is performed over the interval $[4, 8]$. The incompatibility between the values assigned by M and the duration of the second action affects $Mod(M,\Upsilon)$

but not $Mod^{10}(M,\Upsilon)$. ⋈

8.3.3 Intended model set for an IA chronicle

The notions of the previous subsection are immediately generalized to weakened models containing only the M and R components, as in Chapter 1. Therefore, formally,

Definition.
$$\Sigma_{\mathbf{IA}}(\Upsilon) = \mathsf{Ci}(Mod(\Upsilon)),$$
$$\Sigma_{\mathbf{IA}}(M,\Upsilon) = \mathsf{Ci}(Mod(M,\Upsilon)),$$
$$\Sigma_{\mathbf{IA}}^{t}(M,\Upsilon) = \mathsf{Ci}(Mod^{t}(M,\Upsilon)),$$

where in all cases
$$\mathsf{Ci}(W) = \{\langle M, R \rangle \mid \langle \mathcal{B}, M, R, \mathcal{A}, \mathcal{C} \rangle \in W\}. \;\; ⋈$$

The simple properties for Mod transfer trivially to $\Sigma_{\mathbf{IA}}$, so, in particular:
$$\Sigma_{\mathbf{IA}}(\Upsilon) = \bigcup_{M} \Sigma_{\mathbf{IA}}(M,\Upsilon)$$
and
$$\Sigma_{\mathbf{IA}}(M,\Upsilon)_{0:t} \subseteq \Sigma_{\mathbf{IA}}^{t}(M,\Upsilon).$$

8.3.4 Classical models for an IA chronicle

Definition. Let σ, π, and Υ be as in the previous definitions. The classical model set for Υ is written $[\![\Upsilon]\!]$ and is defined as $[\![\Pi(\text{SCD}) \cup \text{OBS}]\!]_{\mathcal{O}}$. ⋈

Comparing the definitions of $\Sigma_{\mathbf{IA}}(\Upsilon)$ and $[\![\Upsilon]\!]$, they differ in particular in how the description of the world is handled. For $\Sigma_{\mathbf{IA}}(\Upsilon)$, the action laws are used to identify a corresponding IDS world $\mathbf{W}$, and one then proceeds to identify all games played by that world against an arbitrary ego. For $[\![\Upsilon]\!]$, on the other hand, one combines the action laws and the schedule to form $\Pi(\text{SCD})$, and identifies the set of all developments that satisfy the resulting formulae.

Lemma 8.3. Let $\mathcal{O}$ be an object domain, let $\langle \text{Infl}, \text{Trajs} \rangle$ be a description of an **IA** world $\mathbf{W}$, and let Π be an FTNF representation of $\langle \text{Infl}, \text{Trajs} \rangle$. Also, let $[s,t]E$ be an action statement, let $\langle M, R \rangle$ be a DFL-1 interpretation based on $\mathcal{O}$, and let $\mathbf{s} = M[s]$, $\mathbf{t} = M[t]$. Then $\Pi[[s,t]E]$ is true in $\langle \mathcal{O}, \langle M, R \rangle \rangle$ according to DFL-1 iff there is a trajectory $v \in$ Trajs$(M[E], R(\mathbf{s}))$ of length $\mathbf{t} - \mathbf{s}$, and $v_j \subseteq R(\mathbf{s}+j)$ for $1 \leq j \leq \mathbf{t} - \mathbf{s}$. This trajectory is unique.

Proof We make the proof for the case of empty $\mathcal{O}$. $\Pi[\texttt{[s,t]}E]$ is true in $\langle\mathcal{O},\langle M,R\rangle,\rangle$ iff $M[\Pi[\texttt{[s,t]}E]]$ is. But
$$M[\Pi[\texttt{[s,t]}E]] \simeq \Pi M[\texttt{[s,t]}E] \simeq \Pi[\texttt{[s,t]}E'],$$
where $\mathbf{s} = M[s]$, $\mathbf{t} = M[t]$, and $E' = M[E]$. By the definition of Π in Chapter 8, subsection 8.1.2, this is equivalent to $pt[M[E], r, \mathbf{s}, \mathbf{t}]$, where $r = R(\mathbf{s})/\emptyset$, since all the other conjuncts in the definition of Π are implications with a false antecedent. This in turn is equivalent to
$$\bigvee[change[r', \mathbf{s}, \mathbf{t}, \texttt{Trajs}(M[E], r)] \mid r' \in last(\texttt{Trajs}(M[E], r))]$$
or, after further expansion,
$$\bigvee[\texttt{[s,t]} \bigwedge[\ldots \mid \ldots] \wedge \psi(\mathbf{s}, \mathbf{t}, resl(\texttt{Trajs}(M[E], r), r')) \mid r' \in \ldots],$$
and the claim follows directly. This concludes the proof for the case of empty $\mathcal{O}$. The proof is easily generalized to the case of nonempty $\mathcal{O}$. □

Lemma 8.4. *Let $\mathcal{O}$ be an object domain, let $\langle\texttt{Infl}, \texttt{Trajs}\rangle$ be a description of an **IA** world $\mathbf{W}$, and let Π be an FTNF representation of $\langle\texttt{Infl}, \texttt{Trajs}\rangle$. Also, let $\langle\mathcal{B}, M, R, \mathcal{A}, \mathcal{C}\rangle$ be a development resulting from a game between $\mathbf{W}$ and some ego, using an initial state over the object domain $\mathcal{O}$, and let $[s_i, t_i]E_i \in \mathcal{A}$. Then $\Pi[[s_i, t_i]E_i]$ is true in $\langle\mathcal{O}, \langle M', R\rangle\rangle$ for any M'.*

Proof Since $\ulcorner[s_i, t_i]E_i\urcorner \in \mathcal{A}$, it follows that $E_i = M'[E_i]$ for any valuation M'. According to the definition of $\mathcal{K}$-**IA** worlds, there is some $v \in \texttt{Trajs}(E_i, R(s_i))$ of length $t_i - s_i$ such that $R_{0:t_i} = R_{0:s_i} \triangleright v$. The result follows from Lemma 8.3. □

8.3.5 Chronicle completion

We are now ready to obtain the precise relationship between the classical and the intended model sets of a chronicle.

Proposition 8.5. *If $\Upsilon = \langle\mathcal{O}, \Pi, \text{SCD}, \text{OBS}\rangle \in \mathcal{K}$-**IA** then*
$$\Sigma_{\mathbf{IA}}(\Upsilon) \subseteq [\![\Upsilon]\!].$$

Proof Recall that $[\![\Upsilon]\!]$ is defined as $[\![\Pi(\text{SCD}) \cup \text{OBS}]\!]_\mathcal{O}$, which equals $[\![\text{OBS}]\!]_\mathcal{O} \cap [\![\Pi(\text{SCD})]\!]$. It is immediately clear that $\Sigma_{\mathbf{IA}}(\Upsilon) \subseteq [\![\text{OBS}]\!]_\mathcal{O}$, since
$$\Sigma_{\mathbf{IA}}(\Upsilon) = \{\langle M, R\rangle \mid \langle\mathcal{B}, M, R, \mathcal{A}, \mathcal{C}\rangle \in Mod(\Upsilon)\},$$
and the definition of $Mod(\Upsilon)$ requires that $\langle M, R\rangle$ be formed using $\mathcal{O}$ and that all members of OBS are true there.

To prove that $\Sigma_{\mathbf{IA}}(\Upsilon) \subseteq [\![\Pi(\text{SCD})]\!]$, let $\langle M, R\rangle$ be an arbitrary member of $\Sigma_{\mathbf{IA}}(\Upsilon)$. There is then some $\langle\mathcal{B}, M, R, \mathcal{A}, \mathcal{C}\rangle \in Mod(\Upsilon)$, and $M[\text{SCD}] = \mathcal{A}$ according to the definition of Mod. Let $[s_i, t_i]E_i$ be an arbitrary member of SCD. It only remains to show that $\Pi[[s_i, t_i]E_i]$ is true in $\langle\mathcal{O}, \langle M, R\rangle\rangle$. But $M[[s_i, t_i]E_i] \in \mathcal{A}$, so by Lemma 8.4, $\Pi M[[s_i, t_i]E_i]$ is true in $\langle\mathcal{O}, \langle M, R\rangle\rangle$. Since the FTNF rewrite does not affect the constant symbols in the expression, $\Pi[[s_i, t_i]E_i]$ is also true in $\langle\mathcal{O}, \langle M, R\rangle\rangle$, which concludes the proof. □

This proposition says that the intended set of models is a subset of the set of classical models. The subset relation is usually strict, since $[\Pi(\text{SCD}) \cup \text{OBS}]$ may also contain models not satisfying inertia, whereas the construction of the intended models guarantees inertia.

The task of chronicle completion is to construct the smaller set $\Sigma_{\text{IA}}(\Upsilon)$ from $[\Upsilon]$ and from operations on the model sets of formula-set partitions in Υ. Several previously proposed methods, in particular those based on circumscription or other minimization techniques, define a selection function ζ and attempt to obtain $\Sigma_{\text{IA}}(\Upsilon)$ as $\zeta[\Upsilon]$. Such an approach can never achieve full generality within $\mathcal{K}$-**IA** according to the following:

Proposition 8.6. *There does not exist any selection function ζ which satisfies $\Sigma_{\text{IA}}(\Upsilon) = \zeta[\Upsilon]$ for all $\Upsilon \in \mathcal{K}$-**IA**.*

Proof It is sufficient to find two chronicles Υ and Υ' where $[\Upsilon] = [\Upsilon']$ but $\Sigma_{\text{IA}}(\Upsilon) \neq \Sigma_{\text{IA}}(\Upsilon')$. Consider the 'toss-coin' action E which has the effect of randomizing the value of a propositional feature p, so that its value after E is T or F regardless of its value when E started. Then let Υ contain the schedule $\{\ulcorner[4,5]E\urcorner\}$, and let Υ' contain the empty schedule. In this case $[\Upsilon] = [\Upsilon']$ is the set of all possible interpretations (arbitrary valuation and arbitrary history), but $\Sigma_{\text{IA}}(\Upsilon) \neq \Sigma_{\text{IA}}(\Upsilon')$. □

More specifically, this proposition suggests that if nondeterministic actions are allowed then partitioning of the premises is necessary in order to identify $\Sigma_{\text{IA}}(\Upsilon)$ precisely. There are two possible courses of action:

(1) To use a wider class of definitions of $\approx$. One possibility is to use filtering techniques with definitions of the form

$$\zeta[\Pi(\text{SCD})]_O \cap [\text{OBS}].$$

(2) To stay with the definitions of the form $\zeta[\Upsilon]$, but to identify a narrower class of chronicles for which the proposed definition can be proved to be always correct.

Both of these approaches will be analyzed in the following chapters. The remainder of the present chapter defines some auxiliary concepts.

8.4 Models over restricted time domains

The remainder of this chapter introduces some technical concepts which will be used in the assessment proofs for chronological entailment methods in Chapters 9 and 10.

8.4.1 Definitions and basic properties

The members of $[\Upsilon]$ are interpretations over the whole time domain $\mathcal{T}$, which may be finite or infinite. In particular the history component R is defined for the whole of the currently used T. However, most of the assessment proofs for chronological entailment methods are induction proofs over time. They need to use interpretations which are defined over a successively extended subset of T, and to compare interpretations which are defined over different subsets of T. The following concepts are introduced for this purpose.

Definition. *If* $\Upsilon = \langle \mathcal{O}, \Pi, \text{SCD}, \text{IOBS} \rangle \in \mathcal{K}\text{p-}\mathbf{IA}$, M *is a good valuation for* SCD, *and* $t \in \text{At}(M, \text{SCD})$, *then* $[\![M, \Upsilon]\!]^t$ *is the set of* **restricted models** *for* M *and* Υ *over* $[0, t]$, *and is defined as the set of all* $\langle M, R \rangle$ *having* $\mathcal{O}$ *as their domain and where* R *is defined over* $[0, t]$ *and all members of* $\Pi(M[\text{SCD}]_{0:t}) \cup M[\text{IOBS}]$ *are true.* ⋈

Notice that the action statements that occur in the interval $[0, t]$ are satisfied in the members of $[\![M, \Upsilon]\!]^t$, after expansion of the abbreviation, while one does not care about action statements at later times. Inertia is not required.

Similarly, the restriction of $[\![\Upsilon]\!]$ to those models containing M will be written $[\![M, \Upsilon]\!]$. The members of $[\![M, \Upsilon]\!]$ are therefore complete models over T. An expression $[\![M, \Upsilon]\!]^u_{0:t}$ is interpreted as $([\![M, \Upsilon]\!]^u)_{0:t}$, that is, one first restricts $M[\text{SCD}]$ to $[0, u]$, then obtains the models, and then truncates the histories to $[0, t]$.

Proposition 8.7. *If* $\Upsilon = \langle \mathcal{O}, \Pi, \text{SCD}, \text{IOBS} \rangle \in \mathcal{K}\text{p-}\mathbf{IA}$, M *is a valuation*, t *and* t' *are members of* $\text{At}(M, \text{SCD})$, *and* $t \leq t'$, *then*
$$[\![M, \Upsilon]\!]_{0:t} \subseteq [\![M, \Upsilon]\!]^{t'}_{0:t} \subseteq [\![M, \Upsilon]\!]^t.$$
However, if t *is* $\geq$ *the ending time for all actions in* $M[\text{SCD}]$, *then*
$$[\![M, \Upsilon]\!]_{0:t} = [\![M, \Upsilon]\!]^{t'}_{0:t} = [\![M, \Upsilon]\!]^t.$$

Proof Only the proof of the second $\subseteq$ relation is given. Let $\langle M, R \rangle$ be an arbitrary member of $[\![M, \Upsilon]\!]^{t'}_{0:t}$. There is then some $\langle M, R' \rangle \in [\![M, \Upsilon]\!]^{t'}$ such that $R = R'_{0:t}$. All members of $M[\text{IOBS}]$ are true in $\langle M, R' \rangle$ and therefore also in $\langle M, R \rangle$. Furthermore, since $M[\text{SCD}]_{0:t} \subseteq M[\text{SCD}]_{0:t'}$ and all members of $\Pi(M[\text{SCD}]_{0:t'})$ are true in $\langle M, R' \rangle$, all members of $\Pi(M[\text{SCD}]_{0:t})$ must be true in $\langle M, R' \rangle$ and therefore also in $\langle M, R \rangle$. Therefore, $\langle M, R \rangle \in [\![M, \Upsilon]\!]^t$. The proof for the first $\subseteq$ relation is similar, and the proof for the equalities follows easily. □

The reason why equality is not always obtained is as was discussed after Lemma 8.2.

Proposition 8.8. *If* $\Upsilon = \langle \mathcal{O}, \Pi, \text{SCD}, \text{IOBS} \rangle \in \mathcal{K}\text{p-}\mathbf{IA}$, M *is a valuation*, *and* t *is a member of* $\text{At}(M, \text{SCD})$, *then*
$$\Sigma^t_{\mathbf{IA}}(M, \Upsilon) \subseteq [\![M, \Upsilon]\!]^t.$$
However, equality is obtained for the special case of $t = 0$,
$$\Sigma^0_{\mathbf{IA}}(M, \Upsilon) = [\![M, \Upsilon]\!]^0.$$

This proposition is analogous to Proposition 8.5 but for restricted intervals.

Proof Let $\langle M, R \rangle \in \Sigma_{\mathbf{IA}}^t(M, \Upsilon)$. It follows at once that all members of IOBS are true in $\langle \mathcal{O}, \langle M, R \rangle \rangle$. Also, there is some $\langle \mathcal{B}, M, R, \mathcal{A}, \mathcal{C} \rangle \in Mod^t(M, \Upsilon)$, where $M[\text{SCD}]_{0:t} = \mathcal{A}$ according to the definition of $Mod^t(M, \Upsilon)$. Let $[s_i, t_i] E_i$ be an arbitrary member of $M[\text{SCD}]_{0:t}$. It only remains to show that $\Pi[[s_i, t_i] E_i]$ is true in $\langle \mathcal{O}, \langle M, R \rangle \rangle$, and this follows from Lemma 8.4, as in the proof of Proposition 8.5.

For the particular case of $t = 0$, $M[\text{SCD}]_{0:t} = \emptyset$. Let $I = \langle M, \{0 \mapsto r_0\} \rangle$ be an arbitrary member of $[\![M, \Upsilon]\!]^0$. Every member of IOBS is true in $\langle \mathcal{O}, I \rangle$, and it follows that $I \in \Sigma_{\mathbf{IA}}^0(M, \Upsilon)$. This concludes the proof. □

The relationship between $\Sigma_{\mathbf{IA}}^t$ and $\Sigma_{\mathbf{IA}}$ is as follows.

Proposition 8.9. *If* $\Upsilon = \langle \mathcal{O}, \Pi, \text{SCD}, \text{IOBS} \rangle \in \mathcal{K}\text{p-}\mathbf{IA}$, *then*
$$\Sigma_{\mathbf{IA}}(\Upsilon) = \{\langle M, R \rangle \mid \langle M, R_{0:t} \rangle \in \Sigma_{\mathbf{IA}}^t(M, \Upsilon) \text{ for all } t \in \text{At}(M, \text{SCD})\}.$$

Proof The definitions and Proposition 8.1 give
$$\Sigma_{\mathbf{IA}}(\Upsilon) = \{\langle M, R \rangle \mid \langle \mathcal{B}, M, R, \mathcal{A}, \mathcal{C} \rangle \in Mod(\Upsilon)\} =$$
$$\{\langle M, R \rangle \mid \langle \mathcal{B}, M, R, \mathcal{A}, \mathcal{C} \rangle \in Mod(M, \Upsilon)\} =$$
$$\{\langle M, R \rangle \mid \exists \langle \mathcal{B}, M, R, \mathcal{A}, \mathcal{C} \rangle [$$
$$\langle \mathcal{B}_{0:t}, M, R_{0:t}, \mathcal{A}_{0:t}, \mathcal{C}_{0:t} \rangle \in Mod^t(M, \Upsilon)$$
$$\text{for all } t \in \text{At}(M, \text{SCD})]\} =$$
$$\{\langle M, R \rangle \mid \langle M, R_{0:t} \rangle \in \Sigma_{\mathbf{IA}}^t(M, \Upsilon) \text{ for all } t \in \text{At}(M, \text{SCD})\}.$$

The last equality in the chain applies because of the construction of Mod^t and, in particular, because $\Sigma_{\mathbf{IA}}^t$ only uses games where the ego is predeterminate. □

8.4.2 Progressive construction of restricted intended models

Assume that $\Upsilon = \langle \mathcal{O}, \Pi, \text{SCD}, \text{IOBS} \rangle \in \mathcal{K}\text{p-}\mathbf{IA}$, and that M is a good valuation for SCD. Consider how the set of restricted intended models $\Sigma_{\mathbf{IA}}^t(M, \Upsilon)$ changes as t ranges from 0 upwards through successive values that are members of $\text{At}(M, \text{SCD})$. In particular, suppose that t and t' are two successive members of $\text{At}(M, \text{SCD})$, where $t < t'$, and $I = \langle M, R \rangle \in [\![M, \Upsilon]\!]^t$. There are two possibilities. First, if $M[\text{SCD}]$ does not contain any action starting in t, so that $t' = t + 1$, it is easily seen that

1. If $I \in \Sigma_{\mathbf{IA}}^t(M, \Upsilon)$ then $\langle M, R \triangleright \langle \emptyset \rangle \rangle \in \Sigma_{\mathbf{IA}}^{t'}(M, \Upsilon)$.

2. If $I' = \langle M, R' \rangle \in \Sigma_{\mathbf{IA}}^{t'}(M, \Upsilon)$ where $R'_{0:t} = R$, then $I \in \Sigma_{\mathbf{IA}}^t(M, \Upsilon)$ and $R' = R \triangleright \langle \emptyset \rangle$.

On the other hand, if $M[\text{SCD}]$ does contain an action $[t, t'] E$ for an action starting in t, then

1. If $I \in \Sigma_{\mathbf{IA}}^t(M, \Upsilon)$ and $v \in \text{Trajs}(E, R(t))$ has length $t' - t$, then $\langle M, R \triangleright v \rangle \in \Sigma_{\mathbf{IA}}^{t'}(M, \Upsilon)$.

2. If $I' = \langle M, R' \rangle \in \Sigma_{\mathbf{IA}}^{t'}(M, \Upsilon)$ where $R'_{0:t} = R$, then $I \in \Sigma_{\mathbf{IA}}^{t}(M, \Upsilon)$ and $R' = R \triangleright v$ for some $v \in \texttt{Trajs}(E, R(t))$.

In particular, if $\texttt{Trajs}(E, R(t))$ does not contain any trajectory of length $t' - t$, then there is no I' in $\Sigma_{\mathbf{IA}}^{t'}(M, \Upsilon)$ such that $I'_{0:t}$ equals the given I. For a given set of intended models over $[0, t]$, some of them may have continuations into intended models over $[0, t']$ and some may not. On the other hand, every intended model over $[0, t']$ is a continuation of some intended model over $[0, t]$. One can therefore think of the sequence of $\Sigma_{\mathbf{IA}}^{t}(M, \Upsilon)$ for increasing t as a tree-growing process. Each interpretation over a particular interval can possibly be extended into several interpretations over a longer period, contributing to the branching of the tree. However, some branches of the tree are pruned when the timing constraints which M imposes on the duration of the actions cannot be satisfied.

These observations can be summarized in a more general and concise form using the following definitions and proposition.

Definition. *Let $\langle \texttt{Infl}, \texttt{Trajs} \rangle$ be a world description, let SCD be a schedule, let M be a good valuation for SCD, let $s \in \texttt{At}(M, \text{SCD})$, and let R be a history over $[0, s]$. The* **candidate trajectory set** $\texttt{Cats}(M, \text{SCD}, R)$ *is defined as* $\texttt{Trajs}(E, R(s))$ *iff there is some action designator $[s, t] E$ in $M[\text{SCD}]$, and as $\{\langle \emptyset \rangle\}$ otherwise.* ⋈

In other words, if an IDS game has progressed up to time s, R is the history of the game up to now, the game has satisfied the schedule SCD under valuation M up to now, and it is supposed to do so in the sequel as well, then $\texttt{Cats}(M, \text{SCD}, R)$ is the set of possible trajectories for the continuation of the history in the next move of the game.

Definition. *Let $\Upsilon = \langle \mathcal{O}, \Pi, \text{SCD}, \text{IOBS} \rangle \in \mathcal{K}\text{p-}\mathbf{IA}$. A function $S^t(M, \Upsilon)$ is* **progressive** *for Υ iff it satisfies the following conditions whenever M is a good valuation for SCD and $t \in \texttt{At}(M, \text{SCD})$:*

1. $S^t(M, \Upsilon) \subseteq [\![M, \Upsilon]\!]^t$.

2. If $I = \langle M, R \rangle \in S^t(M, \Upsilon)$, $v \in \texttt{Cats}(M, \text{SCD}, R)$ has length k, and $t + k$ is the next larger member of $\texttt{At}(M, \text{SCD})$ after t, then $\langle M, R \triangleright v \rangle \in S^{t+k}(M, \Upsilon)$.

3. If $I' = \langle M, R' \rangle \in S^u(M, \Upsilon)$ where u is the next larger member of $\texttt{At}(M, \text{SCD})$ after t, and $R = R'_{0:t}$, then (a) $\langle M, R \rangle \in S^t(M, \Upsilon)$, and (b) $R' = R \triangleright v$ for some $v \in \texttt{Cats}(M, \text{SCD}, R)$. ⋈

Proposition 8.10. $\Sigma_{\mathbf{IA}}^{t}$ *is progressive for any* $\Upsilon \in \mathcal{K}\text{p-}\mathbf{IA}$.

Proof Requirement (1) was obtained in Proposition 8.8. The other two requirements follow directly from the definition of $\Sigma_{\mathbf{IA}}^{t}$ and of the ego-world game. □

Adaptations for branching time 191

The significance of this concept is that some ('chronological') entailment criteria are also progressive, and this structural similarity between the entailment criteria and Σ_{IA}^t will be exploited in their assessment proofs.

8.5 Adaptations for branching time

The presentation in this chapter has assumed the linear time domain $\mathcal{T}\phi$. However, the adaptation to branching time is straightforward.

8.5.1 Chronicle structure

A schedule in a $\mathcal{K}$-**IA** chronicle does not need to specify the order of the actions. In particular, if the schedule consists entirely of action statements of the form $[s_i, t_i] A_i$, then all possible sequential orderings of these actions may be represented by models. Temporal ordering statements, for example of the form $t_i \leq s_j$, may be used to constrain the order of the actions. Similarly for branching time, a schedule consisting only of action statements may allow models where certain actions occur along the same path, as well as models where those same actions occur along divergent paths. Temporal ordering statements may then be used to force certain actions to occur in the same path, or to force them to occur in divergent paths. (Recall that a path in a branching-time tree was defined as a succession of nodes from the root of the tree, choosing one of the successors in each step). Anyway, since each s_i and t_j has one specific timepoint in the tree as its value, each action statement must occur in exactly one position in the tree. Since all developments in $Mod(M, \Upsilon)$ have the same $\mathcal{A}$ component, they assign actions to paths in the same way, but between different M the actions may be assigned differently.

The definition of sequential schedules in subsection 8.3.1 must be modified so that a schedule is sequential if the requirement in the linear-time definition applies within each of its paths. The definition is as follows: If SCD is a schedule in branching time not containing any constant symbols (for example formed as $M[\text{SCD}']$ for an arbitrary schedule SCD'), then it is said to be *sequential* iff all of the following apply.

- $\langle s_i, t_i \rangle$ for every action $[s_i, t_i] E_i$ in SCD is an interval.

- $s_i \notin (s_j, t_j) \land s_j \notin (s_i, t_i)$ for every pair $[s_i, t_i] E_i$ and $[s_j, t_j] E_j$ of distinct actions in SCD.

- If $[s_i, t_i] E$ and $[s_j, t_j] E$ are distinct actions in SCD then $s_i \neq s_j$.

In this way neither of them can start at a timepoint where the other one is going on, but it is possible for them to dovetail or to start at the same time (creating diverging branches). The third condition item rejects schedules such as the one containing

$$[\langle 4,\emptyset\rangle, \langle 6, \{\langle 4, E_2\rangle\}\rangle] E_2$$

and

$$[\langle 4,\emptyset\rangle, \langle 8, \{\langle 4, E_2\rangle\}\rangle] E_2$$

where an E_2 action is claimed to start twice at the same timepoint and with different durations.

The full trajectory normal form for action laws applies equally for branching time. One must only take care that the subexpressions $\psi(\varsigma, \tau, V)$ in the definition of FTNF are written so that any timepoint referred there is guaranteed to be a member of the branching-time interval set $[s, t]$. The definitions for observations continue to apply without modification.

8.5.2 Truth conditions for chronicles

The set $Mod(\Upsilon)$ was defined for sequential time as the set of all complete developments having a certain relation to Υ (subsection 8.3.2). In branching time, complete developments may differ with respect to whether they include (infinite) paths that are not required by Υ. It is natural to adapt the definition so that only complete developments of minimal size are included. This means that in each member $\langle \mathcal{B}, M, \mathcal{R}, \mathcal{A}, \mathcal{C}\rangle$ of $Mod(\Upsilon)$, the tree $\mathcal{B}$ will contain the interval $\langle M[s_i], M[t_i]\rangle$ for each action statement $[s_i, t_i] E_i$ in the schedule component of Υ. It will also contain those 'waiting' intervals which are required in order for $\mathcal{B}$ to be a complete tree, both waiting that leads up to the actions, and waiting after the last action in a path and onwards to infinity. However, it will not contain any intervals formed by an action except those which are required by the action statements. We notice that the $\mathcal{B}$ component is then completely determined by M and the schedule component SCD in Υ, that is, by $M[\text{SCD}]$.

The epistemological assumption $\mathcal{K}$ therefore implies that if M is a good valuation for a chronicle Υ in $\mathcal{K}$-**IA**, then all observations in $M[\text{OBS}]$ must refer to timepoints in the tree defined by $M[\text{SCD}]$. In other words, one does not state observations for timepoints that arise after actions that have not been explicitly stated in SCD.

The definitions and propositions in Section 8.3 were based on the use of an interval $[0, t]$ for some timepoint t in linear time. For branching time, we notice first of all that all developments in $Mod(M, \Upsilon)$ have the same tree component $\mathcal{B}$. Counterparts of $Mod^t(M, \Upsilon)$ and $Mod(M, \Upsilon)_{0:t}$ are now formed as $Mod^B(M, \Upsilon)$ and $Mod(M, \Upsilon)_B$, where B is a subtree of the tree component $\mathcal{B}$ of the members of $Mod(M, \Upsilon)$. B is a subtree of $\mathcal{B}$ iff B is a tree and $B \subseteq \mathcal{B}$. In particular this means that $\mathcal{B}$ and its subtrees have the same root; it is the branches that may be omitted in the subtree.

The other notation such as $R_{0:t}$ is revised accordingly, and the set

At(M, SCD) is replaced by the set of subtrees of the tree defined by M[SCD].

The definition of Cats must be modified to take E as an additional argument, since the choice of E is no longer implicit from the other arguments. All the other definitions as well as the theorems of Sections 8.3-8.4 generalize without difficulty in accordance with the changes which have now been outlined.

8.6 Summary

We have introduced full trajectory normal form (FTNF), whereby action laws are expressed as DFL-1 formulae, but in a way that corresponds directly to the world description according to the trajectory semantics. Using FTNF, we have given a precise definition of the set $\Sigma_{\mathbf{IA}}(\Upsilon)$ of intended models for any chronicle $\Upsilon \in \mathcal{K}\text{-}\mathbf{IA}$.

9
Entailment methods for $\mathcal{K}$-**IA** using DFL-1

It is readily verified that SCD $\subseteq$ SCD$'$ does not imply
$$\Sigma_{\mathbf{IA}}(\langle \mathcal{O}, \Pi, \text{SCD}', \text{OBS}\rangle) \subseteq \Sigma_{\mathbf{IA}}(\langle \mathcal{O}, \Pi, \text{SCD}, \text{OBS}\rangle).$$
Chronicle completion is therefore a non-monotonic logical operation. It relies essentially not only on the propositions in the chronicle itself, but also on closure assumptions for the chronicle, which have the effect that $\Sigma_{\mathbf{IA}}(\Upsilon) \subseteq [\![\Upsilon]\!]$ (Proposition 8.5). Beginning in this chapter I shall define and analyze several proposed non-monotonic entailment methods. I start with some simple methods which are variants of previously published ones, then discuss their limitations, and finally formulate a few methods which are correct for the ontological family $\mathcal{K}$-**IA** or major parts thereof. In this way we obtain an assortment of methods with broad applicability within $\mathcal{K}$-**IA**. In Volume II I shall analyze whether any of these methods are also applicable outside $\mathcal{K}$-**IA**.

Section 8.3 in the previous chapter defined the set of intended models, $\Sigma_{\mathbf{IA}}$, as a mapping from chronicles to model sets, using a time-constructive and simulation-like process. The entailment methods being analyzed here are functions which are restrictedly equivalent to $\Sigma_{\mathbf{IA}}$, that is, for some classes of chronicles they give the same value, but they are defined in terms of operations on classical models and model sets.

The DFL-1 logic will be used throughout this chapter, while the model selection function, and thereby the non-monotonic entailment relation, will be allowed to vary. Several of the definitions and lemmas that are introduced here will be used in later chapters as well.

9.1 Entailment methods

9.1.1 Pretransformations and model selection

The entailment methods that we shall consider in the following chapters are in principle model-theoretic: they are defined in terms of preference relations on models. However, they also impose certain trivial syntactic restrictions, which can be accommodated by simple transformations on the given formulae. Entailment methods will therefore be defined to have a syntactic and a semantic part, according to the following definition.

Definition. *Let $\mathcal{Z}$ be a family of chronicles. A **pretransformation function** resulting in $\mathcal{Z}$ is a function $\mathbf{G}$ from chronicles to chronicles such that $\mathbf{G}(\Upsilon) \in \mathcal{Z}$. A **model selection function** for $\mathcal{Z}$ is a function S from $\mathcal{Z}$ to sets of DFL models, such that*
$$S(\Upsilon) \subseteq [\![\Upsilon]\!]$$
*for any chronicle $\Upsilon \in \mathcal{Z}$. An **entailment method** is a pair $\langle \mathbf{G}, S \rangle$ such that $\mathbf{G}$ is a pretransformation function resulting in a family $\mathcal{Z}$, and S is a model selection function for $\mathcal{Z}$. The **set of selected models** for a chronicle Υ according to an entailment method $\langle \mathbf{G}, S \rangle$ is defined as $S(\mathbf{G}(\Upsilon))$.* ⋈

The term 'selection function' was introduced in Chapter 4, subsection 4.4.1, for a function from a set of models to one of its subsets, and it is here given a second although related meaning. The pretransformation concept will be generalized in the next chapter, but we start with the simple version defined here. Since pretransformation functions will be used for corrections of a 'housekeeping' nature, they will be chosen so that $\mathbf{G}(\Upsilon) = \Upsilon$ for many simple chronicles, whereas $\mathbf{G}(\Upsilon) \neq \Upsilon$ for more complex chronicles, and for chronicles where certain important information has been omitted.

One particular model selection function $\mathbf{G}_{seq}$ will be needed because the present model selection functions are only correct when schedules are *safely sequential* according to the following definition.

Definition. *A schedule SCD is **safely sequential** iff $M[\text{SCD}]$ is sequential for every good valuation M of SCD. Similarly, a $\mathcal{K}$-**IA** chronicle is called safely sequential iff it has the form $\langle \mathcal{O}, \Pi, \text{SCD}, \text{OBS} \rangle$, where SCD is safely sequential.* ⋈

The definition means that for every valuation M where all timing statements in SCD are satisfied, then for linear time it must be the case that:

- $M[s] < M[t]$ for every action statement $[s, t]E$ in SCD, and
- the action statements in SCD are sequential under M, that is, if $[s, t]E$ and $[s', t']E'$ are two different members of SCD, then either $M[t] \leq M[s']$ or $M[t'] \leq M[s]$.

Example. The set
$$\Gamma = \{[s_1, t_1]A_1, t_1 \leq s_2, [s_2, t_2]A_2\}$$
is not safely sequential, but $\Gamma \cup \{s_1 < t_1, s_2 < t_2\}$ is safely sequential. ⋈

In the case of branching time we have a generalized definition of 'sequential', but the definition of 'safely sequential' in terms of 'sequential' applies without changes. If a chronicle Υ is not safely sequential, then it has some good valuations for which its schedule is not sequential, and therefore it is possible that $[\![\Upsilon]\!]$ contains some non-sequential models. Such models will not be included in $\Sigma_{\mathbf{IA}}(\Upsilon)$ since they cannot be generated from a game with an **IA** world, but current model selection functions have no systematic way of rejecting them.

In order to deal with this problem we choose $\mathbf{G}_{seq}$ as the pretransformation function defined by
$$\mathbf{G}_{seq}(\langle \mathcal{O}, \Pi, \text{SCD}, \text{OBS}\rangle) = \langle \mathcal{O}, \Pi, \text{SCD} \cup \Delta, \text{OBS}\rangle,$$
where Δ is chosen as $\emptyset$ if $\langle \mathcal{O}, \Pi, \text{SCD}, \text{OBS}\rangle$ is already safely sequential, and otherwise as the set of $\ulcorner \varsigma_i < \tau_i \urcorner$ and $\ulcorner \tau_i \leq \varsigma_j \vee \tau_j \leq \varsigma_i \urcorner$ for all different i, j of action statements $\ulcorner [\varsigma_i, \tau_i]E_i \urcorner$ in SCD.

Lemma 9.1. *If* $\Upsilon = \langle \mathcal{O}, \Pi, \text{SCD}, \text{OBS}\rangle$, *then* $[\![\mathbf{G}_{seq}(\Upsilon)]\!]$ *is a subset of* $[\![\Upsilon]\!]$, *and consists of exactly those members of* $[\![\Upsilon]\!]$ *which are safely sequential.*

Proof Directly from the definitions. □

We proceed to consider entailment methods of the form $\langle \mathbf{G}_{seq}, S\rangle$. The pretransformation $\mathbf{G}_{seq}$ does not have any effect from the point of view of the intended-models function $\Sigma_{\mathbf{IA}}$: its only purpose is to correct certain chronicles that could otherwise give difficulties for the entailment method. It is therefore natural to introduce the following concept.

Definition. *A pretransformation function* $\mathbf{G}$ *is called* **harmless** *iff* $\Sigma_{\mathbf{IA}}(\Upsilon) = \Sigma_{\mathbf{IA}}(\mathbf{G}(\Upsilon))$ *for every* $\Upsilon \in \mathcal{K}\text{-}\mathbf{IA}$. *Likewise, a restriction on chronicles (represented as a chronicle family)* $\mathcal{Z} \subseteq \mathcal{K}\text{-}\mathbf{IA}$ *is called harmless iff there exists a harmless pretransformation function resulting in* $\mathcal{Z}$. ⋈

From Lemma 9.1 and the observation that all members of $\Sigma_{\mathbf{IA}}(\Upsilon)$ are sequential, it follows that the sequentiality restriction is harmless.

9.1.2 Entailment methods formed using minimization

We shall pay special attention to model selection functions S of the form
$$S(\langle \mathcal{O}, \Pi, \text{SCD}, \text{OBS}\rangle) = S'([\![\Pi(\text{SCD})]\!]_{\mathcal{O}}, [\![\text{OBS}]\!]).$$
Such a function is syntax independent: it has only access to the models of the given premise partitions, and not to the formulae that occur in them. The techniques for non-monotonic logics that were described at the end of Chapter 4 are therefore applicable for defining the function S'. I

will immediately drop the prime, and write $S([\![\Pi(\text{SCD})]\!]_\mathcal{O}, [\![\text{OBS}]\!])$, without danger of confusion.

A model selection function S is said to be *direct* iff it is defined as
$$S(W, W') = S'(W \cap W').$$
It is said to be *formed using filtering* iff it is defined as
$$S(W, W') = S'(W) \cap W'.$$
In both cases S' is called the *essential* selection function. S is *formed using minimization* iff it is direct or it is formed using filtering, and the essential selection function is defined as
$$S'(W) = Min(\ll, W) = \{I \in W \mid \neg\exists I'[I' \in W \wedge I' \ll I]\}$$
for some *preference relation* $\ll$ on interpretations.

Specific entailment methods will be named with multi-letter abbreviations, for example PCM for 'prototypical chronological minimization'. Its main selection function will be written as S_{pcm}, and the corresponding entailment relation as $\models_{pcm}$. If the selection function is formed using minimization, then I will write its preference relation as $\ll_{pcm}$ and its essential selection function as pcm. In the example we have
$$\text{PCM} = \langle \mathbf{G}_{seq}, S_{pcm} \rangle,$$
$$S_{pcm}(\langle \mathcal{O}, \Pi, \text{SCD}, \text{OBS} \rangle) = Min(\ll_{pcm}, [\![\Pi(\text{SCD})]\!]_\mathcal{O} \cap [\![\text{OBS}]\!]),$$
and with other notation,
$$S_{pcm}(\Upsilon) = pcm[\![\Upsilon]\!],$$
$$pcm(W) = Min(\ll_{pcm}, W),$$
$$\Upsilon \models_{pcm} \alpha \text{ iff } S_{pcm}(\Upsilon) \subseteq [\![\alpha]\!].$$
We first consider entailment methods $\langle \mathbf{G}_{seq}, S_x \rangle$ whose model selection function S_x has the form
$$S_x(\Upsilon) = Min(\ll_x, [\![\Upsilon]\!])$$
for some preference relation $\ll_x$ over DFL-1 interpretations. In the above terms, such methods are direct and formed using minimization. Three entailment methods will be defined, namely prototypical chronological minimization of change (PCM), original chronological minimization of change (OCM), and prototypical global minimization of change (PGM). According to Proposition 8.6, such entailment methods cannot possibly be applicable throughout $\mathcal{K}$-**IA**, so the problem is to identify the narrower range of applicability for each of them.

Although their applicability is limited, these entailment methods do have an interest both as simple methods for fairly limited classes of problems, and as starting points for the continued discussion. They also have a historical interest. OCM is essentially Kautz's method from 1986 [38], which was one of the first responses to the Yale shooting scenario. PGM is a possible interpretation of McCarthy's proposal from 1984 [49], which caused the flurry over the Yale shooting scenario. PCM is a modification of OCM with a broader range of applicability.

Table 9.1. Applicability of PCM for representative scenarios

Scen:o	Name	Outcome
YTS	Yale Shooting Scenario	=
HTS	Hiding Turkey Scenario	=
SMM	Stanford Murder Mystery	⊃
FCS	Ferryboat Connection Scenario	=
RTS	Russian Turkey Scenario	⊂
SCS	Stolen Car Scenario	N.A.
TCS	Ticketed Car Scenario	±
FAS	Furniture Assembly Scenario	⊃

9.1.3 Prototypical chronological minimization of change

Let
$$I = \langle M, R \rangle$$
be a DFL-1 interpretation, and let $t \in \mathcal{T}$. The *breakset* of I at time t is defined as
$$breakset(I, t) = \{f_i \mid R(f_i, \theta t) \neq R(f_i, t)\},$$
that is, as the set of all the features whose value according to R has a value change from time θt to time t. In particular, $breakset(I, 0) = \emptyset$ for any I, since $\theta 0 = 0$. If $I = \langle M, R \rangle$ and $I' = \langle M', R' \rangle$ are DFL-1 interpretations for the same vocabulary ν, then I write $I \ll_{pcm} I'$ and say that I is *preferred over* I' (according to PCM) iff $M = M'$ and there is some timepoint t such that

- for all features f in ν, and for all timepoints $t < t$, $R(f, t) = R'(f, t)$

- $breakset(I, t) \subset breakset(I', t)$.

Furthermore, we define $pcm(W) = Min(\ll_{pcm}, W)$, $S_{pcm}(\Upsilon) = pcm[\![\Upsilon]\!]$, and PCM itself as $\langle \mathbf{G}_{seq}, S_{pcm} \rangle$, according to the usual pattern. S_{pcm} is intended to have the same value as $\Sigma_{\mathbf{IA}}$, and it does so, for example, for the Yale shooting scenario with single-timestep actions. However, PCM doesn't always work when there is only partial knowledge about the change. Table 9.1 summarizes, for each of the $\mathcal{K}$-**IA** scenarios given in Chapter 7, section 7.2, whether PCM obtains the intended set of models, assuming single-timestep actions in all cases except the FCS and FAS. For each scenario the table's last column indicates the relationship between the selected and the intended set of models. Thus ⊂ indicates that $S_{pcm}(\Upsilon) \subset \Sigma_{\mathbf{IA}}(\Upsilon)$, and similarly for the other relations. The symbol ± represents 'neither =, ⊂, nor ⊃.' These examples also illustrate the rationale for some aspects of the definition of $\ll_{pcm}$. If the requirement in the first '•' item in the definition of PCM is dropped, then too few models are obtained in the

Hiding turkey scenario. Also, if the requirement $M = M'$ is dropped, then too few models are obtained in the Ferryboat connection scenario.

PCM works correctly for a class of chronicles that includes $\mathcal{K}$p-**IsAn** and $\mathcal{K}$p-**IAte** (Proposition 9.9 below), which means that it may fail for the following cases:

- worlds where there are observations at times later than origo,
- actions which take more than a single timestep,
- actions with random choice between change and nochange in their effects.

In the second case PCM will tend to prefer those developments where all changes take place during the last timestep of the action period, and in the third case PCM will tend to prefer the nochanging outcome over the changing one. The problems in the first case will be discussed below. In addition, of course, PCM is inapplicable for worlds with undescribable trajectory languages, since for such worlds there does not exist a set of action laws that is correct with respect to $\mathcal{K}$-**IA**. However, the same will be true for all the entailment methods that are considered in the sequel as well.

This discussion, and in particular the phrase 'tend to' is deliberately vague, and is only intended to convey the basic intuitions. The precise account of when PCM is correct follows later on in this chapter.

9.1.4 Original chronological minimization of change

The entailment method OCM, for original chronological minimization of change, is similar to PCM but uses the preference relation $\ll_{ocm}$ defined as follows. If $I = \langle M, R \rangle$ and $I' = \langle M', R' \rangle$ are DFL-1 structures for the same vocabulary ν, then I write $I \ll_{ocm} I'$ and say that I is *preferred over* I' (according to OCM) iff $M = M'$ and there is some timepoint t such that

- for all features f in ν, and for all timepoints $t <$ t, $breakset(I, t) = breakset(I', t)$,
- $breakset(I, \text{t}) \subset breakset(I', \text{t})$.

The functions ocm and S_{ocm} are defined in the same way as for PCM, and OCM = $\langle \mathbf{G}_{seq}, S_{ocm} \rangle$.

OCM works correctly for a class that includes $\mathcal{K}$sp-**IAte** (Proposition 9.11 below). The demonstrated range of applicability is strictly smaller than the one for PCM. Table 9.2 summarizes, for each of the $\mathcal{K}$-**IA** scenarios

Table 9.2. Applicability of OCM for representative scenarios

Scen:o	Name	Outcome
YTS	Yale Shooting Scenario	=
HTS	Hiding Turkey Scenario	⊂
SMM	Stanford Murder Mystery	±
FCS	Ferryboat Connection Scenario	=
RTS	Russian Turkey Scenario	⊂
SCS	Stolen Car Scenario	N.A.
TCS	Ticketed Car Scenario	±
FAS	Furniture Assembly Scenario	=

given in Chapter 7, whether OCM obtains the intended set of models, again assuming single-timestep actions in all cases except the FCS and FAS. Notice, for example, that the HTS is correctly handled by PCM but not by OCM, because the initial state is not completely specified by the premises. On the other hand FAS is handled correctly by OCM but not by PCM.

9.1.5 Prototypical global minimization of change

Let
$$I = \langle M, R \rangle$$
be a DFL-1 interpretation. The *changeset* of I is defined as
$$changeset(I) = \{\langle f_i, \mathbf{t}\rangle \mid R(f_i, \theta \mathbf{t}) \neq R(f_i, \mathbf{t})\},$$
that is, as the set of instances where a feature changes its value according to R. If $I = \langle M, R \rangle$ and $I' = \langle M', R' \rangle$ are DFL-1 structures for the same vocabulary ν, then I write $I \ll_{pgm} I'$ and say that I is *preferred over* I' (according to PGM) iff $M = M'$ and $changeset(I) \subset changeset(I')$. Also, $pgm(W) = Min(\ll_{pgm}, W)$, $S_{pgm}(\Upsilon) = pgm[\![\Upsilon]\!]$, and PGM $= \langle \mathbf{G}_{seq}, S_{pgm}\rangle$ as for PCM and OCM.

Prototypical global minimization has a very simple preference relation, but at the expense of limited applicability. In particular it requires that actions 'always change the same features' (in a sense to made precise below), and it also imposes a consistency requirement (Proposition 9.14). For example, it does not work in general if it is conditional on the starting state whether an action affects a certain fluent, such as in the Yale shooting scenario, where the effect of *Fire* on *alive* depends on whether the gun is loaded or not. In such a case, PGM will obtain unintended preferred models. Also, it does not in general work correctly if different members of the same $\mathtt{Trajs}(E, r)$ allow the same feature to change either once or several times within the action period, since PGM will then tend to minimize the number of such changes.

Table 9.3. Applicability of PGM for representative scenarios

Scen:o	Name	Outcome
YSS	Yale Shooting Scenario	⊃
HTS	Hiding Turkey Scenario	⊃
SMM	Stanford Murder Mystery	⊃
FCS	Ferryboat Connection Scenario	⊃
RTS	Russian Turkey Scenario	⊃
SCS	Stolen Car Scenario	N.A.
TCS	Ticketed Car Scenario	=
FAS	Furniture Assembly Scenario	⊃

Table 9.3 gives a summary of how PGM applies for the same scenarios as in the table for PCM above. In particular, there is only one case where PGM obtains the intended set of models, namely the Ticketed car scenario! In all the other cases, a superset of the intended set of models is obtained. However, it is also easy to construct an example where some intended models are not selected, for example a scenario containing a single occurrence of an action in **IA−IAn**.

9.2 Assessments of applicability

This section assesses the applicability of the entailment methods PCM, OCM, and PGM which have now been defined. The basic concepts for correctness are as follows.

Definition. *An entailment method* $\langle \mathbf{G}, S \rangle$, *where S uses DFL-1 interpretations, is said to be* **correctly applicable** *to the ontological family* $\mathcal{Z} \subseteq \mathcal{K}\text{-}\mathbf{IA}$ *iff* $S(\mathbf{G}(\Upsilon)) = \Sigma_{\mathbf{IA}}(\Upsilon)$ *for any chronicle Υ in $\mathcal{Z}$.* ⋈

Definition. *A pretransformation function $\mathbf{G}$* **preserves** *a chronicle family* $\mathcal{Z} \subseteq \mathcal{K}\text{-}\mathbf{IA}$ *iff* $\Upsilon \in \mathcal{Z} \to \mathbf{G}(\Upsilon) \in \mathcal{Z}$. ⋈

Lemma 9.2. *If $\mathbf{G}$ is a harmless pretransformation function which preserves $\mathcal{Z}$ and results in $\mathcal{Z}'$, and the model selection function S satisfies $S(\Upsilon) = \Sigma_{\mathbf{IA}}(\Upsilon)$ for any chronicle Υ in $\mathcal{Z} \cap \mathcal{Z}'$, then the entailment method $\langle \mathbf{G}, S \rangle$ is correctly applicable to $\mathcal{Z}$.*

Proof If $\Upsilon \in \mathcal{Z}$ then $\mathbf{G}(\Upsilon) \in \mathcal{Z}$, since $\mathbf{G}$ preserves $\mathcal{Z}$. Also, $\mathbf{G}(\Upsilon) \in \mathcal{Z}'$ since $\mathbf{G}$ results in $\mathcal{Z}'$. It follows that $S(\mathbf{G}(\Upsilon)) = \Sigma_{\mathbf{IA}}(\mathbf{G}(\Upsilon))$, since S is correct in $\mathcal{Z} \cap \mathcal{Z}'$, and $\Sigma_{\mathbf{IA}}(\mathbf{G}(\Upsilon)) = \Sigma_{\mathbf{IA}}(\Upsilon)$ since $\mathbf{G}$ is harmless. □

In particular, if a model selection function S obtains the correct results for any safely sequential chronicle in some chronicle family $\mathcal{Z}$, then the entailment method $\langle \mathbf{G}_{seq}, S \rangle$ is correctly applicable to $\mathcal{Z}$. This lemma will be used for all the assessments proofs in the sequel. Notice that all ontological restrictions are preserved by $\mathbf{G}_{seq}$, since it only modifies the schedule part of the chronicle. Only for epistemological restrictions need we check whether they are preserved by $\mathbf{G}_{seq}$.

9.2.1 Time-restricted selection functions

The intended models function $\Sigma_{\mathbf{IA}}$ can itself be considered as a selection function. It has already been generalized from models over the whole time domain ($\Sigma_{\mathbf{IA}}(\Upsilon)$) to restricted models over a part $[0,t]$ of $\mathcal{T}$ for a given valuation M ($\Sigma_{\mathbf{IA}}^t(M,\Upsilon)$). We make the same generalization for selection functions in general:

Definition. *A selection function $S^t(M,\Upsilon)$ is a* **correct time-restriction** *of a given selection function S for $\mathcal{K}$-**IA** iff, for any $\Upsilon = \langle \mathcal{O}, \Pi, \text{SCD}, \text{IOBS}\rangle \in \mathcal{K}\text{p-}\mathbf{IA}$ and any good valuation M for Υ,*
 1. *$S^t(M,\Upsilon) \subseteq [\![M,\Upsilon]\!]^t$ for any $t \in \text{At}(M,\text{SCD})$, and*
 2. *$S(\Upsilon) = \{\langle M,R\rangle \mid \langle M,R_{0:t}\rangle \in S^t(M,\Upsilon)$ for all $t \in \text{At}(M,\text{SCD})\}$.* ⋈

It follows at once that

Lemma 9.3. *$\Sigma_{\mathbf{IA}}^t(M,\Upsilon)$ is a correct time-restriction of $\Sigma_{\mathbf{IA}}(\Upsilon)$.*

Proof Follows from Propositions 8.8 and 8.9. □

Functions of the form $S^t(M,\Upsilon)$ have already appeared in the definition of progressive in subsection 8.4.2. The following lemma captures the essential induction step for the assessment proof for PCM:

Lemma 9.4. *Let $S^t(M,\Upsilon)$ be a progressive, correct time-restriction for the function $S(\Upsilon)$ for $\mathcal{K}$-**IA**, and let $\hat{S}^t(M,\Upsilon)$ be similar for $\hat{S}(\Upsilon)$. If $\Upsilon = \langle \mathcal{O}, \Pi, \text{SCD}, \text{IOBS}\rangle \in \mathcal{K}\text{p-}\mathbf{IA}$ and $S^0(M,\Upsilon) = \hat{S}^0(M,\Upsilon)$, then $S(\Upsilon) = \hat{S}(\Upsilon)$.*

The restriction to $\mathcal{K}\text{p-}\mathbf{IA}$ is used throughout when we deal with finite interpretations. Observations for time later than zero may cause trouble from a formal point of view, and the $\mathcal{K}\text{p-}\mathbf{IA}$ case is sufficient for the later use of these concepts and lemmas.

Proof We first prove by induction that $S^t(M,\Upsilon) = \hat{S}^t(M,\Upsilon)$ for all $t \in \text{At}(M,\text{SCD})$. Suppose this equality holds for some $t \in \text{At}(M,\text{SCD})$, and let u be the next larger member of $\text{At}(M,\text{SCD})$. Consider an arbitrary $I' = \langle M,R'\rangle \in S^u(M,\Upsilon)$, and let $I = \langle M,R\rangle = \langle M,R'_{0:t}\rangle$. By the definition of progressive in subsection 8.4.2, $I \in S^t(M,\Upsilon)$ according to (part 3a), and $R' = R \triangleright v$ for some $v \in \text{Cats}(M,\text{SCD},R)$ according to (part 3b). By the induction hypothesis, $I \in \hat{S}^t(M,\Upsilon)$, and by the definition of progressive (part 2), $I' \in \hat{S}^u(M,\Upsilon)$.

Now since $S^0(M,\Upsilon) = \hat{S}^0(M,\Upsilon)$, it follows that $S^t(M,\Upsilon) = \hat{S}^t(M,\Upsilon)$ for all $t \in \text{At}(M,\text{SCD})$. Finally, by the definition of a correct time-restriction,
$$S(\Upsilon) = \{\langle M,R\rangle \mid \langle M,R_{0:t}\rangle \in S^t(M,\Upsilon) \text{ for all } t \in \text{At}(M,\text{SCD})\} =$$

$$\{\langle M, R\rangle \mid \langle M, R_{0:t}\rangle \in \hat{S}^t(M, \Upsilon) \text{ for all } t \in \mathsf{At}(M, \mathrm{SCD})\} = \hat{S}(\Upsilon),$$
which concludes the proof. □

9.2.2 Assessment of PCM

The definition of PCM was that if $\Upsilon = \langle \mathcal{O}, \Pi, \mathrm{SCD}, \mathrm{OBS}\rangle$, then
$$S_{pcm}(\Upsilon) = pcm[\![\Upsilon]\!].$$
For the assessment of PCM we define the auxiliary function
$$S^t_{pcm}(M, \Upsilon) = pcm([\![M, \Upsilon]\!]^t),$$
when M is a good valuation for $\langle \mathcal{O}, \Pi, \mathrm{SCD}, \mathrm{IOBS}\rangle$, and $t \in \mathsf{At}(M, \mathrm{SCD})$. What this means is that for given M, the schedule is restricted to include only those actions that terminate before or at time t, obtain the restricted models over $[0, t]$ for that (restricted) schedule, and then minimize. (The M operation on IOBS to obtain $M[\mathrm{IOBS}]$ is only of interest if IOBS contains object constant symbols, since otherwise $M[\mathrm{IOBS}] = \mathrm{IOBS}$).

It follows immediately that:

Lemma 9.5. $S^0_{pcm}(M, \Upsilon) = \Sigma^0_{\mathbf{IA}}(M, \Upsilon)$ *for any chronicle* $\Upsilon \in \mathcal{K}\mathrm{p\text{-}IA}$.

Proof According to the second clause in Proposition 8.8, $\Sigma^0_{\mathbf{IA}}(M, \Upsilon) = [\![M, \Upsilon]\!]^0$. Since $breakset(I, 0) = \emptyset$ for any I, it is not possible that $I \ll_{pcm} I'$ when I and I' are members of $[\![M, \Upsilon]\!]^0$. It follows that $S^0_{pcm}(M, \Upsilon) = [\![M, \Upsilon]\!]^0$, which completes the proof. □

Several of the assessments will be defined in terms of how the preference relation operates on the trajectory sets $\mathtt{Trajs}(E, r)$. For this purpose we extend previous notation as follows. If r is a state and v is a trajectory, then $r \triangleright v$ is an abbreviation for $\langle r\rangle \triangleright v$, or in other words
$$r \triangleright \langle r_1, r_2, ..., r_k\rangle = \langle r, r{\oplus}r_1, r{\oplus}r_2, ..., r{\oplus}r_k\rangle.$$
Also, if R and R' are histories over the same time domain and $\ll$ is a preference relation which is independent of the valuation, then $R \ll R'$ is defined as $\langle M, R\rangle \ll \langle M, R'\rangle$.

Definition. *A trajectory-semantics world is said to be* **PCM-compatible** *iff* $\neg(r \triangleright v \ll_{pcm} r \triangleright v')$ *when v and v' have equal length and both are members of* $\mathtt{Trajs}(E, r)$. ⋈

We notice that PCM-compatibility is independent of which description $\langle \mathtt{Infl}, \mathtt{Trajs}\rangle$ is chosen for a given world. The concept will be applied both to world descriptions, to the worlds themselves, and to chronicles containing action laws which are formed using such a description.

Lemma 9.6. *Let* $\Upsilon = \langle \mathcal{O}, \Pi, \mathrm{SCD}, \mathrm{IOBS}\rangle$ *be a PCM-compatible, safely sequential chronicle in the $\mathcal{K}\mathrm{p\text{-}IAe}$ family, and let M be a a good valuation for Υ. If t and u are two successive members of $\mathsf{At}(M, \mathrm{SCD})$, and $\langle M, R'\rangle \in S^u_{pcm}(M, \Upsilon)$, then $R' = R'_{0:t} \triangleright v$ for some $v \in \mathsf{Cats}(M, \mathrm{SCD}, R'_{0:t})$.*

Proof Suppose this were not the case. However, $\langle M, R'\rangle \in [\![M, \Upsilon]\!]^u$, so if $M[\textsc{scd}]_{0:u}$ contains an action $[s_i, t_i] E_i$ then $\Pi([s_i, t_i] E_i)$ must be true in $\langle \mathcal{O}, \langle M, R'\rangle\rangle$. Let $R = R'_{0:t}$. According to Lemma 8.3, there must therefore be some $v \in \mathsf{Cats}(M, \textsc{scd}, R)$ such that $v_j \subseteq R'(t+j)$ for $1 \leq j \leq u - t$, that is, all changes dictated by v are performed in R' from $t+1$ to u, although R' may also contain additional changes in that interval. But then $breakset(\langle M, R \triangleright v\rangle, t+j) \subseteq breakset(\langle M, R'\rangle, t+j)$ for all those j. If there is equality for all j then there are no changes in any other feature than those changed according to v, so R' is formed as $R \triangleright v$, and if there is a strict subset relation for some j, then $\langle M, R'\rangle$ cannot be a PCM-selected model. In both cases a contradiction has been obtained. This concludes the proof. □

Lemma 9.7. S^t_{pcm} *is a correct time-restriction of S_{pcm} and is progressive for any PCM-compatible, safely sequential chronicle Υ in the $\mathcal{K}$p-**IAe** family.*

Proof We write S^t for S^t_{pcm}. The first requirement in the definition of a correct time-restriction is the same as the first requirement in the definition of progressive. It is trivial from its definition that S^t satisfies this requirement.

The second requirement for correct time-restriction is that
$$S_{pcm}(\Upsilon) = \{\langle M, R\rangle \mid \langle M, R_{0:t}\rangle \in S^t(M, \Upsilon) \text{ for all } t \in \mathsf{At}(M, \textsc{scd})\}.$$
We prove that each member of the set to the left of the equality sign is also a member of the set to its right, and vice versa. First, suppose $\langle M, R\rangle \in S_{pcm}(\Upsilon)$. It follows that $\langle M, R\rangle \in [\![M, \Upsilon]\!]$, so $\langle M, R_{0:t}\rangle \in [\![M, \Upsilon]\!]^t$ by Proposition 8.7. To prove that it is also PCM-preferred, suppose it were not. Then there is some other $I' = \langle M, R'\rangle \in [\![M, \Upsilon]\!]^t$ which is preferred over $\langle M, R_{0:t}\rangle$. Since Υ is equidurational, there is some $\langle M, R''\rangle$ which is defined over $[0, \infty]$ such that $R''_{0:t} = R'$. Then $\langle M, R''\rangle$ would be PCM-preferred over $\langle M, R\rangle$, which is a contradiction.

On the other hand, assume that $\langle M, R\rangle$ satisfies $\langle M, R_{0:t}\rangle \in S^t(M, \Upsilon)$ for all $t \in \mathsf{At}(M, \textsc{scd})$. Consider, in particular, a choice of t where the ending times of all actions in $M[\textsc{scd}]$ are $\leq t$. Then $R(u) = R(t)$ for all $u > t$, since otherwise $\langle M, R_{0:u}\rangle$ would not be PCM-preferred. It follows that $\langle M, R\rangle$ is also PCM-preferred, that is, $\langle M, R\rangle \in S_{pcm}(\Upsilon)$. This concludes the proof that S^t_{pcm} is a correct time-restriction of S_{pcm}.

Now consider a safely sequential and PCM-compatible chronicle $\Upsilon = \langle \mathcal{O}, \Pi, \textsc{scd}, \textsc{iobs}\rangle \in \mathcal{K}\text{p-}\mathbf{IAe}$, and a valuation M which is a good valuation for Υ. Let $t \in \mathsf{At}(M, \textsc{scd})$, and address in turn the requirements that are mentioned in the definition of progressive.

1. $S^t(M, \Upsilon) \subseteq [\![M, \Upsilon]\!]^t$ follows directly from the definition.

2. If $I = \langle M, R \rangle \in S^t(M, \Upsilon)$, $v \in \text{Cats}(M, \text{SCD}, R)$ has length k, and $t + k$ is the next larger member in $\text{At}(M, \text{SCD})$ after t, we are to show that $\langle M, R \triangleright v \rangle \in S^{t+k}(M, \Upsilon)$. Assume the contrary. Then there is some $\langle M, R'' \rangle \in [M, \Upsilon]^{t+k}$ which is preferred over $\langle M, R \triangleright v \rangle$, and maximally preferred according to PCM. If the timepoint of preferred comparison is $\leq t$, then $\langle M, R''_{0:t} \rangle$ must be preferred over $I = \langle M, R \rangle$, which contradicts the assumption. Otherwise by the definition of $\ll_{pcm}$, $R = R''_{0:t}$. According to Lemma 9.6, $R'' = R \triangleright v''$ for some $v'' \in \text{Cats}(M, \text{SCD}, R)$. But $(R \triangleright v'') \ll_{pcm} (R \triangleright v)$ is only possible if $(R(t) \triangleright v'') \ll_{pcm} (R(t) \triangleright v)$, which contradicts the assumption that Υ is PCM-compatible.

3. If $I' = \langle M, R' \rangle \in S^u(M, \Upsilon)$, where u is the successor of t in $\text{At}(M, \text{SCD})$ and $R'_{0:t} = R$, prove that (a) $I = \langle M, R \rangle \in S^t(M, \Upsilon)$, and (b) $R' = R \triangleright v$ for some $v \in \text{Cats}(M, \text{SCD}, R)$. Suppose that $I \notin S^t(M, \Upsilon)$. Then there is some other $I'' = \langle M, R'' \rangle \in [M, \Upsilon]^t$ which is preferred over I. Since Υ is equidurational there is some $v'' \in \text{Cats}(M, \text{SCD}, R'')$ of length $u - t$. Therefore, $\langle M, R'' \triangleright v'' \rangle \in [M, \Upsilon]^u$, and by the definition of PCM it must be preferred over I', which contradicts the assumption. Therefore, $I \in S^t(M, \Upsilon)$, which proves (a). The (b) part follows directly from Lemma 9.6.

This concludes the proof. □

From this we obtain:

Proposition 9.8. *PCM is correctly applicable to the family of those chronicles in $\mathcal{K}$p-**IAe** which are PCM-compatible.*

Proof This follows directly from Lemmas/Propositions 8.10 and 9.3-9.7, and because $\mathbf{G}_{seq}$ preserves this family of chronicles. □

The notion of PCM-compatible is difficult to grasp intuitively. The following special case is therefore useful:

Proposition 9.9. *PCM is correctly applicable to the $\mathcal{K}$p-**IsAn** and $\mathcal{K}$p-**IAte** ontological families.*

Proof It suffices to show that all IDS worlds in **IsAn** and **IAt** are PCM-compatible, since $\mathcal{K}$p-**IsAn** $\subseteq \mathcal{K}$p-**IsA** $\subseteq \mathcal{K}$p-**IAe** by their definitions. The PCM-compatibility of **IAt** follows trivially from the definition of that subcharacteristic. For **IsAn**, let $\langle \text{Infl}, \text{Trajs} \rangle$ be a description of a world in **IsAn**, and let $\{\langle r' \rangle, \langle r'' \rangle\} \subseteq \text{Trajs}(E, r)$ for some E and r. It follows that $\text{breakset}(r \triangleright \langle r' \rangle, 2) = \text{Infl}(E, r) = \text{breakset}(r \triangleright \langle r'' \rangle, 2)$. Therefore, it cannot be the case that $r \triangleright \langle r' \rangle \ll_{pcm} r \triangleright \langle r'' \rangle$, and the result follows.
□

9.2.3 Assessment of OCM

For OCM we need an analogous compatibility concept as for PCM. However, we define two related concepts, which will allow us to obtain two different assessments which impose different kinds of restrictions.

Definition. *A trajectory-semantics world is said to be* **OCM-compatible** *iff* $\neg(r \triangleright v \ll_{ocm} r \triangleright v')$ *when v and v' have equal length and both are members of* $\text{Trajs}(E, r)$. *It is said to be* **OCM-favorable** *iff it is the case that* $\neg(r \triangleright v \ll_{ocm} r' \triangleright v')$ *when v and v' have equal length, $v \in \text{Trajs}(E, r)$, and $v' \in \text{Trajs}(E, r')$.* ⋈

It follows at once that if a chronicle is OCM-favorable then it is also OCM-compatible, whereas the converse does not hold in general. These concepts apply equally to world-descriptions and to chronicles, like for 'PCM-compatible'.

Proposition 9.10. *OCM is correctly applicable to the family of OCM-compatible chronicles in $\mathcal{K}$sp-**IAde**.*

Proof It is clear at once that $\mathbf{G}_{seq}$ preserves the indicated chronicle family. Consider a chronicle $\Upsilon = \langle \mathcal{O}, \Pi, \text{SCD}, \text{IOBS} \rangle$ in $\mathcal{K}$sp-**IAde** which is safely sequential and OCM-compatible, and a good valuation M for SCD. For given Υ and M, at most one state r satisfies the observations for origo time, according to the epistemological property $\mathcal{K}$s. Since all actions are deterministic, all intended models for the given M have the same $R(t)$ for those timepoints t which are members of $\text{At}(M, \text{SCD})$. Because of the OCM-compatibility, $\ll_{ocm}$ will not prefer one intended model over another.

It remains to show that no unintended model is included in $ocm[\![\Upsilon]\!]$. Suppose that $\Sigma_{\mathbf{IA}}(M, \Upsilon) \neq \emptyset$ but there is some member $I' = \langle M, R' \rangle$ of $[\![M, \Upsilon]\!] - \Sigma_{\mathbf{IA}}(M, \Upsilon)$ such that no member of $\Sigma_{\mathbf{IA}}(M, \Upsilon)$ is preferred over I'. Attempt to construct $I = \langle M, R \rangle \in \Sigma_{\mathbf{IA}}(M, \Upsilon)$ from I' as follows. The construction proceeds over successively extended intervals $[0, s]$, where $s \in \text{At}(M, \text{SCD})$, and halts if it gets to a timepoint where R and R' differ.

First choose $R(0) = R'(0)$. If $R(s)$ has been constructed, $R_{0:s} = R'_{0:s}$, and $M[\text{SCD}]$ does not contain any action starting in s, then define $R(s + 1) = R(s)$. If $R(s + 1) \neq R'(s + 1)$ then halt, otherwise proceed. If the process halts at this point it follows that any $\langle M, R^+ \rangle$ in $\Sigma_{\mathbf{IA}}(M, \Upsilon)$ where $R^+_{0:s+1} = R$ will be OCM-preferred over I'. Also, since Υ is equidurational and R' is a model, we are guaranteed that such an R^+ exists.

If $R(s)$ has been constructed, $R_{0:s} = R'_{0:s}$, and $M[\text{SCD}]$ contains an action $[s, t]E$, then identify the member v of $\text{Trajs}(R(s), E)$ that exists for R' according to Lemma 8.3 since $I' \in [\![M, \Upsilon]\!]$. Extend R to $[0, t]$ by defining $R_{0:t} = R_{0:s} \triangleright v$. It follows that R and R' must agree over $[s+1, t]$ for all those features which are defined in v, and that R does not have any changes in the other features. If $R'_{0:t} = R_{0:t}$ then proceed. In the opposite case, necessarily $breakset(I, u) \subseteq breakset(I', u)$ for all $u \in (s, t]$, with a strict subset relation in at least one timepoint. Again it is possible to construct an R^+ that is OCM-preferred over R'.

This process must halt at some point in time, because the contrary would violate the assumption that $I' \notin \Sigma_{\mathbf{IA}}(M, \Upsilon)$. However, whenever the process halts it is possible to construct an R^+ so that $\langle M, R^+ \rangle$ is OCM-preferred over I', which violates the assumption about I'.

Finally, if the set $\Sigma_{\mathbf{IA}}(M, \Upsilon)$ of intended models with the given M is empty, then the set $[\![M, \Upsilon]\!]$ of classical models is also empty due to the equidurational requirement. This concludes the proof. □

The condition of OCM-compatible is hard to visualize. The following is a special case:

Proposition 9.11. *Corollary: OCM is correctly applicable to* $\mathcal{K}$sp-**IAte**.

Proof The **At** subcharacteristic is defined by the condition that it is not possible for $\mathtt{Trajs}(E, r)$ to have more than one member of the same length, which guarantees OCM-compatibility. □

Comparing the proven ranges of applicability of OCM and PCM, we make the following observations.

Proposition 9.12. *If a chronicle is OCM-compatible then it is PCM-compatible.*

Proof If $I \ll_{pcm} I'$ then $I \ll_{ocm} I'$ according to the definitions. Therefore, if $\neg (I \ll_{ocm} I')$ then $\neg (I \ll_{pcm} I')$. The result follows from the definitions of OCM-compatible and PCM-compatible. □

The other assessment for OCM is:

Proposition 9.13. *OCM is correctly applicable to the family of OCM-favorable chronicles in* $\mathcal{K}$p-**IAe**.

Proof It follows from the previous proposition that PCM is correct for OCM-favorable chronicles in $\mathcal{K}$p-**IAe**, and since $I \ll_{pcm} I' \rightarrow I \ll_{ocm} I'$ it follows that $\Upsilon \in \mathcal{K}$p-**IA** implies
$$ocm[\![\Upsilon]\!] \subseteq pcm[\![\Upsilon]\!] = \Sigma_{\mathbf{IA}}(\Upsilon).$$
It is therefore sufficient to show that $\ll_{ocm}$ does not apply between two members of $\Sigma_{\mathbf{IA}}(\Upsilon)$. Suppose the contrary, that is,
$$I \ll_{ocm} I', I \in \Sigma_{\mathbf{IA}}(\Upsilon), I' \in \Sigma_{\mathbf{IA}}(\Upsilon).$$
But since I and I' are intended models, they can only have changes according to the trajectories of the action in $M[\text{SCD}]$, and these changes cannot achieve $\ll_{ocm}$-preference. □

Both the proven ranges of applicability for OCM are therefore more restrictive than that of PCM. Notice, however, that this refers only to the proven ranges, which constitute lower bounds on the actual range of applicability. Upper bounds will be considered in Chapter 13.

9.2.4 Assessment of PGM

PCM's restriction to $\mathcal{K}p$ does not apply to prototypical global minimization, PGM. However, PGM does also have an epistemological restriction, namely for those chronicles where $\Sigma_{IA}(\Upsilon) = \emptyset$ although $[\![\Upsilon]\!] \neq \emptyset$. In such cases PGM will select a non-empty set of models. The PGM method can therefore only apply for *consistency-retaining* chronicles, that is, for those chronicles Υ which satisfy
$$[\![\Upsilon]\!] \neq \emptyset \to \Sigma_{IA}(\Upsilon) \neq \emptyset.$$
A note on terminology in passing. A model selection function S is said to be *consistency-preserving* iff it satisfies
$$[\![\Upsilon]\!] \neq \emptyset \to S(\Upsilon) \neq \emptyset,$$
so that if the chronicle has some classical models, then it cannot be that the set of selected models is empty. The function Σ_{IA} itself is not consistency-preserving, as illustrated by the stolen car scenario. A consistency preserving model selection function, such as S_{pgm} for example, is therefore definitely restricted to those chronicles Υ that satisfy
$$[\![\Upsilon]\!] \neq \emptyset \to \Sigma_{IA}(\Upsilon) \neq \emptyset,$$
and it is for those chronicles that we use the term consistency-retaining.

Actually, the applicability requirement for consistency-preserving selection functions must be even a bit stronger than this, as is illustrated by the following example chronicle:

obs1 $[0]p$
scd1 $[4,5]A$
scd2 $[9,10]A$
obs2 $[t_1]p$
obs3 $8 \leq t_1$

In this chronicle, A is an action that toggles the value of the propositional feature p. Clearly, this chronicle has some intended models under $\mathcal{K}\text{-}\mathbf{IA}$, namely models where t_1 is 10 or larger. There is no intended model where t_1 is 8 or 9. However, since PGM will not impose a preference between models containing different valuations, it will also consider valuations where $M[t_1]$ is 8, and find some preferred and unintended models, for example, one where p is reset to true at time 7. We introduce therefore the following definition.

Definition. *A chronicle Υ is said to be* **pointwise consistency-retaining** *iff it satisfies*
$$\langle M, R \rangle \in [\![\Upsilon]\!] \to \Sigma_{IA}(M, \Upsilon) \neq \emptyset.$$
The epistemological property symbol $\mathcal{K}r$ is used to represent pointwise consistency-retaining chronicles. ⋈

Following the pattern of the other entailment methods we use compatibility-like concepts for PGM, but again with a new variant:

Definition. A trajectory-semantics world $\langle \text{Infl}, \text{Trajs} \rangle$ is said to be **PGM-perfect** iff it satisfies the following requirement. For any action designator E, if $v \in \text{Trajs}(E, r)$ and $v' \in \text{Trajs}(E, r')$ are two trajectories of equal length, then $\text{changeset}(r \triangleright v) = \text{changeset}(r' \triangleright v')$. It is said to be **PGM-favorable** iff it satisfies the following, weaker requirement. For any action designator E, let $v \in \text{Trajs}(E, r)$ and $v' \in \text{Trajs}(E, r')$ be two trajectories of equal length. Then it must not be the case that $\text{changeset}(r \triangleright v) \subset \text{changeset}(r' \triangleright v')$. ⋈

Thus, PGM-perfect is a stronger concept than PGM-favorable, which in turn is analoguous to OCM-favorable. The concept of PGM-perfect will be used in the range of applicability, and PGM-favorability will be used in Chapter 13 in a result on the upper bound of range of applicability. We obtain

Proposition 9.14. *PGM is correctly applicable for PGM-perfect chronicles in the $\mathcal{K}r$-**IA** ontological family.*

Proof It follows from the definitions that $\mathbf{G}_{seq}$ preserves $\mathcal{K}r$-**IA** and PGM-perfection. Since $\langle M, R \rangle \ll_{pgm} \langle M', R' \rangle \to M = M'$ it is sufficient to consider the case for each M separately. If $\Sigma_{\mathbf{IA}}(M, \Upsilon) = \emptyset$ then there will quite correctly be no member $\langle M, R \rangle$ in $S_{pgm}(\Upsilon)$. Suppose, then, that $\Sigma_{\mathbf{IA}}(M, \Upsilon) \neq \emptyset$ where $\Upsilon \in \mathcal{K}r$-**IA** is PGM-perfect and safely sequential. It follows at once that all members of $\Sigma_{\mathbf{IA}}(M, \Upsilon)$ have the same changeset, which we call C_M, so that no member of $\Sigma_{\mathbf{IA}}(M, \Upsilon)$ is PGM-preferred over another one. It remains to show that for every unintended model I' there is some intended model I that is PGM-preferred over I'. Suppose therefore that $I' = \langle M, R' \rangle \in [\![\Upsilon]\!]$ and $I' \notin \Sigma_{\mathbf{IA}}(\Upsilon)$. Since I' must satisfy the action laws it follows at once that $\text{changeset}(I') \supseteq C_M$, and that equality is impossible since $I' \notin \Sigma_{\mathbf{IA}}(\Upsilon)$. This concludes the proof. □

9.2.5 Discussion of the restrictions

Comparing the range of applicability of PCM and PGM, they are both fairly restrictive. PCM makes the requirement of equidurationality, which is a strong one, and the single-timestep family **IsA** is probably the only interesting part of **IAe**. This suggests that PCM may be restricted to applications where actions are (described as being) performed in a single timestep, and so that the intermediate states during the execution of the action are not being described. PGM does not require equidurationality, but on the other hand the requirement of PGM-perfection is a very strong one indeed. In order to understand it more concretely, consider the restriction of PGM-perfection to **IsA**. It specializes to **IsAun**, that is, it requires that the set of features that are actually changed by an action is the same for all starting states. More generally,

Proposition 9.15. *Every PGM-perfect chronicle is OCM-favorable.*

This follows directly from the definitions. Consequently, PGM-perfect chronicles are also OCM-compatible and PCM-compatible. On the other hand, since the ontological restriction on PGM does not imply equidurationality, neither of the proven ontological restrictions on PCM and PGM implies the other one. On balance, it seems that the restrictions on PGM are significantly more damaging.

The epistemological requirements $\mathcal{K}$p and $\mathcal{K}$r are related according to the following observation.

Proposition 9.16. $\mathcal{K}$p-**IAe** $\subset$ $\mathcal{K}$r-**IAe**

This means that if an equidurational chronicle has only initial-time observations, then it is pointwise consistency-retaining.

Proof Let Υ be an arbitrary member of $\mathcal{K}$p-**IAe**, and let $\langle M, R \rangle \in [\Upsilon]$. We are to show that $\Sigma_{\text{IA}}(M, \Upsilon) \neq \emptyset$. Since there is such an $\langle M, R \rangle$, it follows that $M[\text{SCD}]$ is consistent, and likewise for $M[\text{OBS}]$. Now construct a development with $R(0)$ as the starting state, M as the valuation, and $M[\text{SCD}]$ as the set of actions, by successively choosing a trajectory of the appropriate length for each of the actions in $M[\text{SCD}]$. This is possible since Υ is equidurational. The development so obtained defines a revised interpretation $\langle M, R' \rangle$, which clearly is a member of $\Sigma_{\text{IA}}(\Upsilon)$. Therefore, $\mathcal{K}$p-**IAe** $\subseteq$ $\mathcal{K}$r-**IAe**.

On the other hand, it is trivial to construct a chronicle that is in $\mathcal{K}$r-**IAe** but not in $\mathcal{K}$p-**IAe**. This concludes the proof. $\square$

Previous articles have stated assessments in terms of other ontological restrictions which are however subsumed by the ones reported here. A previously proposed epistemological restriction for PGM in terms of 'linear' chronicles has turned out to be incorrect.

One may wonder whether the assessment of PCM can be strengthened to use $\mathcal{K}$r-**IAe** instead of $\mathcal{K}$p-**IAe**. The following example shows that this is not possible. Let f be a feature whose value domain is $\{\text{R}, \text{Y}, \text{G}\}$, and let A be a non-deterministic action which always takes a single timestep, and which changes f from its present value to either of the other two values. Let the schedule and the observations be

    obs1 $[0] f_1 \hat{=} \text{G}$
    scd1 $[4, 5] A$
    obs2 $[10] f_1 \hat{=} \text{R}$.

It is immediately verified that PCM does not obtain the intended results for this chronicle, and that it is not in $\mathcal{K}$p-**IAe**. It has an intended model, namely where A changes f to R. Since it is entirely independent of the valuation M, it retains consistency even pointwise, so it is in $\mathcal{K}$r-**IAe**.

Note, also, that the result in Proposition 9.16 depends on equidurationality. On the other hand, $\mathcal{K}$p-**IA** $\not\subseteq$ $\mathcal{K}$r-**IA**, as proved by the following

example. Assume that the action A takes one timestep if p is true, and two timesteps if p is false, and consider the chronicle
 obs1 $[0]p$
 scd1 $[1,3]A.$
This chronicle belongs to $\mathcal{K}$p-**IA** and it has classical models but no intended model.

9.3 Discussion of possible improvements

In view of the quite limited range of applicability for the methods that have been investigated so far in this text, it is necessary to discuss the possible ways of changing them to obtain wider applicability. Let us first discuss some general issues in the light of the approaches that have been introduced so far.

9.3.1 Ambiguities

The term 'reasoning with incomplete information' is often used to characterize reasoning with defaults [24, 79]. In fact, however, the very point with almost all practical logical reasoning is to make do with incomplete information; the case of complete information and just one model is the exception rather than the rule.

The real problems arise when several kinds of incompleteness appear together, and in particular when some incompleteness of information is intended to be filled by defaults and some is not [31, 34]. This happens in particular for the following types of situations in the $\mathcal{K}$-**IA** family, which I shall refer to as ambiguities. An *ambiguity* is, then, a situation in which the set of intended models of a chronicle fails to specify feature changes completely. The following types of ambiguity can be observed in various examples of chronicles.

- Initial value ambiguity: The values of fluents at the beginning time of the scenario are not completely specified. Without this ambiguity, the epistemological property $\mathcal{K}$s is obtained.

- New value ambiguity: There are some points in time where a change is known to occur, but the new value is not completely specified.

- Changetime ambiguity: The exact time of change is not completely specified. This case arises for two reasons: (1) the action may have been specified using temporal constant symbols rather than actual timepoints; (2) the action takes place over an extended period of time.

If one has the **Is** class of worlds and no temporal constant symbols (only explicit timepoints in the formulae), then no changetime ambiguity can arise.

- **Change incidence ambiguity:** It is specified that a certain change may take place (overriding the default for no change), but it is not certain that there is a change. Even for one particular development up to the changetime, there may be several continuations with and without change. This case arises for non-deterministic actions, that is, outside $\mathcal{K}$-**IAd** and when the order of the actions is incompletely specified. With an appropriate syntax extension it can also be obtained from expansion of statements of the form 'maybe during the time interval $[s,t]$ the action A takes place'.

The following is an example of change that is specified unambiguously, so that the time of change is specified as a member of $\mathcal{T}$, and the new feature-value is specified as a singleton:
$$[13,14]\, f_1 := \text{R}.$$
The following is an example of changetime ambiguity:
$$10 < t_1 \leq 15 \wedge [\theta t_1, t_1]\, f_1 := \text{R}$$
and the following is an example of new-value ambiguity:
$$[13,14]\, f_1 := \text{RG}.$$
The initial value and new value ambiguities are handled incorrectly by OCM and correctly by PCM. In an ordering where OCM precedes PCM, the step to PCM may therefore be considered as the device that allows the method to deal correctly with these two types of ambiguity.

The fact that some non-monotonic logics have difficulties with ambiguities has also been observed for applications other than temporal reasoning [57, 12]. One reason why it becomes more difficult to find correct formalizations in the presence of ambiguity may be that without ambiguity we expect to have exactly one extension for any reasonable scenario. Therefore, the only thing that can 'go wrong' is that too many extensions are obtained[40], as in the case of the Yale shooting scenario. However, with ambiguity, multiple extensions are expected, which means there is the possibility of both too many and too few extensions.

Ambiguities vs. postdiction

In order to understand how the restrictions on the entailment methods work, consider again the example that we used just after Proposition 9.16, and which was
$$\text{obs1} \quad [0]\, f_1 \hat{=} \text{G}$$

---

[40] Using a non-monotonic logic with the property of consistency preservation.

scd1 [4,5]A
obs2 [10]$f_1\hat{=}$R.

where the action A has the effect of changing f from its previous value to any other value in the domain of f, which is {G, R, Y}. Under PCM entailment, these formulae will have two PCM-selected models: one model where f_1 is G until time 5 and then becomes R, and another model where f_1 is G until time 5 when it becomes Y, and then is Y until time 10 when it becomes R. Only the first one of these is an intended model in $\mathcal{K}$-**IA**, since a game with an **IA** world can only result in a development where the fluent takes the observed value R already at time 5, and not the model where it takes the value Y and later changes value to R. No intended model can have a change at time 10.

Under OCM entailment, only one model is obtained, namely the intended one. The two models selected by PCM have equal breaksets over [0, 9], although their R components differ, and OCM will therefore prefer the model having an empty breakset at time 10. It may be surprising at first that OCM obtains the correct result and PCM does not, since the assessed range of applicability for PCM is broader. However, this chronicle is outside the range of applicability for both PCM and OCM. The premise [10]$f\hat{=}$R violates the $\mathcal{K}$p requirement that both OCM and PCM have. It just so happens that OCM obtains the correct result anyway.

This example also illustrates the futility of using test examples for promoting or comparing entailment methods.

9.4 Entailment methods that use filtering, and their assessments

The technique of filtering was mentioned already at the beginning of this chapter, and it allows us to weaken some of the restrictions on PCM and PGM. We shall now define and assess those variants of entailment methods.

9.4.1 Definitions

Let a chronicle $\Upsilon = \langle \mathcal{O}, \Pi, \text{SCD}, \text{OBS}\rangle$ be given as usual, and consider the entailment method PCMF = $\langle \mathbf{G}_{seq}, S_{pcmf}\rangle$, where the model selection function S_{pcmf} is defined by
$$S_{pcmf}(\Upsilon) = pcm[\![\Pi(\text{SCD})]\!]_{\mathcal{O}} \cap [\![\text{OBS}]\!].$$
It is similar to PCM, but uses filtering rather than direct minimization, since S_{pcm} was defined as
$$S_{pcm}(\Upsilon) = pcm([\![\Pi(\text{SCD})]\!]_{\mathcal{O}} \cap [\![\text{OBS}]\!]).$$
The problem for PCM that we saw in the example at the end of the previous section is eliminated using PCMF. The definition of S_{pcm} may at first seem to be the most natural definition: one just takes all the known facts, lumps

them together, and forms their preferred model set. However, the definition S_{pcmf} is also plausible in its own way: given that $pcm[\![\Pi(\text{SCD})]\!]$ is the set of all possible developments in the world according to the premises in SCD and the given action laws, and regardless of any observations, then it makes sense to first identify that set, and then 'filter' it using the actual observations.

In this particular example it is easily seen that $S_{pcmf}(\Upsilon) \subset S_{pcm}(\Upsilon)$ without equality, and in general $S_{pcmf}(\Upsilon) \subseteq S_{pcm}(\Upsilon)$ in view of the following.

Proposition 9.17. *If $\ll$ is a preference relation and Γ_1 and Γ_2 are sets of formulae, then*
$$Min(\ll, [\![\Gamma_1]\!]) \cap [\![\Gamma_2]\!] \subseteq Min(\ll, [\![\Gamma_1]\!] \cap [\![\Gamma_2]\!]).$$

Proof $Min(\ll, [\![\Gamma_1]\!]) \cap [\![\Gamma_2]\!]$ satisfies the conditions for being a member of $Min(\ll, [\![\Gamma_1]\!] \cap [\![\Gamma_2]\!])$. □

It follows immediately that filter preferential entailment is monotonic with respect to observations, so that if SCD and OBS filter preferentially entail α, and OBS $\subseteq$ OBS', then SCD and OBS' also filter preferentially entail α. Naturally, it is not monotonic with respect to the schedule statements.

With filter preferential entailment it is quite possible that while $[\![\Upsilon]\!]$ is a non-empty set, the selected set of models is anyway empty. Filter preferential entailment therefore does not have the property of consistency preservation, which is exactly what we want since Σ_{IA} does not either. The filtering counterpart of PGM (prototypical global minimization) will be called PGMF and is defined as $\langle \mathbf{G}_{seq}, S_{pgmf} \rangle$, where
$$S_{pgmf} = pgm[\![\Pi(\text{SCD})]\!]_O \cap [\![\text{OBS}]\!].$$
Filter preferential entailment was first described by Sandewall in [72]. Crawford and Etherington later suggested that such a separation of system description formulae and observation formulae is consistent with the working practices of qualitative reasoning [15].

9.4.2 Assessments

First we establish an additional lemma.

Lemma 9.18. *If S is a selection function for DFL-1 models defined as $S(W, W') = S'(W) \cap W'$, and if $S(\Upsilon) = \Sigma_{\text{IA}}(\Upsilon)$ for an observation-free $\Upsilon = \langle \mathcal{O}, \Pi, \text{SCD}, \emptyset \rangle$, and $\Upsilon' = \langle \mathcal{O}, \Pi, \text{SCD}, \text{OBS} \rangle$, then $S(\Upsilon') = \Sigma_{\text{IA}}(\Upsilon')$.*

Proof The assumptions give
$$S(\Upsilon') = S'[\![\Pi(\text{SCD})]\!]_O \cap [\![\text{OBS}]\!] = S(\Upsilon) \cap [\![\text{OBS}]\!] = \Sigma_{\text{IA}}(\Upsilon) \cap [\![\text{OBS}]\!]$$
but by the definition of Σ_{IA},
$$\Sigma_{\text{IA}}(\Upsilon') = \Sigma_{\text{IA}}(\Upsilon) \cap [\![\text{OBS}]\!],$$
and the proof is complete. □

Prototypical chronological minimization (PCM) was defined without filtering with
$$S_{pcm} = pcm[\Upsilon],$$
and has been proved to be correctly applicable for PCM-compatible chronicles in $\mathcal{K}$p-**IAe**, where p represents initial-observation chronicles. The assessment for PCMF follows immediately:

Proposition 9.19. *PCMF is correctly applicable for the family of PCM-compatible chronicles in $\mathcal{K}$-**IAe**.*

Proof If $\Upsilon = \langle \mathcal{O}, \Pi, \text{SCD}, \text{OBS}\} \in \mathcal{K}$-**IAe** then $\Upsilon' = \langle \mathcal{O}, \Pi, \text{SCD}, \emptyset\} \in \mathcal{K}$p-**IAe**. PCM-compatibility is not affected. The result follows from Proposition 9.8 and Lemma 9.18. □

Finally we turn to prototypical global minimization (PGM), which was defined without filtering using
$$S_{pgm} = pgm[\Upsilon],$$
and which has been proved to be correctly applicable for PGM-perfect chronicles in $\mathcal{K}$r-**IA**, where r represents pointwise consistency retention. This epistemological property cannot in general be removed by filtering, since the lack of consistency retention may be due to the schedule part of the chronicle. The removal is only possible in the equidurational subcase:

Proposition 9.20. *PGMF is correctly applicable for the family of PGM-perfect chronicles in $\mathcal{K}$-**IAe**.*

Proof If $\Upsilon = \langle \mathcal{O}, \Pi, \text{SCD}, \text{OBS}\} \in \mathcal{K}$-**IAe** then
$$\Upsilon' = \langle \mathcal{O}, \Pi, \text{SCD}, \emptyset\} \in \mathcal{K}\text{p-}\mathbf{IAe} \subseteq \mathcal{K}\text{r-}\mathbf{IAe},$$
according to Proposition 9.16. PGM-compatibility is not affected. The result follows from the assessment of PGM (Proposition 9.14) and Lemma 9.18. □

In summary, the filtering technique eliminates the epistemological restrictions on PCM, and is somewhat useful even for PGM. It now remains to eliminate the ontological constraints of equidurationality and of PCM-compatibility or PGM-perfection, respectively. However, this requires some more radical changes to the structure of the entailment methods and to the base logic.

9.5 Branching time

It is straightforward to verify that the definitions and proofs in this chapter work equally well for branching time as defined in Sections 6.6 and 8.5.

9.6 Summary

An entailment method has been formally defined as a pair of a pretransformation function and a model selection function. The concept of range of applicability for entailment methods has been defined. A number of entailment methods for DFL-1 have been defined and assessed. The restriction to initial observations can be avoided using a general technique, namely, filtering. The other restrictions have been discussed, but their solution remains to the following chapters.

The detailed summary of assessments for each of the entailment methods in this chapter will follow in the tables at the end of Chapter 11.

10
Duration constraints

The chronological entailment criteria in the previous chapter required equidurationality for correctness. This restriction is quite severe: if actions are allowed to have a duration in time other than 1, and if the result of the action can vary due to the initial conditions, one would certainly expect that the duration can vary as well. The present chapter will provide a method which removes this restriction.

10.1 Chronological assignment of valuation

10.1.1 Discussion

The following is a simple example of a chronicle where PCM obtains an incorrect result because the chronicle is not equidurational. (The Furniture assembly scenario in Section 7.2 was another example of such a problem).

    scd1 $[4, t_1] A$
    obs1 $[0] p \wedge q$
    law2 $[s, t] A \Rightarrow ([s] p \rightarrow [s, t] q := \text{F} \wedge t = s + 2) \wedge$
               $([s] \neg p \rightarrow [s, t] p := \text{T})$

Here the action A is non-equidurational since the set of possible durations of the action is $\{2\}$ in those starting states where p is T, and the set of the positive integers otherwise. The preference orders that have been considered so far will only compare models having the same valuation M, that is, the minimization is performed separately for each M. Now consider PCM for the case where $M[t_1] = 7$. In those interpretations where p is inert from time 0 to time 4, the value of p at time 4 will be T, and clearly no such interpretation can be a model. However, there are in any event some classical models where $M[t_1] = 7$, namely those where p changes from T to F before or at time 4, and PCM will prefer those models where p is T at times 0, 1, 2, and 3 but it is F at time 4. In one of the PCM-selected models, p changes from T to F between times 3 and 4, and back to T from time 6 to time 7. That model is of course not an intended one.

For PGM the problem is similar except that the time where p changes without reason from T to F can be any one of 1, 2, 3, or 4. The problem is not solved merely by using filtering, since the chronicle is already a member of $\mathcal{K}$p-**IA**.

10.1.2 The model selection function for CAMC

One can think of several ways of dealing with the problem that has now been described:

- Change the logical language in such a way that the beginning time and the ending time of an action can be in separate formulae. The formula for the ending time may then be placed among the observations.

- Distinguish between two types of constant symbols, designated by s and t respectively, where the valuation-equality requirement in the preference relation is restricted to s constants.

- Weaken the valuation-equality requirement but using the value of the constant symbols rather than a classification of the symbols themselves.

I shall pursue the third alternative here, simply because it does not require any changes to the base logic, and only a simple change on the meta level. First the required piece of notation. If M is a valuation as before, and s and t are timepoints, then $M_{s:t}$ represents a subset of M consisting of (1) all maplets binding object constant symbols to their values, and (2) those maplets $t_i : t'$ (mapping the constant symbol t_i to the timepoint t') or $s_i : t'$, where $s \leq t' \leq t$. In other words, we retain the bindings where the constant symbol becomes bound to a point within the interval $[s, t]$, and ignore the others.

Now a selection function CAMC (for 'chronological assignment of valuation while minimizing change') is defined using the preference relation $\ll_{camc}$ as follows. Let $I = \langle M, R \rangle$ and $I' = \langle M', R' \rangle$ be DFL interpretations for the same vocabulary ν. I write $I \ll_{camc} I'$ and say that I is *preferred over* I' (according to CAMC) iff all of the following hold for some timepoint **t**:

- $M_{0:\mathbf{t}} = M'_{0:\mathbf{t}}$.

- For all features f in ν, and for all timepoints $t < \mathbf{t}$, $R(f, t) = R'(f, t)$.

- $breakset(I, \mathbf{t}) \subset breakset(I', \mathbf{t})$.

The function *breakset* was defined in subsection 9.1.3. The difference between $\ll_{pcm}$ and $\ll_{camc}$ is that PCM will only compare two interpretations if their entire valuation components are equal, whereas CAMC will compare them as soon as the valuations are equal up to the timepoint of comparison (t in the definition).

By the usual pattern, if W is a set of DFL interpretations, then $camc(W)$ is defined as the set of $\ll_{camc}$-preferred members of W. The set of CAMC-selected models for a chronicle $\Upsilon = \langle \mathcal{O}, \Pi, \text{SCD}, \text{OBS} \rangle$ is defined using filtered preferential model selection as
$$S_{camc}(\Upsilon) = camc[\![\Pi(\text{SCD})]\!]_{\mathcal{O}} \cap [\![\text{OBS}]\!].$$
It is fairly easy to see that CAMC does not suffer from the equidurationality problem in the Furniture assembly scenario or in the introductory example of this chapter. It now remains to asses the range of applicability of CAMC in general terms, and at the same time to identify an appropriate pretransformation function.

10.1.3   Reformulation functions

The concept of pretransformation which was introduced in the last chapter is not sufficient for CAMC or for some of the entailment methods that are to follow, since they need pretransformations that introduce additional constant symbols or feature symbols. We replace it by the following, more general, concept.

Definition. *A **reformulation function** resulting in a family $\mathcal{Z}$ of chronicles is a mapping whose members (maplets) are of the form*
$$\langle \nu, \Upsilon \rangle \mapsto \langle \nu', \Upsilon', K' \rangle$$
taking a vocabulary $\nu = \langle \sigma, \mathcal{O} \rangle$ and a chronicle Υ for ν (also using some occurrence vocabulary π for ν) as arguments. It produces a complementary vocabulary $\nu' = \langle \sigma', \mathcal{O}' \rangle$ which does not use the feature symbols or objects in ν, a modified chronicle $\Upsilon' \in \mathcal{Z}$ for $\langle \sigma \cup \sigma', \mathcal{O} \cup \mathcal{O}' \rangle$, and a set K' of constant symbols which occur in Υ' but not in Υ. ⋈

Definition. *If $\langle M', R' \rangle$ is a DFL-1 interpretation, and ν' and K' are as in the definition of a reformulation function, then $\text{Wdraw}(\langle M', R' \rangle, \langle \sigma', \mathcal{O}' \rangle, K')$ is a modified interpretation $\langle M, R \rangle$ obtained as follows. M is the restriction of M' where maplets assigning values to constant symbols in K' have been removed. R is the restriction of R' which is only defined for those $R(f,t)$ where the function symbol in the feature f is not defined in σ', and no object in $\mathcal{O}'$ occurs among the arguments in f. If W is a set of interpretations, then $\text{Wdraw}(W) = \{\text{Wdraw}(I) \mid I \in W\}$.* ⋈

The relationship between I' and $I = \text{Wdraw}(I', \nu', K')$ can also be understood in terms of conclusions: the set of formulae which are true in I should equal those formulae which are true in I' and which do not use any of the additions to the vocabulary or object domain mentioned in ν', nor any of the new constant symbols in K'.

The concept of an entailment method $\langle \mathbf{G}, S \rangle$ is generalized accordingly, compared to the original definition in Chapter 9 subsection 9.1.1, so that its $\mathbf{G}$ component can be a reformulation function. (The notation $\mathbf{G}$ will be used for both pretransformation functions and reformulation functions). The usefulness of reformulation functions in the present context arises when $\mathbf{G}$ can map a broader set of chronicles into a more limited range where S is correct.

Definition. *If* $\mathbf{G}$ *is a reformulation function, then the entailment method* $\langle \mathbf{G}, S \rangle$ *is* **correct for** *a chronicle* Υ *iff* $\mathbf{G}(\nu, \Upsilon) = \langle \nu', \Upsilon', K' \rangle$ *implies*
$$\Sigma_{\mathbf{IA}}(\Upsilon) = \mathtt{Wdraw}(S(\Upsilon'), \nu', K').$$
The entailment method is correct for a family $\mathcal{Z}$ *of chronicles iff it is correct for every member of the family. The reformulation function* $\mathbf{G}$ *itself is* **correct for** *a chronicle* Υ *iff* $\langle \mathbf{G}, \Sigma_{\mathbf{IA}} \rangle$ *is correct for* Υ. ⋈

This means that $\mathbf{G}$ is correct for Υ iff $\mathbf{G}(\nu, \Upsilon) = \langle \nu', \Upsilon', K' \rangle$ implies
$$\Sigma_{\mathbf{IA}}(\Upsilon) = \mathtt{Wdraw}(\Sigma_{\mathbf{IA}}(\Upsilon'), \nu', K').$$
Reformulation functions may also be useful from other points of view, and for other types of commonsense reasoning than those being considered here. It is well known that shifts of representation may have dramatic effects on the computational properties of problem-solving methods in common-sense domains. Also, the comparison of different logics for common-sense reasoning sometimes depends heavily on how a particular scenario description is represented in the various logics.

Without violating the formal definitions, it is possible to define reformulation functions that even perform the non-monotonic inference. For example, the method of explanation closure may be formulated using a reformulation function, in which case the model selection function is simply $[.]_\mathcal{O}$. However, this is not an intended usage of the construct of reformulation function; it is intended that 'simple' chronicles shall not need to be modified by the reformulation function.

It follows that if the pretransformation $\mathbf{G}$ is harmless, then the reformulation function
$$\{\langle \nu, \Upsilon \rangle \mapsto \langle \langle \emptyset, \emptyset \rangle, \mathbf{G}(\Upsilon), \emptyset \rangle\}$$
is correct for all members of $\mathcal{K}\text{-}\mathbf{IA}$.

10.2 Intended models using executable schedules

The intuitive reason for comparing interpretations the way CAMC does is as follows. The comparison of interpretations in terms of the full valuation does not correspond completely to the assumptions of the underlying semantics. Since the M component of interpretations is selected at the beginning of the IDS game, and the preference relations only compare interpretations with equal M components, not only the starting times but also the termination times of actions are selected when M is chosen. This can be thought of as if the termination times are part of the specification of the ego. However, the underlying semantics allows the ego to decide when an action is to start, and allows the world to decide when the action is to terminate. This mismatch appears to be the underlying reason for the cases where PCM fails to obtain the intended model set.

This observation suggests the alternative definition $\hat{\Sigma}_{IA}^u(M_0, \Upsilon)$, which will be used for the assessment of CAMC. The definition $\hat{\Sigma}_{IA}^u(M_0, \Upsilon)$ differs from $\Sigma_{IA}^t(M, \Upsilon)$ in the treatment of the valuation M. In Σ_{IA}^t the valuation was given in advance, and one identifies those developments and models where the actions happen to have the duration prescribed by the given M. In $\hat{\Sigma}_{IA}^u$, on the other hand, some parts of the valuation are constructed in the course of the ego-world game.

The change of definitions represents as well an improvement in how the ego/world game models the knowledge level of a cognitive robot. In particular, it becomes possible to view the schedule SCD as a 'program' which defines the required behavior of the ego in a step-by-step fashion.

10.2.1 Executable schedules

In order that the gradual construction of M will work, it is necessary to impose certain additional syntactic restrictions on the schedule component of the chronicle:

Definition. *A schedule* SCD *is called* **simply executable** *iff it is safely sequential and it only contains the following three kinds of formulae:*

- *Action statements of the form* $[s_i, t_i]E_i$, *where all* s_i *and* t_i *are unique constant symbols so that each constant symbol occcurs at most once in all the action statements in* SCD.

- *Timing statements of the form* $s_i < t_i$, *where* s_i *and* t_i *occur in some action statement* $[s_i, t_i]E_i$ *in* SCD.

- *Timing statements which are disjunctions of one or more formulae of the form* $t_j \leq s_i$, *where* s_i *and* t_j *occur in different action statements* $[s_i, t_i]E_i$ *and* $[s_j, t_j]E_j$ *in* SCD.

A chronicle Υ *is called simply executable iff its schedule component is simply executable.* ⋈

We define the reformulation function $\mathbf{G}_{seqx}$ through
$$\mathbf{G}_{seqx}(\nu, \langle \mathcal{O}, \Pi, \text{SCD}, \text{OBS} \rangle) = \langle \langle \emptyset, \emptyset \rangle, \mathbf{G}_{seq}(\langle \mathcal{O}, \Pi, \text{SCD}', \text{OBS}' \rangle), K' \rangle$$
where SCD′, OBS′, and K' are obtained as follows. Initialize SCD′ as SCD, OBS′ as OBS, and K' as $\emptyset$. All the timing statements in SCD, except those which are allowed in a simply executable schedule, are included in OBS′ and omitted in SCD′. For every action statement $[s,t]\varepsilon$ in SCD′, if s is not a temporal constant symbol, then introduce a constant symbol s_i which has not been used elsewhere, replace the action statement by $[s_i, t]\varepsilon$ in SCD′, add the formula $s = s_i$ to OBS′, and add the constant symbol s_i to K'. Do likewise for the t subexpression in each $[s,t]\varepsilon$, and for each case where the same constant symbol is used more than once in SCD′. After this process has been concluded, the pretransformation $\mathbf{G}_{seq}$ is applied. It follows at once that $\mathbf{G}_{seqx}$ results in simply executable schedules, and that it preserves $\mathcal{K}$p-**IA**. Correctness is obtained as follows.

Lemma 10.1. *The reformulation function $\mathbf{G}_{seqx}$ is correct for $\mathcal{K}$-**IA**.*

Proof If $\langle \mathcal{O}, \Pi, \text{SCD}, \text{OBS} \rangle$ is a $\mathcal{K}$-**IA** chronicle, and Γ is a set of timing statements, then it follows from the definition of $\Sigma_{\mathbf{IA}}$ that
$$\Sigma_{\mathbf{IA}}(\langle \mathcal{O}, \Pi, \text{SCD} \cup \Gamma, \text{OBS} \rangle) = \Sigma_{\mathbf{IA}}(\langle \mathcal{O}, \Pi, \text{SCD}, \text{OBS} \cup \Gamma \rangle).$$
This means that those formulae that $\mathbf{G}_{seqx}$ added to the set of observations can be moved to the schedule without affecting the set of intended models. The result follows immediately. □

A possible reason in favor of non-linear time for the purpose of commonsense reasoning is that when we reason about a scenario, we seem to refer to timepoints in the scenario such as the starting times and ending times of various occurrences, and the set of such timepoints is not necessarily completely ordered. In so-called non-linear planning [66], for example, one obtains an action plan where the actions are intended to be performed sequentially, but where the exact order of the actions is only partially determined by the planner. The temporal constant symbols in an executable schedule can serve well as the 'timepoints' in such a non-linear time (which I would rather call pseudo-time) and a non-linear plan can be represented as an executable schedule.

The name 'executable schedule' suggests that it can be used as a program. Actually, in such a case one would wish the schedule to contain some more information, such as constraints on the delay from the termination of one action to the start of the next one. A generalization from 'simply executable' to 'executable' is foreseen, therefore. In the remainder of this chapter, only simply executable schedules will be considered, and the acronym SXS will be used for 'simply executable schedule'.

10.2.2 Auxiliary concepts

A few auxiliary concepts will be needed together with executable schedules. Schedules denoted denoted SXS are implicitly declared to be simply executable. A partial valuation M_0 is said to be a *good initial valuation* for a simply executable schedule SXS iff it is defined for all constant symbols (for objects and for timepoints) except those timepoint constants which occur in SXS.

If M_0 is a good initial valuation for SXS, then a valuation $M \supseteq M_0$ is said to be a *correct extension* of M_0 iff $M - M_0$ consists only of maplets $s_i : s$ and $t_i : t$, but always in pairs. In other words, there is never a maplet for s_i without the corresponding t_i, or vice versa.

The *ending time* of a partial or complete valuation M with respect to a schedule SXS is the largest value assigned by M to some temporal constant symbol that occurs in SXS, or -1 if no such symbol is assigned a value in M. A correct extension M of some good initial valuation for a schedule SXS is said to be *consistent for* SXS iff there is some complete valuation $M' \supseteq M$ where all timing statements in SXS are satisfied, and all maplets in $M' - M$ assign values that are $\geq$ the ending time of M w.r.t. SXS.

If M is a correct extension of a good initial valuation for SXS, then $M[\text{SXS}]$ is the set which is formed as follows. If $\ulcorner [s_i, t_i] E_i \urcorner \in \text{SXS}$ and s_i and t_i are defined in M, then $M[[s_i, t_i] E_i]$ is included in $M[\text{SXS}]$. If M is not consistent for SXS, then F is included in $M[\text{SXS}]$, which does not have any additional members besides those just stated.

Let M be a consistent, correct extension of a good initial valuation for a schedule SXS, and let n be the ending time of M with respect to SXS. An action statement $[s_i, t_i] E_i$ in SXS is said to be a *next executable* statement in SXS with respect to M iff s_i and t_i are not defined in M, and $M \cup \{s_i : (n+1), t_i : (n+2)\}$ is consistent for SXS.

Example. Consider the schedule SXS containing the following formulae:

$[s_1, t_1] E_1$
$[s_2, t_2] E_2$
$[s_3, t_3] E_3$
$[s_4, t_4] E_4$
$s_i < t_i$ for $i = 1, 2, 3, 4$
$t_1 \leq s_2$
$t_1 \leq s_3$
$t_2 \leq s_3 \vee t_3 \leq s_2$
$t_2 \leq s_4$
$t_3 \leq s_4$

so that the order betwen E_2 and E_3 is undetermined. Let M_0 be a good initial valuation for SXS, that is, it defines values for all constant symbols except s_i and t_i for i between 1 and 4. Then

$M = M_0 \cup \{s_1 : 4, t_1 : 6, s_2 : 10, t_2 : 12\}$

is a consistent correct extension of M_0. However,
$$M' = M \cup \{s_4 : 13, t_4 : 14\}$$
is not because its ending time is 14 and it is not possible to add maplets for s_3 and t_3 assigning them values ≥ 14 while satisfying the formula $t_3 \leq s_4$ in SXS. Therefore, $[s_4, t_4] E_4$ is not a next executable statement in SXS with respect to M. Informally speaking, the reason is that if E_1 and E_2 have been performed, one has to perform E_3 before E_4. ⋈

It follows easily

Proposition 10.2. *Assume that the time domain of non-negative integers is used. If M is a consistent, correct extension of a good initial valuation M_0 for a simply executable schedule SXS, then either it is a complete valuation, or there is at least one action statement $[s_i, t_i] E_i$ in SXS which is next executable in SXS with respect to M. Furthermore, for every action statement $\ulcorner[s_i, t_i] E_i\urcorner \in$ SXS that is next executable with respect to M, and if $n \leq s < t$ where n is the ending time of M with respect to SXS, then $M \cup \{s_i : s, t_i : t\}$ is also a consistent correct extension of M_0.*

Proof Trivial from the definitions. □

The reason for the first sentence in the proposition is to exclude finite time domains, where the proposition does not hold. The property that has now been shown means that a simply executable schedule can be used for defining the successive actions that an agent is to perform, so it qualifies as a plan in the sense of Chapter 1, subsection 1.2.2.

10.2.3 Performing egos

We previously defined and used the concept of a predeterminate ego as one which does not modify the valuation component M in its argument development. Now we shall define an analogous type of ego which builds up the valuation in a systematic way in order to match the given schedule.

Definition. *Let SXS be a simply executable schedule and let M_0 be a good initial valuation for SXS. A trajectory-semantics ego $\mathbf{K}$ is said to* **perform** *SXS with respect to M_0 iff the following holds. $M' = M$ in those cases where*
$$\mathbf{K}(\langle \mathcal{B}, M, R, \mathcal{A}, \mathcal{C}\rangle) = \langle \mathcal{B}, M', R, \mathcal{A}, \mathcal{C}\rangle.$$
In those cases where
$$\mathbf{K}(\langle \mathcal{B}, M, R, \mathcal{A}, \mathcal{C}\rangle) = \langle \mathcal{B}, M', R, \mathcal{A}, \mathcal{C} \cup \{\langle n_\mathcal{B}, E\rangle\}\rangle,$$
there is some action statement $\ulcorner[s_i, t_i] E_i\urcorner \in$ SXS which is next executable with respect to M, and where $E = M[E_i]$ and $M' = M \cup \{s_i : n_\mathcal{B}\}$. ⋈

In other words, if an ego performs a schedule SXS, then each time it has the move it either does not invoke any action at all, or it invokes a next executable action in SXS, and in the latter case it binds the time-constant for the starting time of that action to the current now-time.

In order for this arrangement to work, there must also be a way of binding the ending time of the action. That binding must be performed by the world, since the IDS world makes the choice of trajectory and, in particular, determines the duration of the action. Therefore, we make a minor modification of the definition of a corresponding world, previously given in Chapter 3, subsection 3.2.4. The modification does not affect the usage of corresponding worlds earlier in this book, and is as follows.

Amendment. *Previously a* **corresponding world** *was required not to change the valuation component M. This restriction is now relaxed, as follows. If the world performs its move at time n_B, and M is the valuation in the development where the move is made, and there is some temporal constant symbol s_i such that $M[s_i] = n_B$ whereas t_i is not assigned any value by M, and if furthermore the world chooses and uses a trajectory $\langle r'_1, r'_2, ..., r'_k \rangle$ in the present move, then it will add the maplet $t_i : n_B + k$ to M in its move. This is the only change in the valuation that the world is allowed to make.* ⋈

In the previous chapters we have only considered the case where the initial valuation M_0 is complete, and such cases are not affected by this amendment.

Definition. *Let a similarity type σ and an occurrence vocabulary π be given, and let $\Upsilon = \langle \mathcal{O}, \Pi, \text{SXS}, \text{OBS} \rangle$ be a simply executable chronicle in $\mathcal{K}$-**IA** over σ and π. The* **executed model set** *of Υ is defined as the set of all complete developments over $\mathcal{O}$, σ, and π,*
$$\langle \mathcal{B}, M, R, \mathcal{A}, \mathcal{C} \rangle,$$
which result from a game between $\mathbf{W}_\Upsilon$ and some trajectory-semantics ego that performs SXS, with arbitrary initial state r_0 and an initial valuation M_0 which is a good initial valuation for SXS, and where M is a complete valuation and $\langle M, R \rangle \in [\![\text{OBS}]\!]_{\mathcal{O}}$. ⋈

Compared to the previous definition of the predetermined model set, we have removed the explicit requirement that $M[\text{SXS}] = \mathcal{A}$. As a consequence of Proposition 10.2, all action statements in SXS must be executed exactly once, since otherwise the resulting M cannot be complete. Therefore, $M[\text{SXS}] = \mathcal{A}$ must hold in every member of the executed model set.

Proposition 10.3. *If $\Upsilon = \langle \mathcal{O}, \Pi, \text{SXS}, \text{OBS} \rangle \in \mathcal{K}$-**IA** is simply executable, then there is some ego that executes SXS, and the executed model set of Υ equals $Mod(\Upsilon)$.*

Proof The existence of the corresponding ego is trivial. As for $Mod(\Upsilon)$:
1. Let $J = \langle \mathcal{B}, M, R, \mathcal{A}, \mathcal{C} \rangle$ be a member of the predetermined model

set of Υ. Then $\mathcal{A} = M[\text{SXS}]$ according to the definition of that set. Select $M_0 \subseteq M$ as a good initial valuation for SXS (which is always possible), select r_0 as $R(0)$, and choose a trajectory-semantics ego **K** which executes SXS in such a way that M is obtained if the world makes the moves specified by J. It is clear that J can be generated from a game between $\mathbf{W}_\Upsilon$ and **K**. Therefore, J is a member of the executed model set of Υ.

2. In a similar fashion it is shown that every member of the executed model set is also a member of the predetermined model set.

Since the predetermined model equals $Mod(\Upsilon)$ according to subsection 8.3.2, the result follows. □

Therefore, we can safely use executed model sets for the assessment of CAMC.

10.2.4 Finite intended model sets

The finite model set $Mod^t(M, \Upsilon)$ was defined using games with predeterminate egos. An analogous function $Xmod^u(M_0, \Upsilon)$ is defined using egos that perform the schedule component of the chronicle Υ:

Definition. *Assume that $\Upsilon = \langle \mathcal{O}, \Pi, \text{SXS}, \text{IOBS} \rangle \in \mathcal{K}\text{p-}\mathbf{IA}$ is simply executable, u is a non-negative integer, and M_0 is a good initial valuation for* SXS.

We write $Xmod^u(M_0, \Upsilon)$ for the set of finite developments over $\mathcal{O}$, σ, and π, of the form $\langle \mathcal{B}, M, R, \mathcal{A}, \mathcal{C} \rangle$ which result from a finite game between $\mathbf{W}_\Upsilon$ and some trajectory-semantics ego which executes SXS*, with an initial r_0 such that* IOBS *is satisfied in $\langle \mathcal{O}, \langle M, \{0 \mapsto r_0\} \rangle \rangle$, with initial valuation M_0 and where the starting time of the last move of the world was u.*

Also, we define $Xmod^{-1}(M_0, \Upsilon)$ as the set of all $\langle \{0\}, M_0, \{0 \mapsto r_0\}, \emptyset, \emptyset \rangle$ such that IOBS *is satisfied in $\langle \mathcal{O}, \langle M_0, \{0 \mapsto r_0\} \rangle \rangle$.*

Finally, we define $\hat{\Sigma}^u_{\mathbf{IA}}(M_0, \Upsilon) = \text{Ci}(Xmod^u(M_0, \Upsilon))$ by analogy with $\Sigma^t_{\mathbf{IA}}(M, \Upsilon)$. ⋈

One difference between $\Sigma^t_{\mathbf{IA}}(M, \Upsilon)$ and $\hat{\Sigma}^u_{\mathbf{IA}}(M_0, \Upsilon)$ is, therefore, that the former uses a fixed valuation M, generates the developments, and then selects those where the actions turned out to have the duration that was ascribed to them by the given M in the first place. $\hat{\Sigma}^u_{\mathbf{IA}}(M_0, \Upsilon)$ is obtained, instead, through games with an ego that 'interprets' the schedule SXS, executes the actions that are listed there, and assigns appropriate additional bindings to the valuation, in interaction with the world, so that the resulting development satisfies $M[\text{SXS}]$. This means in particular that the value of $\hat{\Sigma}^u_{\mathbf{IA}}(M_0, \Upsilon)$ may contain interpretations with different M components, although $M_0 \subseteq M$ for all of them.

It is not required that all action statements in SXS have been performed within the finite game, just that those that have been performed have been chosen in such a way that it would be possible to continue execution and complete the whole schedule within its time constraints.

Another difference is as follows. If
$$\langle \mathcal{B}, M, R, \mathcal{A}, \mathcal{C} \rangle \in Mod^t(M, \Upsilon),$$
then R is defined over $[0, t]$. However, if
$$\langle \mathcal{B}, M, R, \mathcal{A}, \mathcal{C} \rangle \in Xmod^u(M_0, \Upsilon),$$
then R is defined over some $[0, t]$ where $t > u$, and either of the following holds:

(a) there is a member $\langle u, E, t \rangle \in \mathcal{A}$, and $R = R_{0:u} \triangleright v$ for some $v \in \text{Trajs}(E, R(u))$;

(b) $u = t - 1$, $R(t) = R(u)$, and all members of $\mathcal{A}$ end at a time $\leq u$.

In both cases, of course, $\mathcal{C} = \emptyset$.

Proposition 10.4. *If* $\Upsilon = \langle \mathcal{O}, \Pi, \text{SXS}, \text{IOBS} \rangle \in \mathcal{K}\text{p-}\mathbf{IA}$ *is simply executable, then*

1. *If* $\langle M, R \rangle \in \Sigma_{\mathbf{IA}}(\Upsilon)$, $M_0 \subseteq M$ *is a good initial valuation for* Υ, *and u and t are two successive members of* $\{-1\} \cup \text{At}(M, \text{SXS})$, *then* $\langle M', R_{0:t} \rangle \in \hat{\Sigma}^u_{\mathbf{IA}}(M_0, \Upsilon)$ *for some* $M' \subseteq M$.

2. *If* $\langle M, R \rangle \in \hat{\Sigma}^u_{\mathbf{IA}}(M_0, \Upsilon)$ *and R is defined over $[0, t]$, then there is some* $\langle M', R' \rangle \in \Sigma_{\mathbf{IA}}(\Upsilon)$ *such that* $M \subseteq M'$ *and* $R = R'_{0:t}$.

Proof Straightforward from Propositions 10.2 and 10.3. □

In Lemma 8.2 we showed that with a similar choice of t,
$$Mod(M, \Upsilon)_{0:t} \subseteq Mod^t(M, \Upsilon),$$
but without necessarily having equality. Proposition 10.4 shows that $\hat{\Sigma}^u_{\mathbf{IA}}$ has a property which corresponds to having equality in Lemma 8.2.

Example. Consider again the example following Lemma 8.2 in Chapter 8, and let us relate it to item 2 in Proposition 10.4. SXS was chosen to consist of

$[s_1, t_1] E, s_1 < t_1,$
$[s_2, t_2] E, s_2 < t_2,$
$t_1 \leq s_2,$

which clearly is simply executable. We also assumed action laws which require E to always have the duration 4. If $M = M_0 \cup \{s_1 : 4, t_1 : 8\}$ then $\langle M, R \rangle \in \Sigma^3_{\mathbf{IA}}(M_0, \Upsilon)$ for some R, defined over $[0, 8]$, where the first but not the second occurrence of E is executed. One can choose M' for example as $M \cup \{s_3 : 10, t_4 : 14\}$ and R' as

$$R \triangleright \langle \emptyset, \emptyset \rangle \triangleright v \triangleright \langle \emptyset, \emptyset, ... \rangle$$

for some $v \in \text{Trajs}(E, R(8))$, and obtain $\langle M', R' \rangle \in \Sigma_{\mathbf{IA}}(\Upsilon)$. If, instead, the duration of E is dependent on its starting state, then this arrangement makes it possible to choose a suitable M' for each value of $R(8)$.

It follows that $Mod(M, \Upsilon)_{0:10} = \emptyset$ because $Mod(M, \Upsilon) = \emptyset$, but $Mod^{10}(M, \Upsilon)$ consists of all developments over $[0, 10]$ where E is performed over the interval $[4, 8]$. The incompatibility between the values assigned by M and the duration of the second action affects $Mod(M, \Upsilon)$, but not $Mod^{10}(M, \Upsilon)$. ⋈

10.3 Assessment of applicability for CAMC

We have already defined S_{camc} by
$$S_{camc}(\Upsilon) = Min(\ll_{camc}, [\![\Pi(\text{SCD})]\!]_{\mathcal{O}}) \cap [\![\text{OBS}]\!],$$
where the preference relation $\ll_{camc}$ was defined in Section 10.1 above. The entailment method itself can be defined, using reformulation to simply executable schedules, as CAMC $= \langle \mathbf{G}_{seqx}, S_{camc} \rangle$. We now proceed to the assessment of the range of applicability for CAMC. The proof for this case is analogous to the proof for the assessment of PCM, but somewhat shorter because we can use the property shown in Proposition 10.4.

10.3.1 Auxiliary definitions

In order to define the appropriate compatibility requirement for S_{camc} we must first generalize the definition of $\ll_{camc}$ to compare finite interpretations of unequal length.

Definition. *If $I = \langle M, R \rangle$ and $I' = \langle M', R' \rangle$ where R is defined over $[0, k]$ and R' over $[0, k']$ where $k < k'$, then $\ll_{camc}$ is defined as follows. If $I \ll_{camc} I'_{0:k}$ then $I \ll_{camc} I'$. If $I'_{0:k} \ll_{camc} I$ then $I' \ll_{camc} I$. Otherwise, neither $I \ll_{camc} I'$ nor $I' \ll_{camc} I$.* ◻

In particular if $I = I'_{0:k}$ then neither of them is preferred over the other. The assessments of PCM and OCM used the notion of PCM-compatible and OCM-compatible worlds, and an analogous notion will be used for CAMC:

Definition. *A world $\mathbf{W}$ having a description $\langle \texttt{Infl}, \texttt{Trajs} \rangle$ is **CAMC-compatible** iff every trajectory set $V = \texttt{Trajs}(E, r)$ satisfies the following requirement. If v and v' are two members of V then neither $(r \triangleright v \ll_{camc} r \triangleright v')$ nor $(r \triangleright v' \ll_{camc} r \triangleright v)$.* ◻

In other words, if v and v' are different then they must differ in such a way that neither of them is CAMC-preferred over the other. This concept applies to worlds and to chronicles like the previos compatibility concepts.

10.3.2 Assessment

The assessment result is divided into two steps, since the first part will be used separately in Chapter 13.

Proposition 10.5. *$\Upsilon \in \mathcal{K}\text{p-}\mathbf{IA}$ is a simply executable chronicle, then*
$$S_{camc}(\Upsilon) \subseteq \Sigma_{\mathbf{IA}}(\Upsilon).$$

Proof By a generalization of Lemma 9.18 it is sufficient to prove the proposition for Υ of the form $\langle \mathcal{O}, \Pi, \text{sxs}, \emptyset \rangle$. From Proposition 10.2 it then follows that $\Sigma_{\mathbf{IA}}(\Upsilon) \neq \emptyset$ for such Υ.

Suppose the proposition does not hold, that is, that $S_{camc}(\Upsilon) \not\subseteq \Sigma_{IA}(\Upsilon)$, and let $\langle M, R \rangle \in S_{camc}(\Upsilon) - \Sigma_{IA}(\Upsilon)$. By the definition of $\ll_{camc}$, $R(t)$ must be constant when t is $\geq$ the ending time of M with respect to SXS, and the same holds for every member $\langle M', R' \rangle$ of $\Sigma_{IA}(\Upsilon)$. Therefore, one can chose t as the smallest timepoint such that $R_{0:t}$ is not equal to $R'_{0:t}$ for any $\langle M', R' \rangle \in \Sigma_{IA}(\Upsilon)$. For example, the ending time of M w.r.t. SXS has this property.

It is clear that $t > 0$. Let u be the largest member of $\text{At}(M, \text{SXS})$ which satisfies $u < t$. Also, let M_0 be that (unique) good initial valuation for SXS which is $\subseteq M$. Either of the following cases must apply:

(a) There is some $\ulcorner[s_i, t_i] E_i\urcorner \in$ SXS such that $M[s_i] = u$. Since $\langle M, R \rangle \in [\![\Upsilon]\!]$, we can conclude, using Lemma 8.3, that there exists a trajectory v in $\text{Trajs}(M[E_i], R(u))$ of length $k \geq t - u$ such that $v_j \subseteq R(u+j)$ for $1 \leq j \leq k$. Using Proposition 10.4 it follows that $\langle M'', R_{0:u} \triangleright v \rangle \in \hat{\Sigma}^u_{IA}(M_0, \Upsilon)$ for some M'' for every such v. (Actually $M'' = M_{0:u+k}$ but this requires separate verification, and is irrelevant for the present proof). Since $R_{0:t}$ does not agree with any member of $\Sigma_{IA}(\Upsilon)$, $breakset(R, t)$ must have some member that is not included in $\text{Infl}(M[E_i], R(u))$. But this is inconsistent with the assumption that $\langle M, R \rangle \in S_{camc}(\Upsilon)$.

(b) There is no $\ulcorner[s_i, t_i] E_i\urcorner \in$ SXS such that $M[s_i] = u$. In this case $u = t - 1$. By the assumption, $\langle M_{0:u}, R_{0:u} \rangle = \langle M'_{0:u}, R'_{0:u} \rangle$ for some $\langle M', R' \rangle \in \Sigma_{IA}(\Upsilon)$. Therefore, $I' = \langle M_{0:u}, R_{0:u} \triangleright \langle \emptyset \rangle \rangle \in \Sigma^u_{IA}(M_0, \Upsilon)$, and by Proposition 10.4 it has a continuation to ∞ that is a member of $\Sigma_{IA}(\Upsilon)$. Since $\langle M, R \rangle$ does not agree in t with any member of $\Sigma_{IA}(\Upsilon)$ it follows that $R(t) \neq R(u)$. But in that case, $I' \ll_{camc} \langle M, R \rangle$, contradicting the assumption that $\langle M, R \rangle \in S_{camc}(\Upsilon)$. This concludes the proof. $\square$

Proposition 10.6. *CAMC is correctly applicable for CAMC-compatible chronicles in $\mathcal{K}$-IA.*

Proof As in the previous proposition it is sufficient (using Lemmas 9.2, 9.18, and 10.1 as well) to prove that if $\Upsilon = \langle \mathcal{O}, \Pi, \text{SXS}, \emptyset \rangle \in \mathcal{K}\text{p-IA}$ is a simply executable and CAMC-compatible chronicle, then
$$S_{camc}(\Upsilon) = \Sigma_{IA}(\Upsilon).$$
Again, $\Sigma_{IA}(\Upsilon) \neq \emptyset$. The previous proposition proved the $\subseteq$ relation between the two sides of that equality, and it remains to prove that the $\supseteq$ relation also holds.

Suppose $\langle M, R \rangle$ is a member of $\Sigma_{IA}(\Upsilon) - S_{camc}(\Upsilon)$. Since $\Sigma_{IA}(\Upsilon) \subseteq [\![\Upsilon]\!]$ and $S_{camc} = Min(\ll_{camc}, [\![\Upsilon]\!])$ there must be some other model $\langle M', R' \rangle \in S_{camc}(\Upsilon) \subseteq \Sigma_{IA}(\Upsilon)$ which is CAMC-preferred over $\langle M, R \rangle$. Let t be the timepoint of preferential comparison between $\langle M, R \rangle$ and $\langle M', R' \rangle$. Then $R_{0:\theta t} = R'_{0:\theta t}$ and $M_{0:t} = M'_{0:t}$. Therefore, one of the following cases must apply:

(a) $\langle M, R \rangle$ and $\langle M', R' \rangle$ terminate the same action at time t. Let $[s_i, t_i] E_i$ be its action statement, so that $M[t_i] = $ t, and let $M[s_i] = $ s. It follows from Lemma 8.10 that $R_{0:t} = R_{0:s} \triangleright v$ and $R'_{0:t} = R'_{0:s} \triangleright v'$, where v and v' are members of $\text{Trajs}(M[E_i], R(\text{s}))$. Since R' is CAMC-preferred over R in t and $R(\text{s}) = R'(\text{s})$, it follows that $R(\text{s}) \triangleright v' \ll_{camc} R(\text{s}) \triangleright v$, contradicting the assumption that Υ is CAMC-compatible.

(b) These interpretations both execute the same action at time t, where both the ending times are $>$ t but they are not necessarily equal.

(c) Neither of these interpretations execute any action between the time θt and the time t.

Cases (b) and (c) lead to contradictions by arguments similar to the argument for case (a). This concludes the proof. □

Therefore, the modification of the preference relation from PCMF to CAMC has had the intended effect of removing the requirement of equidurationality. It turns out that we have had to pay a certain price with respect to the compatibility requirement, as shown by the following result:

Proposition 10.7. *If a chronicle is CAMC-compatible then it is PCM-compatible.*

Proof If $I \ll_{pcm} I'$ then $I \ll_{camc} I'$ according to the definitions. □

PCM-compatibility is a weaker requirement since $\ll_{pcm}$ only applies to trajectories of equal length. For example, let $r'_\text{T} = \{p_1 : \text{T}\}$, $r'_\text{F} = \{p_1 : \text{F}\}$, $r[p_1] = \text{T}$, and $\text{Trajs}(E, r) = \{\langle r'_\text{T}, r'_\text{F} \rangle, \langle r'_\text{F} \rangle\}$. This world fragment is PCM-compatible but not CAMC-compatible. However, the easily understandable cases for PCM-compatibility are also included in CAMC-compatibility:

Proposition 10.8. *All chronicles in $\mathcal{K}$-IsAn and $\mathcal{K}$-IAt are CAMC-compatible.*

Proof Follows directly from the definitions. □

The remaining task of the next chapter is to find a way of removing the requirement of PCM- or CAMC-compatibility altogether.

10.4 Generalization to branching time

Branching time was defined in Chapter 6, Section 6.6. The definitions and results in the present chapter can be generalized to branching time in a straightforward manner, using the approach that was introduced in Chapter 8, subsection 8.5.2. In particular, the notation $M_{s:t}$ for the restriction of a valuation M to an interval $[s, t]$ must be replaced by the restriction M_B of the valuation M to maplets where the value is either a timepoint in $[B]$ where B is a tree, or an object.

10.5 Summary

One way of avoiding the restriction of equidurationality is to use the preference method of chronological assignment of valuation, which has been defined and assessed here. This method requires the use of a more complex reformulation function, which transforms the given schedule to simply executable form. Both the reformulation and the new preference relation have the disadvantage of fairly complicated definitions. However, this can possibly be balanced by the observation that the restriction to executable schedules is essentially the same restriction as is required if a schedule is to be used as a plan for an agent's behavior.

11
Entailment methods for $\mathcal{K}$-**IA** using occlusion

Compared to the list of requirements for the correct applicability of PCM, we have now removed the requirements of initial-time observations and equidurationality, but the compatibility requirements remain.

In general, when a chronicle Υ is outside the range of applicability of currently available logics and entailment methods, there are several possibilities as to how to proceed:

- Reformulate Υ.
- Construct a new entailment method.
- Change the base logic.

In previous chapters we have used the second of these options and, to a certain extent, the first one. The third option would represent more force than is necessary. For the remaining restrictions it may be possible to use reformulation in some cases, but for full generality it seems necessary to modify the base logic. The present chapter considers these alternatives.

11.1 Discussion

The only essential restriction on CAMC-entailment is that the chronicle must be CAMC-compatible. In terms of the lower-level ambiguity concepts that were introduced in Chapter 9, subsection 9.3.1, it remains to deal with change incidence ambiguity and changetime ambiguity.

11.1.1 Change incidence ambiguity

The essential problem in change incidence ambiguity is that the intended models do not specify, for some point in time, whether a certain feature changes or not. A simple example is the action of tossing a coin. For the examples in this section we let the action symbol A_r represent an action that takes a single timestep and that causes the propositional feature p_1 to obtain either the value T or F regardless of its previous value. This action has been used already in the proof of Proposition 8.6, so it is a well-known cause of trouble. It is easily verified that A_r violates all the compatibility criteria (OCM-, PCM-, and CAMC-compatibility) that have been introduced so far.

It does not seem possible to deal with this type of ambiguity by changing the entailment method, even with the help of partitioning. First we consider a reformulation approach.

Reformulation using oracle features

One possible approach is to replace a non-deterministic action by a deterministic action, whose outcome depends on an additional constant or feature whose value is unknown. In the case of the toss-coin action A_r, let q_1 be a feature symbol that has not previously been used, assign a propositional domain to it, and let A'_r be the action defined by the following action law:
$$[s,t]\, A'_r \Rightarrow ([0]q_1 \leftrightarrow [t]p_1) \wedge s = \theta t.$$
Clearly, A'_r is PCM-compatible. In a schedule containing a single occurrence of A_r one may replace A_r by A'_r and obtain the intended result from PCM even for cases where the original schedule would not. The *oracle feature* q_1 'knows' at time 0 what the outcome of the coin-toss will be.

Unfortunately, this method does not work in a schedule containing more than one occurence of A_r, since the set of intended models (according to $\mathcal{K}$-**IA**) for the revised schedule will then obtain the same outcome each time the coin is tossed within a model. Introducing one oracle feature for every non-deterministic occurrence is a possible but unattractive solution. A more general method is to retain the use of a single oracle feature, but to use its value at the starting time of the action, writing the action law as
$$[s,t]\, A'_r \Rightarrow ([s]q_1 \leftrightarrow [t]p_1) \wedge s = \theta t.$$
This solution will work for this particular example provided that one can exempt the oracle feature from inertia, since otherwise one has the same problem as before. The same method appears to work even for more complex non-deterministic actions, involving possibly correlated changes in several, non-propositional features. Essentially, a non-inert feature can be viewed as a counterpart of a random-number generator, and as such it can be used for defining the set of possible outcomes of the action. I return below to the case of actions with extended duration.

It remains to find a method for exempting oracle features from inertia. Two methods come to mind:

- To use a syntactic distinction where the vocabulary structure is modified so that it indicates which features are assumed or not assumed to have inertia. Such a distinction may be useful anyway for characterizing 'things that change by themselves' [Lifschitz] or for representing secondary features, that is, those whose value is functionally dependent on some other features.

- To extend the base logic so that the fact of whether a feature is inert or not can be expressed in that logical language. Such an extension will be considered below, but as we shall see it allows one to deal with change incidence ambiguity directly and without the use of oracle features.

In summary, the use of a syntactic distinction between inert and non-inert features, together with a reformulation method using oracle features, seems to be adequate for dealing with at least some cases of change incidence ambiguity.

The occlusion predicate

The purpose of the occlusion predicate is to make explicit statements about the inertia of a certain feature at a certain time. This is used for restricting the impact of the preference minimization, so that in some intervals inertia is not assumed, and chronological minimization of change does not therefore apply. The specification of those intervals is often problem dependent, and must therefore be given by the premises. Also, since one wishes to consider inertia as a default and favor it over non-inertia, there has to be a separate default mechanism to that effect.

From the point of view of trajectory semantics, in any interval $[s,t]$ where $[s,t]A$ holds, every $f \in \text{Infl}(A, R(s))$ is considered to be occluded in the timepoints $s+1, s+2, \ldots t-1, t$.

This approach requires the introduction of an additional predicate in the base logic. Therefore, the DFL syntax is extended with one more kind of *elementary fluent formula*, of the form

$\mathbf{X}f$,

where f is a feature expression. It is intended that $\mathbf{X}f$ [41] shall be true at time t iff there is no preference for the fluent f to retain its value in the transition from θt to t. In other words, $\mathbf{X}f$ represents a lack of information about f, whereas $\neg \mathbf{X}f$ means that one 'knows about' f in the sense that if

---

[41] Pronounced 'occluded f', its negation is pronounced 'transparent f'. The term 'occlude' was originally selected for its meaning of 'hide', 'prevent from being seen', which is commonly used, for example, in the computer vision literature. Later on it was discovered that this meaning of the word is not listed in the major English dictionaries, but it was too late to change the terminology then.

f changes its value, then this fact follows from the given premises, or is even included in them. Features should be transparent by default, therefore, so that occlusion is minimized.

For example, the action statement $[4, 5] A_r$ will be translated to
    scd1 $[5] \mathbf{X} p_1$
    obs1 $[5] p_1 \hat{=} \mathtt{TF}$,
where the second statement is of course a tautology. This approach was previously described in [70, 72, 71], and is essentially the same as the use of the *Ends* predicate in [68].

The occlusion predicate can be seen as a special-purpose abnormality predicate, where the reason for the abnormality has been made precise. Notice that occlusion does not imply change; it merely allows it. Occlusion is also related to the causality-based approaches [34, 40] which make a distinction as to whether a particular change has or does not have an identified cause. In our case the primary difference is between occluded and transparent change.

11.1.2 Changetime ambiguity

Another type of incompleteness in the information about change is when the time of change is not known precisely but only constrained. Consider the following set of premises after expansion:
    obs1 $[0] f_1 \hat{=} \mathtt{G}$
    scd1 $[10, 15] f_1 := \mathtt{R}$,
where the second premise results from an encapsulated action, and is intended to say that f_1 obtains the value R some time during the time interval between time 10 and time 15. This intention agrees with how sets of trajectoris were mapped to FTNF formulae in Chapter 8. Such ambiguity regarding the exact time of change is typical, therefore, of what is obtained when action statements are translated using action laws: an action law typically specifies the values of features when the action begins and ends, but it does not have to specify exactly when the changes take place within the action period. In FTNF, if the timing of the changes is known, they have to be specified explicitly.

Under the PCM preference criterion, there will be one preferred model, where the change of f_1 from G to R takes place at time 15, because according to PCM one prefers models where the change takes place later rather than earlier. However, this does not agree with the intended model set $\Sigma_{\mathbf{IA}}(\Upsilon)$: one intends that there is a change between time 10 and time 15 so, in particular, the old value still holds at time 10 and the new value has certainly been adopted at time 15, but between those times one does not know when the changes occur.

In terms of the underlying trajectory semantics, the intended set of trajectories corresponding to $[10,15] f_1 := \text{R}$ has a number of members, allowing all different choices of the time of change for f_1. However, the DFL-1 translation of $[10,15] f_1 := \text{R}$ as $[15] f_1 \hat{=} \text{R}$ does not capture that degree of freedom. The intended set of trajectories is not PCM-compatible or compatible with any other one of the entailment methods.

The two methods which were discussed for change incidence ambiguity can be used for this case as well.

Reformulation using time-valued oracle features

If an action has extended duration but satisfies direct change (**IdA**), that is, each influenced feature has only one change within the action's interval, then one can use an oracle feature that 'knows' in advance when the change will occur. If the zero-time value of the oracle is used, then for every change at any imprecise time greater than zero, one needs a separate, time-valued feature whose value at time zero specifies the time of change. In the example given above, suppose j is the oracle feature specifying the time of change. The relevant premises would be

$[0] 10 < j \leq 15$,

$\forall t [[0] t = j \rightarrow [t] f_1 := \text{R}]$.

In order to work conveniently with multiple occurrences one should use the value of the oracle at the starting time of the action, so that the example becomes

$[10] 10 < j \leq 15$,

$\forall t [[10] t = j \rightarrow [t] f_1 := \text{R}]$.

This approach requires that the logic is extended to allow time-valued features, which is a simple extension within the syntactic level. Possibly the CAMC entailment condition has to be modified, then, so that the value-restriction that it imposes on M is also imposed on time-valued features.

A variant of this method, previously noticed by Kautz [38], is to use time-valued Skolem functions instead of time-valued features as oracles. If the second formula in the example is written as

$10 < t_1 \leq 15 \wedge [t_1] f_1 := \text{R}$,

then the choice of changetime is represented in the M component of an interpretation. By our definition of PCM, interpretations with different M components are not comparable preferentially, so all choices of t_1 are equally preferred.

In this case the Skolem function has no arguments, that is, it is a temporal constant symbol. In the more general case it would have to be a function of the action's starting time, so the example would be

$10 < t_1(10) \leq 15 \wedge [t_1(10)] f_1 := \text{R}$,

so again an extension of the base logic is required.

Regardless of whether features or functions are used as oracles, this method is difficult to use when several changes are allowed within the action interval.

Reformulation using item-valued oracle features

Another possibility is to use the same method as for change incidence ambiguity, that is, to have a non-inert oracle feature that serves as a kind of random-value generator, and to use action laws that use the oracle for defining the possible ways of performing the action. On an informal level no restriction has been found for this approach, but it has not been analyzed systematically.

Occlusion

The occlusion technique is immediately applicable to chronicles with change-time ambiguity. The example above would then be written as follows:

obs1 $[0] f_1 \hat{=} G$
obs2 $[15] f_1 \hat{=} R$
scd1 $(10, 15] \mathbf{X} f_1$.

To understand why the third line is $(10, 15] \mathbf{X} f_1$ and not $[10, 15] \mathbf{X} f_1$, recall that $[\tau] \mathbf{X} f$ expresses occlusion of change in f between timepoint $\theta(\tau)$ and τ. $[10, 15] f_1 := R$ is intended to say that the possible change takes place somewhere between time 10 and time 15, that is, in either of the steps from 10 to 11 (captured by $[11] \mathbf{X} f_1$), from 11 to 12 (captured by $[12] \mathbf{X} f_1$), etc. until the step from 14 to 15 (captured by $[15] \mathbf{X} f_1$). Therefore, $\mathbf{X} f_1$ holds during the interval $(10, 15]$ in this example.

The example as shown above allows any sequence of values for f_1 at times 11, 12, 13, and 14, and only constrains it at time 15. If some constraints are known for the possible sequences of values, then those constraints have to be expressed in the action law. The FTNF is designed to do exactly that.

11.2 Minimization of occlusion

I now proceed to develop the occlusion approach, which is one way of removing the remaining restrictions within $\mathcal{K}$-**IA**. The definition of discrete fluent logic DFL-1 is revised, obtaining DFL-2, which allows the new entailment criteria. The syntax and semantics for DFL-2 will be defined in a cumulative fashion in this chapter, by indicating the changes from DFL-1.

11.2.1 Syntax extensions for DFL-2

Core syntax

The core syntax in Chapter 6 is augmented with one more construct: $\mathbf{X}f$ is an elementary fluent formula when f is a feature expression.

Definition of := for DFL-2

We reconsider the definition of the abbreviation :=. For DFL-1, the abbreviation $[\varsigma, \tau] f := \mathcal{X}$ was defined to expand into $[\tau] f \hat{=} \mathcal{X}$. For DFL-2 we change the definition to the following:

- $[\varsigma, \tau] f := \mathcal{X}$ abbreviates $(\varsigma, \tau] \mathbf{X} f \wedge [\tau] f \hat{=} \mathcal{X}$.

This expansion states that the fluent denoted by the feature f is occluded during the time interval $[\varsigma, \tau]$ and has a value in $\mathcal{X}$ at the end of the interval.

Whenever an action is performed over an interval of time, and the action changes or may change a fluent, it is intended that the fluent shall be occluded over the entire interval. The occluded fluents are selected as the members of $\mathtt{Infl}(E, r)$.

Expressions formed using := are solid expressions, that is, if they apply to an interval one cannot conclude that they apply to arbitrary subintervals or superintervals, except to a right-aligned subinterval.

Most of the concepts and notation that were introduced for DFL-1 can be used without difficulty for DFL-2. The following detail of the previous definitions of abbreviations has to be clarified: in subsection 6.5.3 a conditional reassignment formula was defined to have the syntax
$$[\varsigma] \phi \rightarrow [\varsigma, \tau] \alpha,$$
where ϕ is a fluent formula, which in DFL-1 can only be formed using $\hat{=}$. Since ϕ is intended to characterize the present state of the world, one must now add the statement that ϕ cannot be or contain a formula of the form $\mathbf{X}f$. In other words, the $[\varsigma]\phi$ component must still be a DFL-1 formula. Similarly, the previously established specification that the observation set must consist of DFL-1 formulae is not being changed at this point. Thus the occlusion operator is not allowed among the observations, except in the right-hand side of translations of action statements.

11.2.2 Semantics extensions for DFL-2

An interpretation for DFL-2 is a threetuple
$$\langle M, R, X \rangle,$$

where the first two components are as for DFL-1, the occlusion component X is a function from rigid feature expressions in ν, times $\mathcal{T}$, to truth-values T or F, and where $X(f,0) = \mathrm{F}$ for all features f.

The definition of the value of a formula is completed with the following addition:
$$mval[\mathbf{X}f,s,t] = X(den[f,s],t),$$
$$den[\Lambda f,s] = den[f,s].$$

In line with the notation $\mathrm{F} \prec \mathrm{T}$, I shall write $X \preceq X'$ iff
$$\forall f \forall t [X(f,t) \prec X'(f,t) \vee X(f,t) = X'(f,t)]$$
and $X \prec X'$ iff $X \preceq X' \wedge X \neq X'$. The conversion from DFL-2 to DFL-1 interpretations is performed using the *coercion function* Ci, which is defined so that
$$\mathrm{Ci}(\langle M,R,X \rangle) = \langle M,R \rangle,$$
$$\mathrm{Ci}(\langle M,R \rangle) = \langle M,R \rangle,$$
and for sets of interpretations
$$\mathrm{Ci}(W) = \{\mathrm{Ci}(I) \mid I \in W\}.$$
Since a given chronicle has both a set of DFL-1 models and a set of DFL-2 models, I will write DFL1:$[\![\Upsilon]\!]$ and DFL2:$[\![\Upsilon]\!]$ to distinguish the two options. In this chapter $[\![\Upsilon]\!]$ without prefix is understood as DFL2:$[\![\Upsilon]\!]$. From the structure of FTNF and the definitions of := it follows that
$$\text{DFL1:}[\![\Upsilon]\!] = \mathrm{Ci}(\text{DFL2:}[\![\Upsilon]\!]).$$

11.2.3 Correctness condition

In the context of DFL-1, the correctness condition for an entailment method $\langle \mathbf{G}, S \rangle$ was defined by $S(\mathbf{G}(\Upsilon)) = \Sigma_{\mathbf{IA}}(\Upsilon)$, where Υ is a chronicle and $\Sigma_{\mathbf{IA}}(\Upsilon)$ is a set of DFL-1 interpretations of the form $\langle M, R \rangle$. Since DFL-2 interpretations have the structure $\langle M, R, X \rangle$ it is necessary to modify the definition. We choose to generalize it to the requirement $\mathrm{Ci}(S(\mathbf{G}(\Upsilon))) = \Sigma_{\mathbf{IA}}(\Upsilon)$. The correctness condition is therefore not concerned with the correctness of conclusions involving occlusion. Occlusion is seen as a technical device in the logic, not as a property in the world being described.

11.3 Chronological entailment methods with occlusion

The introduction of occlusion makes a range of new entailment methods possible, differing in the following aspects:

- Global, chronological, or pointwise minimization of occlusion? (Pointwise minimization means to minimize X separately for each choice of R).

- How to perform the model reduction for unoccluded change: semantically or syntactically? In the former case one uses preference relations, in the latter case one uses additional premises that restrict change relative to occlusion.

- In the former case, by chronological or global minimization of unoccluded changes?

I shall define a few of these at once, and then proceed to their assessments. I begin with the semantical approaches.

11.3.1 Chronological minimization of occlusion and change

Chronological minimization of occlusion and change (CMOC) is the most straightforward way of using occlusion preferentially. The preference relation $\ll_{cmoc}$ for chronological minimization of occlusion is defined as follows. Let $I = \langle M, R, X \rangle$ and $I' = \langle M', R', X' \rangle$ be DFL-2 interpretations for the same vocabulary ν. The definition of *breakset* in subsection 9.1.3 is revised, obtaining
$$brs(I, \mathsf{t}) = \{f_i \mid R(f_i, \theta\mathsf{t}) \neq R(f_i, \mathsf{t}) \land \neg X(f_i, \mathsf{t})\}.$$
Thus $brs(I, \mathsf{t})$ is the set of features with transparent ($=$ unoccluded) changes at time t, so $brs(\langle M, R, X \rangle, \mathsf{t}) \subseteq breakset(\langle M, R \rangle, \mathsf{t})$.

I write $I \ll_{cmoc} I'$ and say that I is *preferred over* I' (according to CMOC) iff $M = M'$ and for some timepoint t both of the following hold:

- for all timepoints $t, t < \mathsf{t} \to R(t) = R'(t) \land X(t) = X'(t)$;

- either of the following holds:

    * $X(\mathsf{t}) \prec X'(\mathsf{t})$;
    * $X(\mathsf{t}) = X'(\mathsf{t}) \land brs(I, \mathsf{t}) \subset brs(I', \mathsf{t})$.

In this way, one first minimizes the occlusion predicate and then, for a given occlusion predicate, one minimizes the set of transparently changing features. All this is done chronologically and in the same preference relation, that is, in the same 'sweep' of the time line. The latter minimization should always obtain an empty set of changes.

The remaining definitions follow the usual pattern. If W is a set of DFL-2 interpretations, then $cmoc(W)$ is defined as the set of $\ll_{cmoc}$-preferred members of W. The selection function is defined using filtered preferential selection as
$$S_{cmoc}(\langle \mathcal{O}, \Pi, \text{SCD}, \text{OBS} \rangle) = cmoc[\![\Pi(\text{SCD})]\!]_{\mathcal{O}} \cap [\![\text{OBS}]\!],$$
and the entailment method is defined as
$$\text{CMOC} = \langle \mathbf{G}_{seq}, S_{cmoc} \rangle.$$

11.3.2 Chronological assignment of valuation and minimization of occlusion and change

Chronological assignment of valuation and minimization of occlusion and change (CAMOC) combines the features of CAMC which was introduced in the previous chapter, and CMOC which was defined in the previous subsection. The preference relation $\ll_{camoc}$ is defined as follows. Let $I = \langle M, R, X \rangle$ and $I' = \langle M', R', X' \rangle$ be DFL-2 interpretations for the same vocabulary ν. I write $I \ll_{camoc} I'$ and say that I is *preferred over* I' (according to CAMOC) iff there is some timepoint t where all of the following hold:

- $M_{0:\mathbf{t}} = M'_{0:\mathbf{t}}$;
- for all timepoints t, $t < \mathbf{t} \rightarrow R(t) = R'(t) \wedge X(t) = X'(t)$;
- either of the following holds:
  * $X(\mathbf{t}) \prec X'(\mathbf{t})$;
  * $X(\mathbf{t}) = X'(\mathbf{t}) \wedge brs(I, \mathbf{t}) \subset brs(I', \mathbf{t})$.

The remaining definitions are as usual:
$S_{camoc}(\Upsilon) = Min(\ll_{camoc}, [\![\Pi(\text{SCD})]\!]_O) \cap [\![\text{OBS}]\!]$,
$\text{CAMOC} = \langle \mathbf{G}_{seqx}, S_{camoc} \rangle$.

11.4 Syntactical approaches with occlusion

We proceed now to 'syntactical' approaches which use occlusion, but where additional premises are introduced to replace the chronological minimization.

11.4.1 Schema of nochange premises

We use the following schema for *nochange premises*. Sometimes the term 'law of inertia' is used for this schema.

Definition. *For the given vocabulary ν, the set of **nochange formulae** for ν, written NCH_ν, is defined as the set of all logical formulae of the form*
$$\Box[\forall o_1...\forall o_k[f(\vec{o}) \neq \Lambda f(\vec{o}) \rightarrow \mathbf{X}f(\vec{o})]]$$
for all choices of feature symbol f and a corresponding list of arguments according to ν, and where $(\vec{o})$ stands for $(o_1, ...o_k)$. ⋈

For example, feature symbols f_0 and f_1 with zero and one argument position, respectively, have the nochange formulae

$\Box[f_0 \neq \Lambda f_0 \to \mathbf{X} f_0]],$
$\Box[\forall o[f_1(o) \neq \Lambda f_1(o) \to \mathbf{X} f_1(o)]].$

These nochange formulae say that the fluent named f_0 or f_1 can only have changed from its previous value if it is occluded. In other words, they are a kind of frame axioms. Notice, however, that the implication is just one-way: the presence of occlusion $\mathbf{X} f$ does not imply $f \neq \Lambda f$, and its intended usage is not restricted to those situations where change occurs.

The section on non-monotonic logics in Chapter 4 defined the concept of premise integration, which is a syntactic transformation on a given set of premises obtaining a stronger set of premises. However, nochange premises are not an example of premise integration. For a given similarity type one obtains the corresponding nochange premises, but if the scenario-specific premises change (action laws, observations, change statements) then there is no corresponding modification in the nochange premises. Nochange premises constitute a schema in the same sense as the schema for equality axioms in first-order logic.

11.4.2 Global minimization of occlusion with nochange premises

Prototypical global minimization of change (PGM) has been introduced in Chapter 9. A similar method using occlusion will be called GMON (global minimization of occlusion using nochange premises) with the following definitions. Let $I = \langle M, R, X \rangle$ and $I' = \langle M', R', X' \rangle$ be DFL-2 interpretations for the same vocabulary ν. Then I write $I \ll_{gmon} I'$ and say that I is *preferred over* I' (according to *gmon*) iff $M = M'$ and $X \prec X'$.

If W is a set of DFL-2 interpretations, then $gmon(W)$ is defined as the set of $\ll_{gmon}$-preferred members of W. The set of GMON-selected models for a chronicle $\Upsilon = \langle \mathcal{O}, \Pi, \text{SCD}, \text{OBS} \rangle$ is defined accordingly as
$$S_{gmon}(\Upsilon) = gmon(\llbracket \Upsilon \rrbracket \cap \llbracket \text{NCH}_\nu \rrbracket).$$
If filtered preferential entailment is used one obtains similarly GMONF defined by
$$S_{gmonf}(\Upsilon) = gmon\llbracket \Pi(\text{SCD}) \rrbracket_\mathcal{O} \cap \llbracket \text{OBS} \rrbracket \cap \llbracket \text{NCH}_\nu \rrbracket.$$
Both entailment methods use $\mathbf{G}_{seq}$ as their pretransformation.

GMON is a possible interpretation of John McCarthy's 1984 proposal for how to solve the frame problem for a simple blocks world [49]. In this proposal, which Hanks and McDermott used for their Yale shooting scenario [32], action laws are supposed to specify the implied abnormality explicitly: if the gun is loaded and one fires it, then the liveness of Fred is abnormal *and* it becomes false. This differs from the formalization used for example by Kautz, and which has later been the basis of all theory-update approaches, where one merely specifies the new facts: if the gun is loaded and one fires it, then Fred is dead in the resulting situation. The old facts are suppressed implicitly in order to avoid a contradiction.

We have already observed that Kautz's approach corresponds almost precisely to OCM. With respect to the approach of McCarthy there is a question of interpretation: if one assumes (as Hanks and McDermott apparently did) that abnormality is only present when a change actually occurs, and if one writes the action laws accordingly, then one obtains an instance of the PGM entailment method. However, if one permits the implication from 'no abnormality' to 'no change' to be one-way rather than two-way, so that the action laws allow the case of abnormality but no change, then one obtains an instance of GMON.

11.4.3 Pointwise minimization of occlusion with nochange premises and filtering

Pointwise minimization of occlusion using nochange premises and filtering, PMON, is defined as follows. Let $I = \langle M, R, X \rangle$ and $I' = \langle M', R', X' \rangle$ be DFL-2 interpretations for the same vocabulary ν. Then I write $I \ll_{pmon} I'$ and say that I is *preferred over* I' (according to *pmon*) iff $M = M'$, $R = R'$, and $X \prec X'$. The selection function is defined accordingly, using filtered preferential entailment, as
$$S_{pmon}(\langle \mathcal{O}, \Pi, \text{SCD}, \text{OBS} \rangle) = pmon[\![\Pi(\text{SCD})]\!]_{\mathcal{O}} \cap [\![\text{OBS}]\!] \cap [\![\text{NCH}_\nu]\!],$$
and the entailment method is defined by
$$\text{PMON} = \langle \mathbf{G}_{seq}, S_{pmon} \rangle.$$
The definition of PMON requires the minimization to be applied 'before' the nochange premises. For the observations it does not matter; the definition of the selection function can equivalently be written as
$$S_{pmon}(\langle \mathcal{O}, \Pi, \text{SCD}, \text{OBS} \rangle) = pmon[\![\Pi(\text{SCD}) \cup \text{OBS}]\!]_{\mathcal{O}} \cap [\![\text{NCH}_\nu]\!].$$
Notice that in the definitions of $\ll_{gmon}$ and $\ll_{pmon}$ there is no chronological component. The occlusion predicate is minimized non-chronologically. There is no need to minimize unoccluded changes since such changes are not allowed at all, and interpretations with unoccluded changes are rejected by the nochange premises.

PMON is defined using filtered preferential entailment. The case without filtering of $[\![\text{NCH}_\nu]\!]$ lacks interest, since if one uses the $\ll_{pmon}$ preference relation without filtering then one will obtain many unintended models: for any intended model $\langle M, R, X \rangle$ one can insert additional unintended changes in R and extend X accordingly so that all the new changes are occluded, thus obtaining additional *pmon*-preferred models.

Doherty has reported a method [20] for reducing the circumscription version of PMON, previously described in [23], to a corresponding monotonic theory.

11.5 Two-stage minimization of occlusion and change

Two-stage minimization of occlusion and unoccluded change (TMOC) was proposed by Persson and Staflin [56], and uses two separate preference relations which are minimized in succession. The first relation minimizes occlusion pointwise, the second relation minimizes change globally. Nochange premises are not used, but the $\ll_{pmon}$ preference relation can be used for the first (inner) minimization. For the second (outer) one we need a new preference relation, using the following auxiliary definition. The definition of *changeset* in subsection 9.1.5 is revised to a definition of the set of unoccluded changes, obtaining
$$chs(\langle M, R, X \rangle) = \{\langle f_i, t \rangle \mid R(f_i, \theta t) \neq R(f_i, t) \land \neg X(f_i, t)\}.$$
It follows that $chs(\langle M, R, X \rangle) \subseteq changeset(\langle M, R \rangle)$.

The outer preference relation $\ll_{tmoc}$ is then defined as follows. Let $I = \langle M, R, X \rangle$ and $I' = \langle M', R', X' \rangle$ be DFL-2 interpretations for the same vocabulary ν. I write $I \ll_{tmoc} I'$ and say that I is *preferred over I'* (according to TMOC) iff $M = M'$ and $chs(I) \subset chs(I')$. Then $tmoc(W)$ is defined as the set of $\ll_{tmoc}$-minimal members of W, as usual. Finally the set of TMOC-selected models for a chronicle Υ is defined by
$$S_{tmoc}(\Upsilon) = tmoc(pmon[\![\Upsilon]\!])$$
$$\text{TMOC} = \langle \mathbf{G}_{seq}, S_{tmoc} \rangle.$$

11.6 Assessments of occlusion-based approaches

The strategy will be to perform the assessment for CAMOC first, and then to obtain the assessment for CMOC as a special case of it. After that we asses the methods PMON, GMON, and GMONF that use nochange premises, and finally we proceed to the two-stage method TMOC.

11.6.1 Assessment of CAMOC

Lemma 8.3 is adapted to DFL-2 as follows.

Lemma 11.1. *Let $\mathcal{O}$ be an object domain, let $\langle \text{Infl}, \text{Trajs} \rangle$ be a description of an IA world $\mathbf{W}$, and let Π be an FTNF representation of $\langle \text{Infl}, \text{Trajs} \rangle$. Also let $[s, t]E$ be an action statement, let $\langle M, R, X \rangle$ be a DFL-2 interpretation based on $\mathcal{O}$, and let $\mathbf{s} = M[s]$, $\mathbf{t} = M[t]$. Then $\Pi([s, t]E)$ is true in $\langle \mathcal{O}, \langle M, R, X \rangle \rangle$ according to DFL-2 iff there is a trajectory $v \in \text{Trajs}(M[E], R(\mathbf{s}))$ of length $\mathbf{t} - \mathbf{s}$, $v_j \subseteq R(\mathbf{s} + j)$ for $1 \leq j \leq \mathbf{t} - \mathbf{s}$, and*
$$f \in \text{Infl}(M[E], R(\mathbf{s})) \land \mathbf{s} < t \leq \mathbf{t} \to X(f, t) = \text{T}.$$
This trajectory is unique.

Proof The proof is similar to the proof for Lemma 8.3. The only change is in the requirement on X. □

From this we obtain the assessment at once.

Proposition 11.2. *CAMOC is correctly applicable to the chronicle family $\mathcal{K}$-IA.*

Proof The proof consists of two parts which are similar to the proofs for Propositions 10.5 and 10.6 respectively. In the first part there are only minor changes. However, the whole proof is replicated here for ease of reading.

It has been verified in subsection 10.2.1 that $\mathbf{G}_{gseqx}$ is correct, harmless, and preserves the property of being simply executable. Therefore, by Lemmas 9.2 and 9.18 it is sufficient to prove that $\text{Ci}(S_{camoc}(\Upsilon)) = \Sigma_{\mathbf{IA}}(\Upsilon)$ for all simply executable Υ of the form $\langle \mathcal{O}, \Pi, \text{SXS}, \emptyset \rangle$. For an arbitrary Υ having those properties, it follows from Proposition 10.2 that $\Sigma_{\mathbf{IA}}(\Upsilon) \neq \emptyset$. We first prove $\text{Ci}(S_{camoc}(\Upsilon)) \subseteq \Sigma_{\mathbf{IA}}(\Upsilon)$ and then $\Sigma_{\mathbf{IA}}(\Upsilon) \subseteq \text{Ci}(S_{camoc}(\Upsilon))$.

1. Suppose that $\text{Ci}(S_{camoc}(\Upsilon)) \not\subseteq \Sigma_{\mathbf{IA}}(\Upsilon)$, let $\langle M, R \rangle$ be a member of $\text{Ci}(S_{camoc}(\Upsilon)) - \Sigma_{\mathbf{IA}}(\Upsilon)$, and choose X so that $\langle M, R, X \rangle \in S_{camoc}(\Upsilon)$. By the definition of $\ll_{camoc}$, $R(t)$ must be constant when t is $\geq$ the ending time of M with respect to SCD, and the same holds for every member $\langle M', R' \rangle$ of $\Sigma_{\mathbf{IA}}(\Upsilon)$. Therefore, one can choose t as the smallest timepoint such that $R_{0:t}$ is not equal to $R'_{0:t}$ for any $\langle M', R' \rangle \in \Sigma_{\mathbf{IA}}(\Upsilon)$. (For example, the ending time of M w.r.t. SXS has this property).

Clearly $t > 0$. Let u be the largest member of of $\text{At}(M, \text{SCD})$ which satisfies $u < t$. Also let M_0 be that (unique) good initial valuation for SXS which is $\subseteq M$. Either of the following cases must apply:

(a) There is some $\ulcorner [s_i, t_i] E_i \urcorner \in \text{SCD}$ such that $M[s_i] = u$. Since $\langle M, R \rangle \in [\![\Upsilon]\!]$ and by Lemma 8.3 there is some trajectory v of length $k \geq t - u$ which is a member of $\text{Trajs}(M[E_i], R(u))$, and such that $v_j \subseteq R(u+j)$ for $1 \leq j \leq k$. Using Proposition 10.4 it follows that $\langle M'', R_{0:u} \triangleright v \rangle \in \hat{\Sigma}^u_{\mathbf{IA}}(M_0, \Upsilon)$ for some M'' for every such v. (Actually $M'' = M_{0:u+k}$ but this requires separate verification, and is irrelevant for the present proof). Since $R_{0:t}$ does not agree with any member of $\Sigma_{\mathbf{IA}}(\Upsilon)$, it must be different from $(R_{0:u} \triangleright v)_{0:t}$, which means that some $R(t')$ for $u < t' \leq t$ must also differ from $R(u)$ in a feature outside those defined in v. Therefore either $\mathbf{X}$ is unnecessarily large, or $brs(R, t') \neq \emptyset$ for some t'. But this is incompatible with the assumption that $\langle M, R \rangle \in \text{Ci}(S_{camoc}(\Upsilon))$.

(b) There is no $\ulcorner [s_i, t_i] E_i \urcorner \in \text{SCD}$ such that $M[s_i] = u$. In this case $u = t - 1$. By the assumption, $\langle M_{0:u}, R_{0:u} \rangle = \langle M'_{0:u}, R'_{0:u} \rangle$ for some $\langle M', R' \rangle \in \Sigma_{\mathbf{IA}}(\Upsilon)$. Therefore $I' = \langle M_{0:u}, R_{0:u} \triangleright \langle \emptyset \rangle \rangle \in \Sigma^u_{\mathbf{IA}}(M_0, \Upsilon)$, and by Proposition 10.4 it has a continuation to ∞ that is a member of $\Sigma_{\mathbf{IA}}(\Upsilon)$. Since $\langle M, R \rangle$ does not agree in t with any member of $\Sigma_{\mathbf{IA}}(\Upsilon)$ it follows that $R(t) \neq R(u)$. But in that case $I' \ll_{camc} \langle M, R \rangle$, contradicting the assumption that $\langle M, R \rangle \in \text{Ci}(S_{camoc}(\Upsilon))$. This concludes this part of the proof.

2. Suppose that $\langle M, R \rangle \in \Sigma_{\mathbf{IA}}(\Upsilon) - \mathrm{Ci}(S_{camoc}(\Upsilon))$. It follows that $\langle M, R \rangle \in \mathrm{DFL1}{:}[\![\Upsilon]\!] = \mathrm{Ci}(\mathrm{DFL2}{:}[\![\Upsilon]\!])$. Let $I = \langle M, R, X \rangle$ be that member of $\mathrm{DFL2}{:}[\![\Upsilon]\!]$ that minimizes X w.r.t. $\prec$. (X is uniquely determined). Since $\langle M, R \rangle \notin \mathrm{Ci}(S_{camoc}(\Upsilon))$ it follows $\langle M, R, X \rangle \notin S_{camoc}(\Upsilon)$. There must therefore be some model $I' = \langle M', R', X' \rangle \in \mathrm{DFL2}{:}[\![\Upsilon]\!]$ which is CAMOC-preferred over $\langle M, R, X \rangle$. It is no restriction to assume that it is CAMOC-minimal in $\mathrm{DFL2}{:}[\![\Upsilon]\!]$. In other words, some unintended model is selected after being preferred over some intended one.

Let $\mathbf{t}$ be the timepoint of preferential comparison between $\langle M, R, X \rangle$ and $\langle M', R', X' \rangle$. Then $R_{0:\theta\mathbf{t}} = R'_{0:\theta\mathbf{t}}$, $X_{0:\theta\mathbf{t}} = X'_{0:\theta\mathbf{t}}$, and $M_{0:\mathbf{t}} = M'_{0:\mathbf{t}}$. Therefore one of the following cases must apply:

(a) $\langle M, R, X \rangle$ and $\langle M', R', X' \rangle$ terminate the same action at time $\mathbf{t}$. Let $[s_i, t_i] E_i$ be its action statement, so that $M[t_i] = \mathbf{t}$, and let $M[s_i] = \mathbf{s}$. Since X is minimal and $\langle M', R', X' \rangle$ is $\ll_{camoc}$-minimal in $[\![\Upsilon]\!]$, it follows that $X(\mathbf{t}) = X'(\mathbf{t})$. Therefore since X' is preferred over X, it follows $brs(I', \mathbf{t}) \subset brs(I, \mathbf{t})$.

However, since $\langle M, R \rangle \in \Sigma_{\mathbf{IA}}(\Upsilon)$, it follows $R_{0:\mathbf{t}} = R_{0:\mathbf{s}} \triangleright v$ for some $v \in \mathrm{Trajs}(E_i, R(\mathbf{s}))$ according to Proposition 8.10. By Lemma 11.1 the minimal X over $[\mathbf{s}, \mathbf{t}]$ is obtained by selecting $X(f) = \mathbf{T}$ exactly when $f \in \mathrm{Infl}(E_i, R(\mathbf{s}))$. It follows that $brs(I, \mathbf{t}) = \emptyset$, contradicting $brs(I', \mathbf{t}) \subset brs(I, \mathbf{t})$.

(b) These interpretations are both executing the same action at time $\mathbf{t}$, where both the ending times are $> \mathbf{t}$ but they are not necessarily equal.

(c) Neither of these interpretations is executing any action between $\theta\mathbf{t}$ and $\mathbf{t}$.

Cases (b) and (c) lead to contradictions by arguments similar to the argument for case (a). This concludes the proof. □

11.6.2 Assessment of CMOC

The result for CMOC follows as a special case:

Proposition 11.3. *CMOC is correctly applicable for the chronicle family $\mathcal{K}$-IAe.*

Proof Since CMOC is formed using filtering, it is sufficient to prove that if $\Upsilon \in \mathcal{K}\mathbf{p}\text{-}\mathbf{IAe}$ is simply executable, then
$$\mathrm{Ci}(Min(\ll_{cmoc}, [\![\Upsilon]\!])) = \Sigma_{\mathbf{IA}}(\Upsilon).$$
Let $W = [\![\Upsilon]\!]$. If $W = \emptyset$ then the result is trivial. Also, since $I \ll_{cmoc} I' \to I \ll_{camoc} I'$, which is easily obtained from the definitions, it follows that
$$\Sigma_{\mathbf{IA}}(\Upsilon) = \mathrm{Ci}(Min(\ll_{camoc}, W)) \subseteq \mathrm{Ci}(Min(\ll_{cmoc}, W)),$$
using Proposition 11.2. We now have to strengthen this subset relationship to equality using $\Upsilon \in \mathcal{K}\text{-}\mathbf{IAe}$. Suppose the contrary holds, so that there is some $\langle M, R, X \rangle \in Min(\ll_{cmoc}, W)$ and $\langle M, R \rangle \notin \Sigma_{\mathbf{IA}}(\Upsilon)$. Then a history

R' is constructed from R by proceeding incrementally over time as follows. First of all, $R'(0) = R(0)$. Then, at any timepoint s such that $R'_{0:s}$ has already been defined, if no action in $M[\text{SCD}]$ starts in s then $R'(s+1) = R'(s)$. On the other hand, if an action $\ulcorner[s,t]E\urcorner \in M[\text{SCD}]$ starts in s, then by Lemma 8.3 there is a unique $v \in \text{Trajs}(E, R(s))$ such that $v_i \subseteq R(s+i)$, where i ranges from 1 to the length k of v. If $R(s) = R'(s)$ then choose $v' = v$, otherwise choose v' as an arbitrary member of $\text{Trajs}(E, R'(s))$ having length k. We are guaranteed that there is such a member since Υ is equidurational. In either case define R' so that $R'_{0:s+k} = R'_{0:s} \triangleright v'$. Finally, choose a $\prec$-minimal X' such that $\Pi(M[\text{SCD}])$ is satisfied in $\langle M, R', X' \rangle$.

Clearly, $\langle M, R' \rangle \in \Sigma_{\mathbf{IA}}(\Upsilon)$. With this construction, either $R' = R$, which contradicts the assumption that $\langle M, R \rangle \notin \Sigma_{\mathbf{IA}}(\Upsilon)$, or using the first timepoint of disagreement between R and R' as the time of preferential comparison it follows that $\langle M, R', X' \rangle \ll_{cmoc} \langle M, R, X \rangle$, which also contradicts the assumptions. This concludes the proof. □

Since the proof is based on the assessment of CAMOC it inherits its restriction to simply executable chronicles. It seems likely that this restriction can be removed, but it is of minor importance.

11.6.3 Assessments of methods that use nochange premises

We proceed now to the assessments of those methods that use nochange premises. These assessments are easy.

Lemma 11.4. *Let* $\Upsilon = \langle \mathcal{O}, \Pi, \text{SCD}, \text{OBS} \rangle$ *as usual, let* $\langle M, R \rangle \in \Sigma_{\mathbf{IA}}(\Upsilon)$, *and choose X so that $X(f, t)$ is true iff there is some action* $\ulcorner[s_i, t_i]E_i\urcorner \in M[\text{SCD}]$ *such that $s_i < t \leq t_i$ and $f \in \text{Infl}(E_i, R[s_i])$. Then*
$$\langle M, R, X \rangle \in pmon[\![\Pi(\text{SCD})]\!]_\mathcal{O} \cap [\![\text{OBS}]\!] \cap [\![\text{NCH}_\nu]\!].$$

Proof $\langle M, R, X \rangle \in [\![\Pi(\text{SCD})]\!]_\mathcal{O} \cap [\![\text{OBS}]\!]$ follows immediately from Proposition 8.5 and Lemma 11.1. Furthermore it follows that the constructed X is minimal in $[\![\Pi(\text{SCD})]\!]_\mathcal{O}$ for the given M and R. Finally it follows using the construction of NCH_ν that $\langle M, R, X \rangle \in [\![\text{NCH}_\nu]\!]$. □

Proposition 11.5. *PMON is correctly applicable for the $\mathcal{K}$-IA family of chronicles.*

Proof We are to prove that $\Sigma_{\mathbf{IA}}(\Upsilon) = \text{Ci}(S_{pmon}(\Upsilon))$ for safely sequential chronicles Υ. Using Lemma 9.18 it is sufficient to show this for chronicles with an empty set of observations. Recall the definition
$$S_{pmon}(\Upsilon) = pmon[\![\Pi(\text{SCD})]\!]_\mathcal{O} \cap [\![\text{OBS}]\!] \cap [\![\text{NCH}_\nu]\!].$$
Let $\Upsilon = \langle \mathcal{O}, \Pi, \text{SCD}, \emptyset \rangle \in \mathcal{K}_{\mathbf{p}}\text{-}\mathbf{IA}$. (1) Suppose $\langle M, R \rangle \in \Sigma_{\mathbf{IA}}(\Upsilon)$, so that it is a member of DFL1:$[\![\Upsilon]\!]$. Construct X like in step (2) of the proof of Proposition 11.2, so that $\langle M, R, X \rangle \in$ DFL2:$[\![\langle \mathcal{O}, \Pi, \text{SCD}, \emptyset \rangle]\!]$ and X is minimal for the given M and R. This means that $\langle M, R, X \rangle$ is PMON-minimal. According to Lemmas 11.1 and 11.4 it follows that $\langle M, R, X \rangle \in$

$pmon[\![\Pi(\text{SCD})]\!]_\mathcal{O}$. It also follows from the construction of $\Sigma_{\mathbf{IA}}(\Upsilon)$ that $\langle M, R, X \rangle$ satisfies the nochange premises NCH_ν, so it is a member of $S_{pmon}(\Upsilon)$.

(2) Suppose that $\langle M, R, X \rangle \in S_{pmon}(\Upsilon)$. It is easily verified that $\langle M, R, X \rangle$ satisfies the conditions for being obtained by the construction process for $\Sigma_{\mathbf{IA}}(\Upsilon)$, since all actions are correctly performed, and the form of the nochange premises guarantee that there are no additional changes in feature values. □

The assessment for GMONF uses a similar proof but requires a restriction.

Definition. *A world description $\langle \texttt{Infl}, \texttt{Trajs} \rangle$ is said to be* **GMON-favorable** *iff*
$$\forall r \forall r' \forall E[\texttt{Infl}(E, r) \subset \texttt{Infl}(E, r') \to \texttt{Pdur}(E, r) \mathbin{\Uparrow} \texttt{Pdur}(E, r')].$$
It is said to be **GMON-perfect** *iff*
$$\forall r \forall r' \forall E[\texttt{Infl}(E, r) \neq \texttt{Infl}(E, r') \to \texttt{Pdur}(E, r) \mathbin{\Uparrow} \texttt{Pdur}(E, r')].$$
⋈

In other words, if for some event designator E there is some trajectory following a state r and some trajectory of the same length following a state r' (the same state or different from r), then the set of influenced features in r must not be a strict subset of the set of influenced features in r' (GMON-favorable) or the two sets of influenced features must be equal (GMON-perfect). The name for this restriction was chosen because it corresponds directly to the GMON preference relation. It follows at once that all chronicles in $\mathcal{K}$-**IAu** are GMON-perfect.

Lemma 11.6. *If $\Upsilon \in \mathcal{K}$-**IA** is GMON-favorable, then*
$$gmon[\![\Pi(\text{SCD})]\!]_\mathcal{O} = pmon[\![\Pi(\text{SCD})]\!]_\mathcal{O}.$$

Proof Clearly $I \ll_{pmon} I' \to I \ll_{gmon} I'$ according to the definitions, so
$$gmon[\![\Pi(\text{SCD})]\!]_\mathcal{O} \subseteq pmon[\![\Pi(\text{SCD})]\!]_\mathcal{O}.$$
It therefore suffices to show that every PMON-minimal model is also GMON-minimal. Consider two arbitrary $I = \langle M, R, X \rangle$ and $I' = \langle M, R', X' \rangle$ in $pmon[\![\Pi(\text{SCD})]\!]_\mathcal{O}$. We are to show $I \not\ll_{gmon} I'$. From Proposition 11.5 it follows $\langle M, R \rangle \in \Sigma_{\mathbf{IA}}(\langle \mathcal{O}, \Pi, \text{SCD}, \emptyset \rangle)$, and from Lemma 11.4 and the definition of $\ll_{pmon}$ it then follows that X is chosen as in Lemma 11.4. The same applies for $\langle M, R', X' \rangle$. Since the valuation component is the same in both interpretations it follows that each action statement in SCD obtains the same duration in $\langle M, R, X \rangle$ and in $\langle M, R', X' \rangle$. Since Π is GMON-favorable it then follows that $X \not\prec X'$, so $\langle M, R, X \rangle \not\ll_{gmon} \langle M, R', X' \rangle$. □

Proposition 11.7. *GMONF is correctly applicable for GMON-favorable chronicles in $\mathcal{K}$-**IA**.*

Proof This follows directly from Proposition 11.5 and Lemma 11.6. □

The step 'down' from GMONF to GMON requires a considerable strengthening of the restrictions:

Proposition 11.8. *GMON is correctly applicable for GMON-perfect chronicles in $\mathcal{K}$r-**IA**.*

Proof Recall the definition, which can be written as
$S_{gmon}(\Upsilon) = gmon([\![\Pi(\text{SCD})]\!]_O \cap [\![\text{OBS}]\!] \cap [\![\text{NCH}_\nu]\!])$.
It follows from Proposition 9.17 that $S_{gmonf}(\Upsilon) \subseteq S_{gmon}(\Upsilon)$. It is therefore sufficient to show that if $\Upsilon \in \mathcal{K}$r-**IA** is GMON-perfect, then every member of $S_{gmon}(\Upsilon)$ is also a member of $S_{gmonf}(\Upsilon)$. Suppose for some GMON-perfect $\Upsilon \in \mathcal{K}$r-**IA** that $\langle M, R, X \rangle$ is a member of $S_{gmon}(\Upsilon)$. Clearly $\langle M, R, X \rangle \in [\![\Pi(\text{SCD})]\!]_O \cap [\![\text{OBS}]\!]$, so $[\![M, \Upsilon]\!] \neq \emptyset$. Therefore, since $\Upsilon \in \mathcal{K}$r-**IA** and $S_{gmonf}(\Upsilon) = \Sigma_{\mathbf{IA}}(\Upsilon)$ it follows that $S_{gmonf}(\Upsilon)$ has at least some member $\langle M, R', X' \rangle$.

Now consider an arbitrary $\langle M, R', X' \rangle \in gmon[\![\Pi(\text{SCD})]\!]_O$. The minimal X' depends only on M and not on R because Υ is GMON-perfect. Since there is some member of $[\![\Pi(\text{SCD})]\!]_O$ with minimal X', it follows from the definition of $\ll_{gmon}$ that all members of $gmon[\![\Pi(\text{SCD})]\!]_O$ with the given M have the same X, that is, they have the form $\langle M, R_i, X \rangle$ for varying R_i but fixed X. Therefore $\langle M, R_i, X \rangle \not\ll_{gmon} \langle M, R, X \rangle$ for any of them, whence it follows $\langle M, R, X \rangle \in S_{gmonf}(\Upsilon)$. This concludes the proof. □

Notice the similarity between the assessments for PGM and GMON: they are correct for PGM-perfect and for GMON-perfect chronicles in $\mathcal{K}$r-**IA**, respectively. For PGMF and GMONF there is no similar correspondence. Notice also that unlike the ontological properties that have been used in the previous assessments, GMON-perfection depends on which of several equivalent world descriptions has been chosen as the basis for the action laws. In particular, since it is always possible to choose a uniform world description, GMONF can be used throughout $\mathcal{K}$-**IA** although at a certain price in terms of the size of the axioms. For example, the law for *Fire* in the Hanks-McDermott formalization of the Yale shooting scenario says 'if the gun is loaded and it is fired, then the aliveness of Fred is abnormal and Fred is not alive after the action'. If the uniform world description is used, then the law becomes instead as follows: 'If the gun is fired then: 1. The aliveness of Fred is abnormal. 2. If the gun was loaded then Fred is not alive after the action. 3. If the gun was not loaded then Fred is alive after the action iff he was alive before the action.' It follows from Proposition 11.8 that GMON is correct for YSS with the latter formulation. The problems that Hanks and McDermott had with the Yale shooting scenario can therefore be ascribed to their particular choice of formalization.

11.6.4 Assessments of a two-stage entailment method

The assessment of TMOC also uses the previous result on the assessment of PMON:

Proposition 11.9. *TMOC is correctly applicable for the $\mathcal{K}$r-**IA** ontological family.*

Proof The definition of TMOC above is equivalent to
$$S_{tmoc}(\Upsilon) = tmoc(pmon[\![\Pi(\text{SCD})]\!]_\mathcal{O} \cap [\![\text{OBS}]\!]).$$
We use the selection function of PMON which is
$$S_{pmon}(\Upsilon) = pmon[\![\Pi(\text{SCD})]\!]_\mathcal{O} \cap [\![\text{OBS}]\!] \cap [\![\text{NCH}_\nu]\!].$$
These definitions can be rewritten in comparable form as
$$S_{tmoc}(\Upsilon) = Min(\ll_{tmoc}, W),$$
$$S_{pmon}(\Upsilon) = W \cap [\![\text{NCH}_\nu]\!],$$
where
$$W = pmon[\![\Pi(\text{SCD})]\!]_\mathcal{O} \cap [\![\text{OBS}]\!].$$
Let W_M be the subset of W consisting of those of its members where the valuation component is M. Suppose first that $W_M \cap [\![\text{NCH}_\nu]\!] = \emptyset$. Then
$$\Sigma_{\mathbf{IA}}(M,\Upsilon) = Ci(W_M \cap [\![\text{NCH}_\nu]\!]) = \emptyset,$$
and therefore $[\![M,\Upsilon]\!] = \emptyset$ since Υ is pointwise consistency retaining. Since $W \subseteq [\![\Upsilon]\!]$ it follows $W_M \subseteq [\![M,\Upsilon]\!] = \emptyset$, and therefore $Min(\ll_{tmoc}, W) = \emptyset$.

On the other hand if $W_M \cap [\![\text{NCH}_\nu]\!] \neq \emptyset$, we observe that an interpretation I is a member of $[\![\text{NCH}_\nu]\!]$ iff $chs(I) = \emptyset$. Those members of W_M which satisfy NCH_ν are therefore the same as those which are $\ll_{tmoc}$-minimal.

We have therefore shown equality of the restrictions of $S_{tmoc}(\Upsilon)$ and $S_{pmon}(\Upsilon)$ to a particular M. The equality for the entire sets is obtained as follows.
$$S_{tmoc}(\Upsilon) = tmoc(W) = tmoc \bigcup_M W_M = \bigcup_M tmoc(W_M) =$$
$$= \bigcup_M (W_M \cap [\![\text{NCH}_\nu]\!]) = \left(\bigcup_M W_M\right) \cap [\![\text{NCH}_\nu]\!] = S_{pmon}(\Upsilon),$$
and the result follows from Proposition 11.5. $\square$

Although TMOC has a somewhat restricted range of applicability with respect to $\mathcal{K}$-**IA**, this may actually be an advantage rather than a handicap in a wider perspective. One of the natural next steps after investigating simple inertia $\mathcal{K}$-**IA**, is to proceed to the $\mathcal{K}$-**IAS** family where surprises are allowed. Those chronicles Υ where there is a disagreement, so that $\Sigma_{\mathbf{IA}}(\Upsilon) \neq \Sigma_{\mathbf{IAS}}(\Upsilon)$, arise exactly when Υ does not retain consistency. What comes out as a restriction with respect to $\mathcal{K}$-**IA** may therefore be a strength with respect to $\mathcal{K}$-**IAS**. In such a case it remains to be considered, however, whether to keep or to drop the condition $M = M'$ in the definition of $\ll_{tmoc}$.

11.6.5 Summary

We have now assessed the range of applicability of twelve different entailment methods. The assessment results for all of them are summarized in Table 11.1. For each entailment method it indicates the current assessment, whether the entailment method uses reformulation to executable schedules, and the full name of the assessed method. The compatibility conditions have been coded so that [PCM] means PCM-compatible, [PGM$^+$] means PGM-favorable, and [PGM*] means PGM-perfect, and similarly for the other methods besides PGM.

Doherty and Łukaszewicz [23] have provided circumscriptive versions of the majority of these entailment methods.

11.7 Oracle features

We proceed now to an assessment of the approach of item-valued oracle features which was introduced at the beginning of this chapter. Essentially, this method is based on a rewriting function where a world with nondeterministic actions is converted into one where actions are deterministic, but inertia only applies for some of the features.

11.7.1 The oracle reformulation function

The reformulation function $\mathbf{G}_{ivo}$ for item-valued oracle features is defined as follows.

$$\mathbf{G}_{ivo}(\langle \sigma, \mathcal{O}\rangle, \langle \mathcal{O}, \Pi, \text{SCD}, \text{OBS}\rangle) = \langle\langle \sigma', \emptyset\rangle, \mathbf{G}_{seq}(\langle \mathcal{O}, \Pi', \text{SCD}, \text{OBS}\rangle), \emptyset\rangle,$$

where σ' and Π' remain to be defined. σ' is obtained from σ by introducing one oracle feature $\hat{f}$ corresponding to each feature f in σ. Features in σ will be called *main features* to distinguish them from the oracle features. The argument structure and value domain of $\hat{f}$ shall be the same as for f. Π' is to be defined in such a way that for every development $J = \langle \mathcal{B}, \mathcal{M}, R, \mathcal{A}, \mathcal{C}\rangle$ of $\Upsilon = \langle \mathcal{O}, \Pi, \text{SCD}, \text{OBS}\rangle$, there is a set of corresponding developments $J' = \langle \mathcal{B}, \mathcal{M}, R', \mathcal{A}, \mathcal{C}\rangle$ of $\Upsilon' = \langle \mathcal{O}, \Pi', \text{SCD}, \text{OBS}\rangle$, each of which has the following properties. For each action $\langle s, E, t\rangle \in \mathcal{A}$ and for each feature f which is influenced during that action, $R(f, u) = R(\hat{f}, \theta u)$ whenever $s < u \leq t$. In other words, the value of $\hat{f}$ at time θu serves as oracle for the value of f at time u. This means that within the duration of each action, each oracle feature varies in the same way as the corresponding main feature, provided that the latter is influenced by the action, and except in the last timestep of the action where it is allowed to change freely.

A set of action laws Π' with those properties can be obtained as follows. Let $\langle \texttt{Infl}, \texttt{Trajs}\rangle$ be the world description corresponding to Π.

Table 11.1. Summary of assessments

Entailment criterion	Assessment	Req execut schedule?	Full name
PGM	$\mathcal{K}$r-**IA**[PGM*]	No	Prototypical global minimization
PGMF	$\mathcal{K}$-**IAe**[PGM*]	No	PGM with filtering
OCM	$\mathcal{K}$sp-**IAde**[OCM]	No	Original chronological minimization
	$\mathcal{K}$p-**IAe**[OCM$^+$]		
PCM	$\mathcal{K}$p-**IAe**[PCM]	No	Prototypical chronological minimization
PCMF	$\mathcal{K}$-**IAe**[PCM]	No	PCM with filtering
CAMC	$\mathcal{K}$-**IA**[CAMC]	Yes	Chronological assignment and minimization of change
CMOC	$\mathcal{K}$-**IAe**	(Yes)	Chronological minimization of occlusion and change
CAMOC	$\mathcal{K}$-**IA**	Yes	Chronological assignment and minimization of occlusion and change
PMON	$\mathcal{K}$-**IA**	No	Pointwise minimization of occlusion with nochange premises
GMON	$\mathcal{K}$r-**IA**[GMON*]	No	Global minimization of occlusion with nochange premises
GMONF	$\mathcal{K}$-**IA**[GMON$^+$]	No	GMON with filtering
TMOC	$\mathcal{K}$r-**IA**	No	Two-stage minimization of occlusion and change

We construct a corresponding world description $\langle \text{Infl}', \text{Trajs}' \rangle$ containing the oracle features, and obtain Π' from it in the standard fashion. Note that $\langle \text{Infl}, \text{Trajs} \rangle$ is constructed using $\langle \sigma, \mathcal{O} \rangle$, and $\langle \text{Infl}', \text{Trajs}' \rangle$ is constructed using $\langle \sigma \cup \sigma', \mathcal{O} \rangle$. $\text{Infl}'(E, r')$ is defined as $\{f, \hat{f} \mid f \in \text{Infl}(E, r)\}$, where r is the restriction of r' to $\langle \sigma, \mathcal{O} \rangle$. Suppose $v = \langle \bar{r}_1, \bar{r}_2, ..., \bar{r}_k \rangle \in \text{Trajs}(E, r)$. Let $\hat{r}_i = \{\hat{f} : v \mid (f : v) \in \bar{r}_i\}$. The trajectory v shall have a number of counterparts, all of which are members of $\text{Trajs}'(E, (r \cup \hat{r}) \oplus \hat{r}_1)$, where $\hat{r}$ is an arbitrary state defined over $\langle \sigma', \mathcal{O} \rangle$. These counterparts are formed as $\langle \bar{r}_1 \cup \hat{r}_2, \bar{r}_2 \cup \hat{r}_3, ..., \bar{r}_{k-1} \cup \hat{r}_k, \bar{r}_k \cup \hat{r}_0 \rangle$, where $\hat{r}_0$ is an arbitrary partial state defined over the same features as the other $\hat{r}_i$.

This defines $\text{Trajs}'(E, r)$ for some, but not necessarily for all the argument combinations that are required by the vocabulary $\langle \sigma \cup \sigma', \mathcal{O} \rangle$. The remaining argument combinations are chosen so that $\text{Trajs}'(E, r'') = \text{Trajs}'(E, r')$ for some r' where a definition has been obtained, and where the restrictions of r'' and r' to $\langle \sigma, \mathcal{O} \rangle$ are equal. For example, suppose the feature f has three possible values 1, 2, and 3, and the first element in the trajectories in $\text{Trajs}(E, r)$, for a certain choice of E and r, assign the value 1 or 2 to f, but none of them assigns the value 3. Within Trajs' one will have $\text{Trajs}'(E, r')$ where $r'[\hat{f}] = 1$, and this set of trajectories will contain counterparts of those trajectories in $\text{Trajs}(E, r)$ where the first element contains $f : 1$. Similarly, those $\text{Trajs}'(E, r')$ where $r'[\hat{f}] = 2$ will contain counterparts of those trajectories in $\text{Trajs}(E, r)$ where the first element contains $f : 2$. Those $\text{Trajs}'(E, r')$ where $r'[\hat{f}] = 3$ is not assigned any members by correspondence, but is then chosen arbitrarily as one of the other cases, let us say as the case where $r'[\hat{f}] = 1$. Therefore, if the last step of the previous action that is executed in the schedule sets $\hat{f}$ to 1 or 3, then the action E will let f be 1 in the first timestep within the action, and if the previous action sets $\hat{f}$ to 2, then the action E will let f be 2 in the first timestep within the action.

This arrangement satisfies the requirements for allowing PCM and similar entailment methods to work correctly even for chronicles which are not PCM-compatible: if v and v' are two members of $\text{Infl}(E, r)$ and $r \triangleright v \ll_{pcm} r \triangleright v'$, then consider the timepoint of comparison $\mathbf{t}$ in their duration. The corresponding oracle features differ in $\theta \mathbf{t}$, and therefore the corresponding trajectories also containing the oracle features will not be PCM-comparable. It is clear that no additional constant symbols or objects need to be introduced by $\mathbf{G}_{ivo}$.

11.7.2 Entailment methods using oracles

Given the original motivation for the use of oracle features, one would expect PCM, at least, to be correct even for chronicles that are not PCM-compatible, if it is applied after reformulation using $\mathbf{G}_{ivo}$. However, the entailment method $\langle \mathbf{G}_{ivo}, S_{pcm} \rangle$ fails to have the intended, broader range of applicability, since it imposes chronological minimization on oracle features as well. The entailment method RPCM (oracle prototypical chronological minimization) is therefore defined as $\langle \mathbf{G}_{ivo}, S_{rpcm} \rangle$, where $S_{rpcm}(\Upsilon) = Min(\ll_{rpcm}, [\![\Upsilon]\!])$, and $\langle M, R \rangle \ll_{rpcm} \langle M', R' \rangle$ iff $M = M'$ and there is some timepoint t such that

- for all features f constructed using $\langle \sigma \cup \sigma', \mathcal{O} \rangle$, and for all timepoints $t < \mathbf{t}$, $R(f,t) = R'(f,t)$;
- $obrs(R, \mathbf{t}) \subset obrs(R', \mathbf{t})$,

where
$$obrs(R, \mathbf{t}) = \{f_i \mid R(f_i, \theta \mathbf{t}) \neq R(f_i, \mathbf{t}) \land (f \text{ is defined in } \sigma)\}.$$
This means that RPCM applies the oracle rewriting function, obtains the classical models for the rewritten chronicle, minimizes chronologically in a PCM-like fashion but only using oracle features for comparing past history, not for identifying the set of changes, and then discarding the oracle features.

Analogous oracle-based variants can be defined for the other entailment methods that have been introduced. A brief look at the assessment results in Table 11.1 suggests, however, that this is only meaningful for PCM, PCMF, and CAMC. For those methods where the assessment contains a requirement of favorability or perfection it is pointless to introduce oracles, since such restrictions apply across $\text{Trajs}(E, r)$ or $\text{Infl}(E, r)$ for several r anyway. The first assessment of OCM is also not going to apply, since its proof depended crucially on the fact that all intended models for the given chronicle are equal except possibly inside actions.

We use the codes RPCM, RPCMF, and RCAMC with the obvious meanings and pronounciations for the latter two. It is clear that the use of item-valued oracles does not relieve the $\mathcal{K}$p-**IAe** restriction for RPCM, and that the use of filtering relieves the initial-observation restriction as usual. It is therefore likely that RPCM is correct for $\mathcal{K}$p-**IAe**, RPCMF for $\mathcal{K}$-**IAe**, and RCAMC for $\mathcal{K}$-**IA**. However, the detailed formal verification has not been made.

11.7.3 Discussion

Assuming that these tentative assessments are correct, then RPCMF has the same range of applicability as CMOC, and RCAMC the same range of applicability as CAMOC. There is in fact a direct correspondence since, compared to the DFL-1 version of the chronicle, one adds either $\mathbf{X}f$ for occlusion or $\hat{f}$ for the oracle feature. They are different since $\mathbf{X}f$ has a truthvalue and $\hat{f}$ has an item value, and the values are used differently, but it should not be a surprise that they are exchangable.

The presentation in this chapter has considered the case of occlusion first, since it is considerably less complicated in its general formulation; it corresponds more directly to the underlying semantics; and it is relevant for a larger number of preference relations. From the point of view of an application, there are additional factors to be considered. The use of oracle features forces one to introduce additional constraints relating the value of the main feature and the value of the oracle. When occlusion is used, the corresponding constraints are taken care of once and for all as occlusion is used in the preference relations. By comparison, the oracle rewriting function results in a relatively complex action law in the most general case.

On the other hand, consider a very simple case where an application contains only one nondeterministic action, which in turn only affects one feature, and where all other actions are deterministic. In favor of oracle features, one may argue that the oracle reformulation can be performed manually, in such a fashion that only the nondeterministically affected feature needs an oracle. One may also consider whether such a 'leaner' oracle reformulation function could be given a systematic definition. CMOC and the other occlusion-based entailment methods assume of course that occlusion is defined and used for all features. On the other hand, in this case one may also consider an intermediate method between PCMF and CMOC, namely, a method which only uses occlusion for those features that may be affected nondeterministically.

In summary, it seems to me that the use of oracle features may be useful in order to rescue a practical situation where an application involves nondeterministic actions, and a chronological-minimization implementation without occlusion is to be used, but that it does not have any significant advantages from the point of view of a systematic theory.

11.8 A perspective on the chronicle description language $\mathcal{A}$

At this point it is appropriate to return to the chronicle description language $\mathcal{A}$ which has been introduced by Gelfond and Lifschitz, and which was already mentioned in Chapter 3, subsection 3.1.7. What is the relationship between the language $\mathcal{A}$, with its syntax and semantics, and the languages that are used in the 'Features and fluents' approach?

In our terms, the original $\mathcal{A}$ language allows one to express certain very simple chronicles in situation calculus, that is, over Herbrand time. Corresponding to the OBS component of a schecule, there are constructs for stating the initial values of features, for example,

Initially $\neg Loaded$.

There are also constructs for specifying simple action laws, corresponding to the Π component of the schedule, for example,

Shoot **causes** ¬*Alive* **if** *Loaded*.

Non-initial timepoints are specified in terms of the actions that led to them, and properties of later timepoints are stated as in the following example.

¬ *Alive* **after** *Load; Wait; Shoot*.

In the 1994 version of the language [37], such expressions are used both for non-initial observations and for stating conclusions from a given chronicle.

The $\mathcal{A}$ language has both a syntax and a semantics. The expressivity of original $\mathcal{A}$ [26] equals $\mathcal{K}$-**IbsAd** [80], and in particular, its semantics defines the effects of actions in terms of strict inertia. Later extensions allow nondeterministic actions, concurrent actions, and indirect effects. The most significant results based on $\mathcal{A}$ and its dialects relate to their translation into circumscription.

In our framework, there are two possible ways of viewing the chronicle description language $\mathcal{A}$ and the results that have been obtained using it. One may view $\mathcal{A}$ as an alternative way of defining the underlying semantics, instead of the trajectory semantics being used here. Or, one may view $\mathcal{A}$ as an entailment method, and ask how it can be translated to circumscription or other realizations.

If the first perspective is adopted, then there are two differences between $\mathcal{A}$ and trajectory semantics: (1) $\mathcal{A}$ has both a syntax and a semantics, whereas the trajectory semantics manages without a syntax; (2) the trajectory semantics is considerably more expressive, especially compared with the original $\mathcal{A}$ language. The latter perspective may also be appropriate, particularly for the extended variant $\mathcal{AR}_0$ which was recently proposed by Kartha and Lifschitz in [37]. This extended language addresses non-deterministic actions and limited cases of ramification. The effects of non-deterministic actions are defined using an additional operator in the language which is called **releases**, and which is the same as occlusion. The semantics of ramification is defined using minimal change in unoccluded (unreleased) features. The semantics of $\mathcal{AR}_0$ is therefore quite similar to the entailment method CMOC.

From my point of view, the construct of occlusion is not sufficiently self-evident to simply be postulated. It is a formal technique that should be subjected to assessment according to some underlying notions, and whose range of applicability is not necessarily universal. The principle of minimal change during update is another such technique which one may or may not choose to postulate. It is usually taken for granted, in the present KR research literature on ramification, but I would actually prefer to treat it as a derived technique as well. For these reasons, I prefer to view the $\mathcal{A}$-based languages as comparable with entailment methods. The methods for translating them to circumscription and to extended logic programming make good sense in this context.

11.9 Interpretation of assessment results

11.9.1 General remarks

We now return to Table 11.1 which summarizes the assessment results. Some observations strike one immediately when reading this table. Except for the OCM method (original chronological minimization), only two epistemological properties are being used, namely "p" for initial-observation chronicles and "r" for consistency-retaining chronicles. It was proved above that "p" is a strictly stronger requirement than "r" for equidurational chronicles, and of course the "sp" requirement that is used by OCM (initial-observation, initial state completely determined) is even stronger. There are three pairs consisting of methods without and with filtering, and the filtering removes the epistemological restriction in all three cases. This does not mean that filtering is always necessary, since one can easily imagine cases where the restrictions are satisfied.

The ontological requirements are **IAe** (equidurational change) and the various requirements for compatibility, favorability, and perfection. **IAe** is a quite strong requirement. It is trivially satisfied if all actions take a single timestep, according to the chosen representation, and for chronicles in $\mathcal{K}$-**I**. It is difficult to believe that it will be satisfied in any other practical cases. Methods which are constrained by the **IAe** requirement are not suitable, therefore, when one needs to reason about intermediate steps during context-dependent actions.

11.9.2 The compatibility requirements

Requirements of compatibility, favorability, and perfection have been introduced as we went along with the assessment of the various methods. The following are the inclusion relations which have been obtained beetween those families.

[PGM*] $\subseteq$ [PGM$^+$]
[OCM$^+$] $\subseteq$ [OCM]
[GMON*] $\subseteq$ [GMON$^+$]
[OCM] $\subseteq$ [PCM]   (Proposition 9.12)
[PGM*] $\subseteq$ [OCM$^+$]   (Proposition 9.15)
[CAMC] $\subseteq$ [PCM]   (Proposition 10.7)
IAt $\subseteq$ [OCM]
IsAn $\subseteq$ [CAMC]   (Proposition 10.8)
IAt $\subseteq$ [CAMC]   (Proposition 10.8)

IAu ⊆ [GMON*]   (Subsection 11.6.3)

IsAd = IsAt   (From definitions)

A few other relationships which have also been specified in the text are omitted here because they follow from the ones listed. Figure 11.1 illustrates these relationships in graphical form, and includes also those additional families for single-timestep actions that are introduced in the following subsection.

11.9.3 The special case of single-step actions

It may be interesting to consider the specialization of Table 11.1 to the case of single-timestep actions. The restrictions are then a bit easier to grasp, and single-timestep actions have been extensively studied in the form of situation calculus. The derived assessments for this special case are summarized in Table 11.2, using the following simplified ontological properties. Notice, for the definition, that if r is a state and r' is a partial state, then $r' - r$ is the set of feature-value pairs in r' where it differs from r, since states are represented as sets of feature-value pairs (maplets). Therefore the domain of $r' - r$ is the set of those features for which r' differs from r.

Definition. *Let* ⟨Infl, Trajs⟩ *be a trajectory-semantics world in* **IsA**. *This world is said to be* **minimal-change-compatible** *iff the domain of* $r' - r$ *is never a strict subset of the domain of* $r'' - r$, *when* ⟨r'⟩ *and* ⟨r''⟩ *are two members of* Trajs(E, r). *The world is said to be* **minimal-change-favorable** *iff the domain of* $r' - r$ *is never a strict subset of the domain of* $\bar{r}' - \bar{r}$, *when* ⟨r'⟩ *is a member of* Trajs(E, r) *and* ⟨$\bar{r}'$⟩ *is a member of* Trajs$(E, \bar{r})$. *The world is said to be* **minimal-change-perfect** *iff the domain of* $r' - r$ *equals the domain of* $\bar{r}' - \bar{r}$, *when* ⟨r'⟩ *is a member of* Trajs(E, r) *and* ⟨$\bar{r}'$⟩ *is a member of* Trajs$(E, \bar{r})$. *The world is said to be* **minimal-occlusion-favorable** *iff* Infl$(E, r) \not\subseteq$ Infl$(E, \bar{r})$. *The world is said to be* **minimal-occlusion-perfect** *iff* Infl$(E, r) = $ Infl$(E, \bar{r})$. *In all cases this must apply for arbitrary choices of* E, r, *and* $\bar{r}$. ⋈

The ontological properties defined here will be denoted as [MC], [MC⁺], [MC*], [MO⁺], and [MO*] in the obvious fashion. The assessments follow easily and are summarized in the table. We see how a major part of the restrictions can be expressed in terms of the relatively simple criteria that were just defined. It follows that **IsAun** is equivalent to [MC*] (every world in one is equivalent to some world in the other), which is indicated by the double line in Figure 11.1.

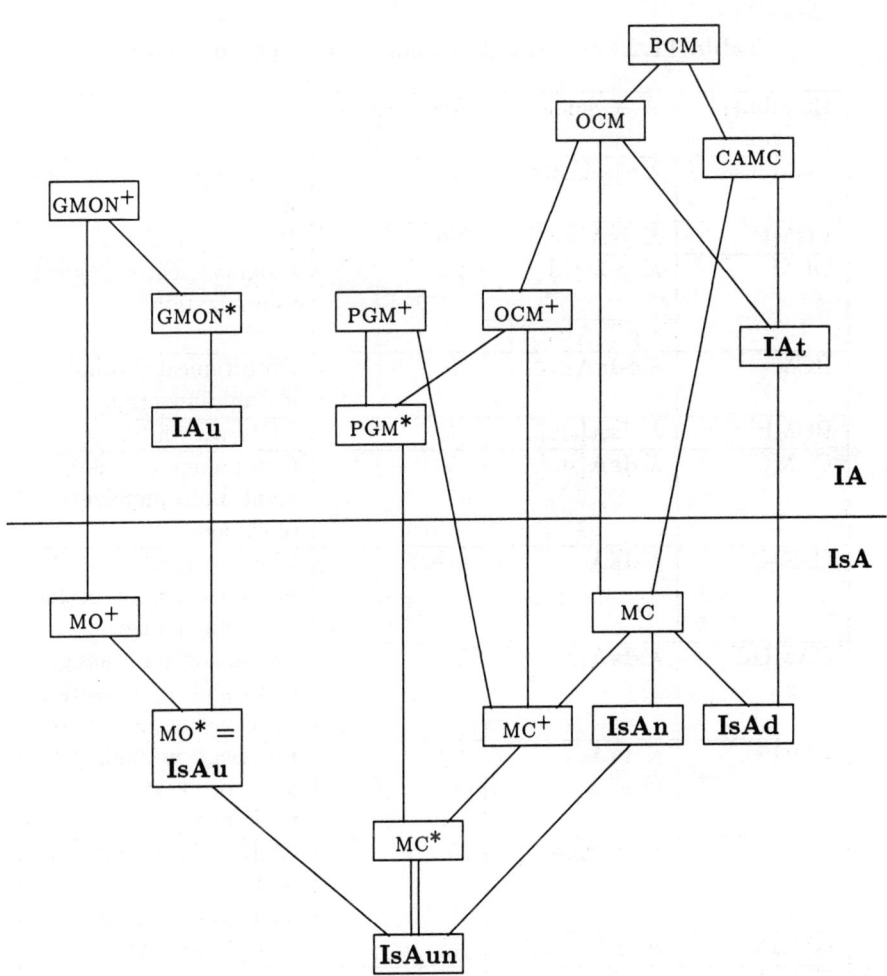

Fig. 11.1. Inclusion diagram for trajectory families

Table 11.2. Assessment summary for single-timestep case

Entailment criterion	Assessment	Req execut schedule?	Full name
PGM	$\mathcal{K}$r-**IsA**[MC*]	No	Prototypical global minimization
PGMF	$\mathcal{K}$-**IsA**[MC*]	No	PGM with filtering
OCM	$\mathcal{K}$sp-**IsAd**	No	Original chronological minimization
	$\mathcal{K}$p-**IsA**[MC$^+$]		
PCM	$\mathcal{K}$p-**IsA**[MC]	No	Prototypical chronological minimization
PCMF	$\mathcal{K}$-**IsA**[MC]	No	PCM with filtering
CAMC	$\mathcal{K}$-**IsA**[MC]	Yes	Chronological assignment and minimization of change
CMOC	$\mathcal{K}$-**IsA**	(Yes)	Chronological minimization of occlusion and change
CAMOC	$\mathcal{K}$-**IsA**	Yes	Chronological assignment and minimization of occlusion and change
PMON	$\mathcal{K}$-**IsA**	No	Pointwise minimization of occlusion with nochange premises
GMON	$\mathcal{K}$r-**IsA**[MO*]	No	Global minimization of occlusion with nochange premises
GMONF	$\mathcal{K}$-**IsA**[MO$^+$]	No	GMON with filtering
TMOC	$\mathcal{K}$r-**IsA**	No	Two-stage minimization of occlusion and change

12
Composite actions

Chapter 6 defined an occurrence language which only allows elementary occurrence expressions, so that each action statement refers to a single action with its timing information and choice of arguments. An eventual generalization to the case of composite action statements was mentioned, and it is the topic of the present chapter. We shall define the language for composite actions, and extend the entailment methods and their assessments to the broader range of action statements.

12.1 Composite actions in schedules

We address the case where composite schedule statements are formed using the classical operations of sequencing, conditional expressions, and loops. Notice that we only consider the case where it is the schedule that contains such composite expressions. The case where the action law for one action is expressed by composition of other actions is a related but possibly a more complex topic.

The distinction between epistemological and ontological characteristics is difficult to apply in the present case. The addition of composite action statements in the schedule is in principle an epistemological one, since it only affects the description of the ego-world game and not the game itself. The $\mathcal{A}$ component of IDS developments will still only contain elementary actions, although one action statement in the schedule may correspond to several members of $\mathcal{A}$. However, if one allows the action law for one action to be expressed in terms of other and more elementary actions, then one has arguably crossed over to an ontological issue. Furthermore, the technique that will be used for analyzing entailment from composite action statements actually integrates the epistemological and the ontological aspects. For these reasons I refrain from assigning any characteristics or subcharacteristics to the case of composite action statements, and continue to refer to it by name.

12.1.1 Syntax extension

The extended syntax is introduced by a modification of the definition of occurrence expressions in Chapter 6, subsection 6.5.2. Previously defined as only being an elementary occurrence expression, it is now generalized so that an occurrence expression can be formed recursively as well. We first define a *condition* as a quantifier-free logic formula in LFL. This means that it can be formed from formulae of the form $f \hat{=} \mathcal{X}$ and $\omega = \omega'$ by composition using logical connectives but not using quantifiers, and provided that f, ω, and ω' may contain object constants and object names, but no variables. A *rigid condition* is a condition which additionally does not contain any constant symbols (only object names).

An *occurrence expression* is defined recursively to be either an elementary occurrence expression, or a *sequential expression* of the form

$\varepsilon_1; \varepsilon_2,$

or a *conditional expression* of the form

if φ then ε_1 else $\varepsilon_2,$

or finally a *loop expression* of the form

do ε until $\varphi,$

where ε_1, ε_2, and ε are occurrence expressions, and φ is a condition. The latter three kinds of expressions will jointly be called *composite occurrence expressions*. It is assumed that parentheses are used to resolve any syntactic ambiguity.

As the range of occurrence expressions is extended in this way, the range of actions statements, schedules, and other syntactic constructs which build on the definition of occurrence expressions is extended automatically as well. The wordings in the definitions of those higher level constructs do not require any changes.

For the remainder of this section (but not after that) we assume that the extended syntax is used unless otherwise noted.

12.1.2 Intended models

The definition of intended models $\Sigma_{\mathbf{IA}}(\Upsilon)$ must be extended to the case where the schedule component of Υ contains composite action statements. The definition of *Mod* is the essential one, whereas the definition of $\Sigma_{\mathbf{IA}}(\Upsilon)$ in terms of *Mod* can remain as it is.

The value $val[\varphi, s]$ of a logic formula in LFL was defined in Chapter 5, subsection 5.2.4, for the case where s is a structure $\langle \mathcal{O}, \langle M, r \rangle \rangle$. In the special case where φ is a rigid condition the value does not depend on M (since there are no constant symbols in φ) nor on $\mathcal{O}$ since there is no quantification over objects. In such cases, we use the notation $val[\varphi, r]$ to stand for $val[\varphi, \langle \mathcal{O}, \langle M, r \rangle \rangle]$ for arbitrary M, and arbitrary $\mathcal{O}$ containing all the object names that occur in φ. Thus $val[\varphi, r]$ is simply the value of the formula φ in the state r by evaluation as usual.

The definition in Chapter 8, subsection 8.3.2, defined $Mod(\Upsilon)$ as a set of developments $\langle \mathcal{B}, M, R, \mathcal{A}, \mathcal{C}\rangle$, where one of the requirements was that $M[\text{SCD}] = \mathcal{A}$. This requirement must now be changed, since any composite occurrence expression in an action statement in SCD will be retained in $M[\text{SCD}]$. All constant symbols will have been removed, of course, but conditional and loop expressions can not be simplified without recourse to the history component R of an interpretation.

In the revised definition, each statement in $M[\text{SCD}]$ corresponds to a set of contiguous actions in $\mathcal{A}$. Such a set will be called an agglomeration. The following definitions are used.

Definition. *A nonempty set of occurrences*
$$\{\langle s_1, E_1, t_1\rangle, \langle s_2, E_2, t_2\rangle, ..., \langle s_k, E_k, t_k\rangle\}$$
is called an **agglomeration over** $[s_1, t_k]$ *iff* $t_{i-1} = s_i$ *for* $i = 2, ..., k$ *(with some ordering of the set).* ⋈

Definition. *An agglomeration G over $[s, t]$ is said to* **realize** *a rigid occurrence expression ε* **in** *a history R iff either of the following cases holds:*

- *ε is an elementary occurrence expression and $G = \{\langle s, \varepsilon, t\rangle\}$.*

- *$\varepsilon \equiv \varepsilon_1; \varepsilon_2$ and $G = G_1 \cup G_2$ where G_1 and G_2 are disjoint sets, G_1 is an agglomeration over $[s, u]$ which realizes ε_1 in R, and G_2 is an agglomeration over $[u, t]$ which realizes ε_2 in R, for some u such that $s < u < t$.*

- *$\varepsilon \equiv$ if φ then ε_1 else ε_2, and either $val[\varphi, R(s)] = \text{T}$ and G realizes ε_1 in R, or $val[\varphi, R(s)] = \text{F}$ and G realizes ε_2 in R.*

- *$\varepsilon \equiv$ do ε_1 until φ, and G can be written as the union of one or more disjoint sets (but a finite number of them),*
  $$G = G_1 \cup G_2 \cup ... \cup G_k,$$
  where each G_i is an agglomeration over $[s_i, t_i]$ and $t_{i-1} = s_i$, each G_i realizes ε_1 in R, $val[\varphi, R(t_i)] = \text{F}$ for $i = 1, 2, ..., k-1$, and $val[\varphi, R(t_k)] = \text{T}$. ⋈

The definition for the loop expression requires the body of the loop, ε_1, to be performed at least once. This is necessary since we do not allow actions with zero duration. The effects of a 'while' expression where the test is done before commencing the first cycle can be obtained using a combination of conditional and loop expressions. However, the rewrite is a bit clumsy in the general case.

After these preliminaries, the definition of *Mod* is modified with respect to the requirement that $M[\text{SCD}] = \mathcal{A}$. The revised definition reads in its entirety:

Definition. *Let a similarity type σ and an occurrence vocabulary π be given, and let $\Upsilon = \langle \mathcal{O}, \Pi, \text{SCD}, \text{OBS} \rangle$ be a chronicle in $\mathcal{K}$-**IA** over σ and π. The* **model set** *$Mod(\Upsilon)$ is defined as the set of all complete developments over $\mathcal{O}$, σ, and π,*

$$\langle \mathcal{B}, M, R, \mathcal{A}, \mathcal{C} \rangle$$

which result from a game between $\mathbf{W}_\Upsilon$ and some trajectory-semantics ego, with arbitrary r_0 and M_0, where $\langle M, R \rangle \in [\![\text{OBS}]\!]$, and where $M[\text{SCD}]$ does not contain F and corresponds to $\mathcal{A}$ in the following way. There must exist a partitioning of $\mathcal{A}$ into disjoint subsets, and a one-to-one correspondence between those subsets and the members of $M[\text{SCD}]$, such that for each $[s_i, t_i]\varepsilon_i \in M[\text{SCD}]$, the corresponding subset is an agglomeration over $[s_i, t_i]$ which realizes ε_i in R. ⋈

For the purpose of $\text{At}(M, \text{SCD})$ it is only the starting time and the ending time of each member of $M[\text{SCD}]$ that is used, not the starting time and ending time of the constituent elementary actions. For example, if $M[\text{SCD}]$ contains the statement ⌜$[5, 12]A_2; A_6$⌝ then 5 and 12 are members of $\text{At}(M, \text{SCD})$ but the intervening numbers are not. In this way, the definition of Mod^t continues to work without problems; one must only replace the condition $M[\text{SCD}] = \mathcal{A}$ in the same way as in the definition of *Mod*.

12.1.3 Direct translation of composite action statements

The approach of the present volume has been to consider action statements as a side language, which is translated into the main language of the logic using the translation function Π, mapping an elementary action statement $[s, t]\varepsilon$ to a corresponding logic formula in DFL-1 or DFL-2. That translation function must now be extended to the case where the action statement is a composite one.

We shall consider two such translations, where the first one is simple and straightforward, but does not correspond so well to the underlying semantics in terms of trajectories and world descriptions. The second translation remedies that problem, at the expense of simplicity.

Definition. *If Π is a translation function for elementary action statements, then the extended translation function Π^k ('k' for 'composite'; 'c' is being used for 'concurrent) is defined as follows:*

1. If ε is an elementary occurrence expression, then $\Pi^k[[\varsigma, \tau]\varepsilon] = \Pi[[\varsigma, \tau]\varepsilon]$.

2. $\Pi^k[[\varsigma, \tau] \text{if } \varphi \text{ then } \varepsilon_1 \text{ else } \varepsilon_2] =$

3. $\Pi^k[[\varsigma,\tau]\varepsilon_1] = \ulcorner[\varsigma]\varphi \to \underline{\Pi}^k[[\varsigma,\tau]\varepsilon_1] \land \neg[\varsigma]\varphi \to \underline{\Pi}^k[[\varsigma,\tau]\varepsilon_2]\urcorner$.
3. $\Pi^k[[\varsigma,\tau]\varepsilon_1;\varepsilon_2] = \ulcorner\exists u[\underline{\Pi}^k[[\varsigma,u]\varepsilon_1] \land \underline{\Pi}^k[[u,\tau]\varepsilon_2]]\urcorner$.
4. $\Pi^k[[\varsigma,\tau]\,\text{do}\,\varepsilon\,\text{until}\,\varphi] = \ulcorner\exists u[\underline{\Pi}^k[[\varsigma,u]\varepsilon] \land \underline{esig}[\varsigma,u,\tau]] \land$
$\forall u[\varsigma < u \leq \tau \land [u]p \to$
$([u]\varphi \to u = \tau) \land$
$(\neg[u]\varphi \to \exists u'[\underline{\Pi}^k[[u,u']\varepsilon] \land \underline{esig}[u,u',\tau]])]\urcorner$,

where p is a propositional feature symbol without arguments called a **signal feature** which is not used for any other purpose, the temporal variables u and u' are assumed chosen so that they do not occur in the expressions ς or τ, and

$esig[\varsigma,v,\tau] = \ulcorner\varsigma < v \leq \tau \land [\varsigma,v]p := \mathsf{T} \land (\varsigma,v)\neg p\urcorner$. ⋈

The first three items should be transparent. Recall that the underlining of a function symbol inside Quine quotes means that the function is to be applied when the expression is constructed. For example if $\phi_1 = \Pi^k[[s,u]\varepsilon_1]$ and $\phi_2 = \Pi^k[[u,t]\varepsilon_2]$, then $\Pi^k[[s,t]\varepsilon_1;\varepsilon_2] = \ulcorner\exists u[\phi_1 \land \phi_2]\urcorner$. The translation of loop statements in the fourth item uses the propositional feature p to indicate that a cycle of the loop has been concluded. The existentially quantified subexpression ensures that the body of the loop statement is executed at least once, and that p is false while the body is being executed but true at the end of the cycle. The universally quantified subexpression ensures that whenever a cycle of the loop has been concluded, either φ is false, the ending time t of the composite statement has not been reached, and the loop is repeated (with the same effect on p as in the first cycle), or φ is true and the ending time t has been reached.

The phrase 'not used for any other purpose' refers at least to the action statement being translated. If that statement is presented as part of a schedule or chronicle, the restriction refers to its entirety. In particular, a feature that occurs in the observation partition of a chronicle Υ must not be used as a signal feature for the schedule of Υ. Different loop expressions that occur in sequence may use the same signal feature p, but nested loop expressions must evidently use different signal features. $\Pi^k(\text{SCD})$ is defined in a similar way as for $\Pi(\text{SCD})$.

These translations correspond in an informal fashion to the intended meanings of the composite action statements. Using them, it is well defined how one obtains the classical model set $[\![\Pi^k(\text{SCD})]\!]$ for a given trajectory. Therefore, we can define $[\![\Upsilon]\!]^k$ as follows:

Definition. *The classical model set $[\![\langle\mathcal{O},\Pi,\text{SCD},\text{OBS}\rangle]\!]^k$ for a chronicle in $\mathcal{K}$-**IA** is the set of models obtained from*

$[\![\Pi^k(\text{SCD}) \cup \text{OBS}]\!]_{\mathcal{O}}$

by removing, throughout the history component of each member model, all signal features that were introduced in the reformulation of the loop statements. ⋈

It follows at once that
$$[\langle \mathcal{O}, \Pi, \text{SCD}, \text{OBS}\rangle]^k = [\langle \mathcal{O}, \Pi, \text{SCD}, \emptyset\rangle]^k \cap [\text{OBS}],$$
and that $[\Upsilon]^k = [\Upsilon]$ if all action statements in Υ are elementary. We also obtain the following generalization of Proposition 8.5.

Proposition 12.1. *If* $\Upsilon \in \mathcal{K}\text{-}\mathbf{IA}$ *then* $\Sigma_{\mathbf{IA}}(\Upsilon) \subseteq [\Upsilon]^k$.

Proof Let $\Upsilon = \langle \mathcal{O}, \Pi, \text{SCD}, \text{OBS}\rangle \in \mathcal{K}\text{-}\mathbf{IA}$, let $\langle M, R\rangle \in \Sigma_{\mathbf{IA}}(\Upsilon)$, and let $J = \langle \mathcal{B}, M, R, \mathcal{A}, \mathcal{C}\rangle$ be a corresponding member of $Mod(\Upsilon)$. Make a choice of signal features for SCD, and construct $\Pi^k(\text{SCD})$ accordingly. Construct $J' = \langle \mathcal{B}, M, R', \mathcal{A}, \mathcal{C}\rangle$ from J by introducing values for signal features using the agglomeration of J for Υ. It is straightforward that all members of $[\Pi^k(\text{SCD}) \cup \text{OBS}]_\mathcal{O}$ are satisfied in J', and the result follows. □

With these definitions, it is well defined how to use the various entailment methods which have been defined and assessed in earlier chapters. However, at this point we have no guarantee that an entailment method which is correct for chronicles with elementary action statements continues to be so with respect to the extended definition of intended models in the previous subsection.

12.1.4 World descriptions for composite action statements

In order to analyze the correctness of entailment methods for composite action statements, the following approach will be used. A $\mathcal{K}\text{-}\mathbf{IA}$ chronicle Υ whose schedule SCD contains composite action statements will be related to a modified chronicle $\mathbf{G}_{agr}^{kw}(\Upsilon)$ where all action statements are elementary, and which is constructed as follows using *aggregation*. Each composite action statement $[s,t]\varepsilon$ is replaced by an elementary action statement $[s,t]A'(\omega_1, ..., \omega_k)$, where A' is an action symbol that is not used elsewhere in the schedule of $\mathbf{G}_{agr}^{kw}(\Upsilon)$. It will be called the *aggregated* action symbol. Similarly the set of action laws in $\mathbf{G}_{agr}^{kw}(\Upsilon)$ is obtained from the set of action laws in Υ by adding definitions for the aggregated action symbols. Every such action symbol A' is defined by the FTNF corresponding to the trajectory set $\lambda r.\text{Trajs}(\varepsilon, r)$. The superindex kw will indicate how the action law for the aggregated action symbol is obtained.

The full definition must deal correctly with the possible arguments of the aggregated action symbol, but first of all it requires that the functions Infl and Trajs in a world description are generalized so that they are also defined with composite occurrence expressions as their first arguments.

We shall show that $\Upsilon \cong_{\text{RM}} \Upsilon'$, or in other words that $Mod(\Upsilon, \text{RM}) = Mod(\Upsilon', \text{RM})$, where $\Upsilon' = \mathbf{G}_{agr}^{kw}(\Upsilon)$. This means that after one has removed the information about the actual actions that have been executed, and as one only retains the valuation and the history, the transformation from Υ

to Υ' does not change the set of intended models. It is not realistic to actually construct and use the FTNF action law for the new action symbols A' in the specific cases. However, if it can be shown that the translation Π^k for a composite action statement $[s,t]\varepsilon$ is classically equivalent to the translation by Π' (the action laws in Υ') of $[s,t]A'(\omega_1,...,\omega_k)$, then everything would fit together correctly. Unfortunately, this is only the case for conditional statements, and not for the other two types. Therefore, we define a modified translation function Π^{kw} (for composite statements, with widening) which works for all cases.

In order to carry out this strategy, the first step is to generalize the two components of a world description $\langle \text{Infl}, \text{Trajs} \rangle$. We define some auxiliary functions. The expression $r \dagger F$ is defined as the partial state which is obtained as the restriction of the state r to those features which are members of the set F. We define
$$widen(\langle r'_1, r'_2, ..., r'_k \rangle, r) = \langle r \oplus r'_1, r \oplus r'_2, ..., r \oplus r'_k \rangle,$$
where r is a partial state. Finally, an expression $v \triangleright v'$, where v and v' are trajectories, is defined by noticing that the existing definition of $R \triangleright v$ (subsection 3.2.4) is equally applicable if the first argument is a trajectory, provided only that the elements of v are defined for a set of features that includes ($\supseteq$) those where the elements of v' are defined. In other words, $v \triangleright v'$ equals the concatenation of v and $widen(v', last(v))$. The generalized world description is then defined as follows.

Definition. *Let $\langle \text{Infl}, \text{Trajs} \rangle$ be a world description for a $\mathcal{K}$-**IA** world. Its two components are generalized as follows to the use of composite action expressions as the first argument.*

$\text{Infl}(\ulcorner \text{if } \varphi \text{ then } \varepsilon_1 \text{ else } \varepsilon_2 \urcorner, r) = \text{Infl}(\text{if } val[\varphi, r] = \text{T then } \varepsilon_1 \text{ else } \varepsilon_2, r)$

$\text{Trajs}(\ulcorner \text{if } \varphi \text{ then } \varepsilon_1 \text{ else } \varepsilon_2 \urcorner, r) = \text{Trajs}(\text{if } val[\varphi, r] = \text{T then } \varepsilon_1 \text{ else } \varepsilon_2, r)$

$\text{Infl}(\ulcorner \varepsilon_1; \varepsilon_2 \urcorner, r) = \text{Infl}(\varepsilon_1, r) \cup \bigcup [\text{Infl}(\varepsilon_2, r \oplus last(v)) \mid v \in \text{Trajs}(\varepsilon_1, r)]$

$\text{Trajs}(\ulcorner \varepsilon_1; \varepsilon_2 \urcorner, r) = \{widen(v_1, r \dagger \text{Infl}(\ulcorner \varepsilon_1; \varepsilon_2 \urcorner, r)) \triangleright v_2 \mid$
$\quad v_1 \in \text{Trajs}(\varepsilon_1, r) \wedge v_2 \in \text{Trajs}(\varepsilon_2, r \oplus last(v_1))\}$

$\text{Infl}(\ulcorner \text{do } \varepsilon \text{ until } \varphi \urcorner, r) = \bigcup [\text{Infl}(\varepsilon, r') \mid r' \in reast(\varepsilon, r, \varphi)]$

$\text{Trajs}(\ulcorner \text{do } \varepsilon \text{ until } \varphi \urcorner, r) =$
$\quad \{widen(v_1, r_0 \dagger F) \triangleright widen(v_2, r_1 \dagger F) \triangleright ..., widen(v_k, r_{k-1} \dagger F) \mid$
$\quad r_0 = r \wedge F = \text{Infl}(\ulcorner \text{do } \varepsilon \text{ until } \varphi \urcorner, r) \wedge$
$\quad v_i \in \text{Trajs}(\varepsilon, r_{i-1}) \wedge r_i = r_{i-1} \oplus last(v_i) \wedge val[\varphi, r_k] = \text{T} \wedge$
$\quad (1 \le i < k \to val[\varphi, r_i] = \text{F})\},$

where $reast(\varepsilon, r, \varphi)$ is the set of states that are reachable from r using trajectories in $\text{Trajs}(\varepsilon)$ while not passing through a state where φ is true. It is defined as the smallest set Q of states such that $r \in Q$ and
$r' \in Q \wedge v \in \text{Trajs}(\varepsilon, r') \wedge r'' = r' \oplus last(v) \wedge val[\varphi, r''] = \text{F} \to$

$r'' \in Q$.

When it is necessary to distinguish the present generalization from others, it will be designated as $\langle \texttt{Infl}^{kw}, \texttt{Trajs}^{kw} \rangle$. ⋈

The part of the definition for conditional statements is straightforward. For sequence and loop statements the complexity of the definition is due to the requirement to have a single set $\texttt{Infl}(\varepsilon, r)$ of occluded features, and for all elements of a trajectory to be defined throughout $\texttt{Infl}(\varepsilon, r)$. That is why it is not possible to simply concatenate the trajectories in these two cases; they must first be 'widened' so that all elements of the trajectory are defined over the same set of features.

Actually, there are also other ways of defining $\langle \texttt{Infl}, \texttt{Trajs} \rangle$ which obtain widening-equivalent but wider $\texttt{Trajs}(\varepsilon, r)$. Each such definition corresponds to its own variant of Π.

12.1.5 Aggregation

The reformulation $\mathbf{G}_{agr}^{kw}$ extends only the occurrence vocabulary, but not the feature vocabulary or the set of constant symbols. Therefore, it can be formulated as a pretransformation function as introduced in Chapter 9, subsection 9.1.1.

Definition. *If $\Upsilon = \langle \mathcal{O}, \Pi, \textsc{scd}, \textsc{obs} \rangle$ is a chronicle in $\mathcal{K}$-IA using the occurrence vocabulary π, then the **aggregation** of Υ is written $\mathbf{G}_{agr}^{kw}(\Upsilon)$ and is a modified chronicle $\langle \mathcal{O}, \Pi', \textsc{scd}', \textsc{obs} \rangle$ over an extended occurrence vocabulary $\pi' \supseteq \pi$ which is obtained as follows. It is assumed that Π corresponds to the world description $\langle \texttt{Infl}, \texttt{Trajs} \rangle$, and extensions $\texttt{Infl}'$ and $\texttt{Trajs}'$ are part of the construction. For each nonelementary action statement $[s, t]\varepsilon$ in $\textsc{scd}$,*

1. Introduce a previously unused action symbol A' to be used for the corresponding aggregated action.

2. Let $\tau_1, \tau_2, ..., \tau_k$ be an ordering of all the object expressions that occur in ε.

3. Corresponding to $[s, t]\varepsilon$, construct a contribution to $\textsc{scd}'$ of the form $[s, t]A'(\tau_1, \tau_2, ..., \tau_k)$.

4. Let $\langle o_1, o_2, ..., o_k \rangle$ be a sequence of k objects (members of Ω), and construct $\varepsilon' = \varepsilon_{\tau_i}^{o_i}$, that is, the modified expression where each subexpression τ_i has been replaced by the corresponding object o_i. Notice that it is not required for all the o_i to be different.

5. Define $\texttt{Infl}'$ and $\texttt{Trajs}'$ for the aggregated action symbol A'. The definitions are chosen so that

$\texttt{Infl}[A'(o_1, ..., o_k), r] = \texttt{Infl}^{kw}(\varepsilon', r),$

$\texttt{Trajs}[A'(o_1, ..., o_k), r] = \texttt{Trajs}^{kw}(\varepsilon', r),$

for every r and for every possible choice of ε' in item 4.

6. *Construct the FTNF contribution to Π' for A' from the contributions to $\langle \mathtt{Infl'}, \mathtt{Trajs'} \rangle$ for A'.* ⋈

It is easily verified that if $\langle \mathtt{Infl}, \mathtt{Trajs} \rangle$ satisfies the restrictions of **IA** compared to **IAD**, then the extension $\langle \mathtt{Infl'}, \mathtt{Trajs'} \rangle$ does so as well. It is also verified that if each loop statement in the schedule possibly terminates for each choice of starting state r, then the extension satisfies the restriction for **IAD**. We obtain

Proposition 12.2. *If $\Upsilon \in \mathcal{K}\text{-}\mathbf{IA}$ then $\Sigma_{\mathbf{IA}}(\Upsilon) = \Sigma_{\mathbf{IA}}(\mathbf{G}_{agr}^{kw}(\Upsilon))$.*

Remember that $\Sigma_{\mathbf{IA}}(\Upsilon) = Mod(\Upsilon, \mathrm{RM})$, and notice that the extension of π does not matter after the reduction of models to their R and M components.

Proof (Outline). Let $\Upsilon = \langle \mathcal{O}, \Pi, \mathrm{SCD}, \mathrm{OBS} \rangle$. It is clearly sufficient to prove this result for the case where SCD consists of one single action statement, and the observation set is empty. Let $[\varsigma, \tau]\varepsilon$ be the action statement in SCD, and consider some member $\langle M, R \rangle$ of $Mod(\Upsilon, \mathrm{RM})$. It corresponds to some $J = \langle \mathcal{B}, M, R, \mathcal{A}, \mathcal{C} \rangle$ of $Mod(\Upsilon)$. $\mathcal{A}$ is an agglomeration, according to the extended definition of Mod. Consider the sequence of trajectories for the members of $\mathcal{A}$ that were used in the game resulting in J. Widen them to $\mathtt{Infl}(M[\varepsilon], R(M[\varsigma]))$, and concatenate them. It is easily seen that their concatenation is widening-equivalent to a member of $\mathtt{Trajs}(M[\varepsilon], R(M[\varsigma]))$. Conversely, each member of $\mathtt{Trajs}(M[\varepsilon], R(M[\varsigma]))$ can be decomposed into a sequence of subtrajectories, each of which is widening-equivalent to some member of the trajectory set for the corresponding action in $\mathcal{A}$. □

In this way the connection between intended models and the trajectory semantics has been extended up to the level of composite action statements. Composite action statements can be safely analyzed in terms of their trajectory representation, therefore.

12.1.6 Widened translation of composite action statements

The construction of $\mathtt{Trajs}(\varepsilon, r)$ for composite ε encountered the need for widening the constituent trajectories in order to accomplish the necessary uniformity. This is a technical requirement in the trajectory semantics, and one might consider a modified semantics where the set of features that are defined in a trajectory may vary between its elements. However, some of the entailment methods would be quick to exclude the use of such a generalization.

Given that the trajectory semantics is defined in its present way, it is natural to define the extension of the translation function Π in the same manner. For simplicity we consider only the case of actions without object arguments. The definition of FTNF in Chapter 8 was as follows, using an assumed function $\psi(\varsigma, \tau, V)$ where ς and τ are timepoint expressions characterizing the starting time and the ending time of an action described by the set of trajectories V.

$resl(V, r) = \{v \mid v \in V \wedge last(v) = r\}$
$assign(r) \equiv \bigwedge[(f_i \doteq x_i) \mid (f_i : x_i) \in r]$
$change[r', \varsigma, \tau, V] \equiv$
$\quad ([\varsigma, \tau] \bigwedge[f_i := x_i \mid (f_i : x_i) \in r']) \wedge \psi(\varsigma, \tau, resl(V, r'))$
$pt[\varepsilon, r, \varsigma, \tau] \equiv \bigvee[change[r', \varsigma, \tau, \text{Trajs}(\varepsilon, r)] \mid r' \in last(\text{Trajs}(\varepsilon, r))]$
$\Pi[[\varsigma, \tau]\varepsilon] \equiv \bigwedge[[\varsigma]assign(r) \rightarrow pt[\varepsilon, r, \varsigma, \tau] \mid r \in \mathcal{R}_0]$.

The revised definition retains the first of these functions, but introduces modified versions of *change* and *pt*. The essence of the modification is to introduce a feature set F as an additional argument, which is used for forcing each feature in F to be occluded and specified by the action law. For example if the two actions E_1 and E_2 influence different sets of features for a starting state r, then $\text{Infl}(\ulcorner E_1; E_2\urcorner, r)$ is the the union of those sets for E_1 starting in r and for E_2 in all states where E_1 can end if it starts in r. The corresponding action law is set up so that every feature in $\text{Infl}(\ulcorner E_1; E_2\urcorner, r) - \text{Infl}(E_1, r)$ is held fixed over the duration of E_1, and similarly for E_2. The F argument in the revised definition accomplishes this. The revised auxiliary functions are as follows. The function symbols Infl and Trajs refer to Infl^{kw} and Trajs^{kw}.

$chaw[r, r', \varsigma, \tau, V, F] \equiv$
$\quad change[r', \varsigma, \tau, V] \wedge \bigwedge[(\varsigma, \tau]\mathbf{X}f \wedge f \doteq \underline{r}(f) \mid f \in F]$
$pw[\varepsilon, r, \varsigma, \tau, F] \equiv$
$\quad \bigvee[chaw[r, r', \varsigma, \tau, \text{Trajs}(\varepsilon, r), nin(F, \varepsilon, r)] \mid r' \in last(\text{Trajs}(\varepsilon, r))]$
where ε is an elementary occurrence expression, and $nin(F, \varepsilon, r) = F - \text{Infl}(\varepsilon, r)$;
$atr[[\varsigma, \tau]\varepsilon, Q, H] \equiv \bigwedge[[\varsigma]assign(r) \rightarrow pw[\varepsilon, r, \varsigma, \tau, H(r)] \mid r \in Q]$
$\Pi^{kw}[[\varsigma, \tau]\varepsilon] \equiv atr[[\varsigma, \tau]\varepsilon, \mathcal{R}_0, \lambda r.\text{Infl}(\varepsilon, r)]$.

Here H is a function from states to sets of features. It is easily seen that if ε is an elementary occurrence expression then the last argument of *chaw* equals the empty set, and $\Pi^{kw}[[\varsigma, \tau]\varepsilon] = \Pi[[\varsigma, \tau]\varepsilon]$. The definitions for composite action statements are obtained by complementary definitions for *pw*, as follows.

$pw(\ulcorner\text{if }\varphi\text{ then }\varepsilon_1\text{ else }\varepsilon_2\urcorner, r, \varsigma, \tau, F) =$
$\quad pw(\text{if }val[\varphi, r] = \mathbf{T}\text{ then }\varepsilon_1\text{ else }\varepsilon_2, r, \varsigma, \tau, F)$
$pw(\ulcorner\varepsilon_1; \varepsilon_2\urcorner, r, \varsigma, \tau, F) \equiv$
$\quad \exists u[\bigvee_{r'}(chaw[r, r', \varsigma, u, \text{Trajs}(\varepsilon_1, r), nin(F, \varepsilon_1, r)] \wedge$
$\quad\quad \bigvee_{r''} chaw[r', r'', u, \tau, \text{Trajs}(\varepsilon_2, r'), nin(F, \varepsilon_2, r')])]$

where r' and r'' range over $\mathcal{R}$. However, the component of $chaw[...]$ which is obtained from $\psi(...)$ will often be F, which in practice reduces the ranges of r' and r''.
$$pw(\ulcorner do\ \varepsilon\ until\ \varphi\urcorner, r, \varsigma, \tau, F) \equiv$$
$$\exists u[pw[\varepsilon, r, \varsigma, u, F] \wedge \underline{esig}[\varsigma, u, \tau]] \wedge$$
$$\forall u[\varsigma < u \leq \tau \wedge [u]p \rightarrow$$
$$([u]\varphi \rightarrow u = \tau) \wedge$$
$$(\neg[u]\varphi \rightarrow \exists u'[atr[[u, u']\varepsilon, reast(\varepsilon, r, \varphi), \lambda r.F] \wedge$$
$$\underline{esig}[u, u', \tau]])].$$

Definition. *If Π is a translation function for elementary action statements, then the extended translation function Π^{kw} ('kw' for 'composite with widening') is defined as*
$$\Pi^{kw}[[\varsigma, \tau]\varepsilon] \equiv atr[[\varsigma, \tau]\varepsilon, \mathcal{R}_0, \lambda r.\mathtt{Infl}^{kw}(\varepsilon, r)]$$
for the case where ε is an occurrence expression which may be composite, but where none of the actions in it has any object arguments. ⋈

The case of actions with object arguments is obtained by generalization in the same fashion as in Chapter 8, subsection 8.1.3. It should be clear how this definition has been obtained as a modification of the definition for Π^k. The size of the definition is only marginally larger, but it is probably harder to read because it uses lower level functions. The notation $[\![\Upsilon]\!]^{kw}$ is defined similar to $[\![\Upsilon]\!]^k$. We obtain:

Proposition 12.3. *If $\Upsilon = \langle \mathcal{O}, \Pi, \mathrm{SCD}, \mathrm{OBS}\rangle \in \mathcal{K}$-**IA**, and all composite action statements in SCD are formed using action symbols without arguments, then*
$$[\![\Upsilon]\!]^{kw} = [\![\mathbf{G}_{agr}^{kw}(\Upsilon)]\!].$$

Proof It is sufficient to prove
$$\Pi^{kw}[[\varsigma, \tau]\varepsilon] \simeq \Pi'[[\varsigma, \tau]E]$$
where E equals ε if the latter is an elementary action expression, and is the aggregated action symbol corresponding to ε if the latter is a composite action expression. Regardless of the structure of ε, we have
$$\Pi^{kw}[[\varsigma, \tau]\varepsilon] = atr[[\varsigma, \tau]\varepsilon, \mathcal{R}_0, \lambda r.\mathtt{Infl}(\varepsilon, r)],$$
which equals
$$\bigwedge\{[s]assign(r) \rightarrow pw[\varepsilon, r, \varsigma, \tau, \mathtt{Infl}(\varepsilon, r)] \mid r \in \mathcal{R}_0\}.$$
Also,
$$\Pi'[[\varsigma, \tau]E] = \bigwedge\{[s]assign(r) \rightarrow pt[E, r, \varsigma, \tau] \mid r \in \mathcal{R}_0\}.$$
Therefore, it is sufficient to prove
$$pw[\varepsilon, r, \varsigma, \tau, \mathtt{Infl}(\varepsilon, r)] \simeq pt[E, r, \varsigma, \tau]$$
for all r, ς, and τ, and where E equals ε if the latter is an elementary action expression, and is the aggregated action corresponding to ε if the latter is a composite action expression. This proof proceeds by induction over the

syntactic structure of ε. The cases where ε is an elementary or conditional occurrence expression are trivial. For the other two cases we first prove the following observations.

1. If F is a set of features, V is a set of trajectories defined over F, r is a total state, $F^+ \supseteq F$, r' is defined over F^+, and
$$V^+ = \{widen(v, r \restriction F^+) \mid v \in V\},$$
then
$$change(r', \varsigma, \tau, V^+) \sim chaw(r, r' \restriction F, \varsigma, \tau, V, F^+ - F),$$
where ς and τ are arbitrary temporal expressions.

2. If F is a set of features, r is a state, V is a set of trajectories over F, $V_{r'}$ is also a set of trajectories defined over F for every $r' \in last(V)$, and
$$V'' = \{v \triangleright v' \mid v \in V \land v' \in V_{last(v)}\},$$
then
$$change(r'', \varsigma, \tau, V'') \sim \exists u[$$
$$\bigvee [change(r', \varsigma, u, V) \land change(r'', u, \tau, V_{r'}) \mid r' \in last(V)]].$$

The proofs for these two observations and for the main proposition are straightforward using the definitions of the recursive functions involved. □

Proposition 12.4. *If* $\Upsilon = \langle \mathcal{O}, \Pi, \text{SCD}, \text{OBS} \rangle \in \mathcal{K}\text{-}\mathbf{IA}$ *and all composite action statements in* SCD *are formed using action symbols without arguments, then*
$$\Sigma_{\mathbf{IA}}(\Upsilon) \subseteq [\![\Upsilon]\!]^{kw} \subseteq [\![\Upsilon]\!]^k.$$

Compare Propositions 8.5 and 12.1. Actually, the proof of Proposition 12.1 is redundant since it is obtained as a side result of the present proof.

Proof $\Sigma_{\mathbf{IA}}(\Upsilon) = \Sigma_{\mathbf{IA}}(\mathbf{G}_{agr}^{kw}(\Upsilon))$ by Proposition 12.2.

$\Sigma_{\mathbf{IA}}(\mathbf{G}_{agr}^{kw}(\Upsilon)) \subseteq [\![\mathbf{G}_{agr}^{kw}(\Upsilon)]\!]$ by Proposition 8.5, since $\mathbf{G}_{agr}^{kw}(\Upsilon)$ does not contain any composite action statements.

$[\![\mathbf{G}_{agr}^{kw}(\Upsilon)]\!] = [\![\Upsilon]\!]^{kw}$ by Proposition 12.3.

$[\![\Upsilon]\!]^{kw} \subseteq [\![\Upsilon]\!]^k$ because the translation according to Π^{kw} implies the translation according to Π^k. □

Informally, the reason why the kw model set is smaller is because the inertia for some of the features may have been encoded into the action laws as a result of the widening operation. Anyway, the importance of this proposition is that it allows us to use aggregation and the aggregated actions A' for assessment purposes, but without having to use them for actual computation. One may attempt to obtain $\Sigma_{\mathbf{IA}}(\Upsilon)$ by applying a model selection function on either of the 'classical' model sets $[\![\Upsilon]\!]^{kw}$ or $[\![\Upsilon]\!]^k$. The correctness of that operation can be determined, for the case of selection from $[\![\Upsilon]\!]^{kw}$, by verifying whether the entailment method in question is correct for $\mathbf{G}_{agr}^{kw}(\Upsilon)$.

12.2 Entailment methods for chronicles containing composite action statements

In Chapter 9, subsection 9.1.2, we addressed *model selection functions* of the form
$$S(\langle \mathcal{O}, \Pi, \text{SCD}, \text{OBS}\rangle) = S'([\![\Pi(\text{SCD})]\!]_\mathcal{O}, [\![\text{OBS}]\!]),$$
where SCD only contains elementary action statements. We now consider their extension to the case of composite action statements in SCD defined by
$$S(\langle \mathcal{O}, \Pi, \text{SCD}, \text{OBS}\rangle) = S'([\![\Pi^{kw}(\text{SCD})]\!]_\mathcal{O}, [\![\text{OBS}]\!]).$$

12.2.1 Requirements for the generalized applicability of entailment methods

The following proposition provides a checklist for whether an entailment method is correctly applicable to chronicles with composite action statements.

Proposition 12.5. *If*
  1. $\Upsilon = \langle \mathcal{O}, \Pi, \text{SCD}, \text{OBS}\rangle \in \mathcal{Z} \subseteq \mathcal{K}\text{-}\mathbf{IA}$, *where all composite action statements in* SCD *are formed using action symbols without arguments,*
  2. $\mathbf{G}^{kw}_{agr}$ *preserves* $\mathcal{Z}$,
  3. *An entailment method* $\langle \mathbf{G}, S\rangle$ *is correctly applicable to* $\mathcal{Z}$ *with elementary action statements, and is direct or formed using filtering,*
  4. $\mathbf{G}(\mathbf{G}^{kw}_{agr}(\Upsilon)) = \mathbf{G}^{kw}_{agr}(\mathbf{G}(\Upsilon))$ *for all* $\Upsilon \in \mathcal{Z}$,
  5. $\mathbf{G}$ *does not depend on or change the set of observations in its argument chronicle,*

then $\langle \mathbf{G}, S\rangle$ *is correctly applicable to* Υ.

Proof Consider the case where S is formed using filtering, with S' as the essential selection function, and let $\Upsilon = \langle \mathcal{O}, \Pi, \text{SCD}, \text{OBS}\rangle$ and $\Upsilon' = \langle \mathcal{O}, \Pi, \text{SCD}, \emptyset\rangle$. Since $\mathbf{G}^{kw}_{agr}$ does not affect OBS and is not affected by it, it follows

$$\Sigma_{\mathbf{IA}}(\Upsilon) = \Sigma_{\mathbf{IA}}(\mathbf{G}^{kw}_{agr}(\Upsilon)) = S(\mathbf{G}(\mathbf{G}^{kw}_{agr}(\Upsilon))) =$$
$$= S(\mathbf{G}(\mathbf{G}^{kw}_{agr}(\Upsilon'))) \cap [\![\text{OBS}]\!] =$$
$$= S(\mathbf{G}^{kw}_{agr}(\mathbf{G}(\Upsilon'))) \cap [\![\text{OBS}]\!] =$$
$$= S'([\![\mathbf{G}^{kw}_{agr}(\mathbf{G}(\Upsilon'))]\!], [\![\text{OBS}]\!]) =$$
$$= S'([\![\mathbf{G}(\Upsilon')]\!]^{kw}, [\![\text{OBS}]\!]) =$$
$$= S[\mathbf{G}(\Upsilon)]^{kw} = S(\mathbf{G}(\Upsilon)).$$

But $\Sigma_{\mathbf{IA}}(\Upsilon) = S(\mathbf{G}(\Upsilon))$ is the definition for $\langle \mathbf{G}, S\rangle$ being correctly applicable to Υ, which was to be proved. The case where S is formed using direct minimization has a similar but simpler proof, where the model set [OBS] does not have to be separated. □

12.2.2 Entailment methods using Π^{kw}

Referring back to the entailment methods that have been defined and assessed in earlier chapters, we notice that all of them are formed using direct model selection or filtering, and that they use the pretransformations $\mathbf{G}_{seq}$ and $\mathbf{G}_{seqx}$. The condition $\mathbf{G}(\mathbf{G}_{agr}^{kw}(\Upsilon)) = \mathbf{G}_{agr}^{kw}(\mathbf{G}(\Upsilon))$ is satisfied when $\mathbf{G}$ is either of those. However, $\mathbf{G}_{seqx}$ does change the observation set.

The remaining question is whether $\Upsilon \in \mathcal{Z}$ implies $\mathbf{G}_{agr}^{kw}(\Upsilon) \in \mathcal{Z}$. Or, more precisely, if Υ satisfies the epistemological constraints in $\mathcal{Z}$, its action laws represent a world that satisfies the ontological constraints in $\mathcal{Z}$, and the schedule in Υ also contains composite action statements – is it then the case that $\mathbf{G}_{agr}^{kw}(\Upsilon) \in \mathcal{Z}$? If this is the case, then any entailment method that is correct for $\mathcal{Z}$ according to the assessments in previous chapters is correct for Υ as well.

Consider some simple cases where the ontological constraints are or are not preserved. Determinism is preserved – if each action in a world is deterministic, then composite actions formed using sequences of actions, conditional expressions, or loops are clearly also deterministic. (But notice that if a condition is allowed to contain the operator Λ then this preservation property is lost). The property of single-timestep actions is preserved under formation of conditional expressions, but not under sequencing or loop formation. The property of equidurationality is not preserved for conditional expressions, since different branches of the conditional may use different action symbols. It is preserved for sequencing of actions and obviously not for loops. Uniformity (**IAu**), finally, is preserved under sequencing but not under the other two composition operations.

The preservation of $\mathcal{K}$-**IA** itself under combination of occurrence expressions is guaranteed as long as each loop statement possibly terminates for each starting state.

In general, it seems that among all the various ontological characteristics and subcharacteristics that have been obtained in the assessments, determinism and strong determinism are the only ones that are preserved under all three composition operations. This is true not only for the 'first generation' subcharacteristics that have just been discussed, but as well for the compatibility-type subcharacteristics. For example, [GMON*] is obviously not preserved for conditional expressions.

The only entailment methods in Table 11.1 that satisfy the criteria in Proposition 12.5 are PMON and TMOC, therefore. CAMOC fails on requirement number 5 in the proposition, and the others fail on requirement number 2 if $\mathcal{Z}$ is chosen as the assessed range of applicability of the respective method. On the other hand, any entailment method in Table 11.1 which is correct for a chronicle family $\mathcal{Z}$ without composite action statements, and which uses $\mathbf{G}_{seq}$, is also correct for $\mathcal{Z} \cap \mathbf{IsAd}$ with composite action statements.

Notice, however, that the requirements in Proposition 12.5 are sufficient but have not been proven to be necessary, and requirement number 5, in particular, may be unnecessarily strict. Notice, also, that there are many applications that do not require the use of composite action statements. The assessments at this point are therefore not sufficient reasons to disconsider all the other entailment methods.

12.2.3 Entailment methods using Π^k

It is clear that the translation function Π^k is more convenient than Π^{kw}. It is also easy to find examples to show that chronological entailment methods, with or without occlusion, do not retain their domain of applicability if extended to composite action statements using Π^k. As for PMON and TMOC, it seems plausible that they should be correctly applicable to $\mathcal{K}$-**IA** or $\mathcal{K}r$-**IA** if extended using Π^k. A formal proof for such a proposition can not be obtained in the framework defined here, however, since Π^k does not have any direct counterpart in terms of the trajectory semantics.

12.3 Summary

Composite action statements using sequences, conditional expressions, and loops have been introduced by a straightforward extension of the syntax for the occurrence language. The set of intended models for chronicles containing composite actions was defined on the level of the ego-world game, and without reference to the underlying trajectory semantics. Thereafter, both the notion of a world description ⟨Infl, Trajs⟩ and of a translation function Π from the occurrence language to logic formulae were extended to composite occurrence expressions. The most straightforward extension of the translation function Π does not correspond well to the trajectory semantics, so a modified and slightly more complex translation had to be introduced.

The entailment methods that have been introduced for elementary action statements have been generalized, using this approach, to composite action statements without object arguments. The further generalization to allow object arguments does not appear to offer any difficulties. The entailment methods PMON and TMOC retain their previous ranges of applicability. Most of the others continue to apply, but for a more restricted range of applicability where deterministic actions are required.

13
Upper applicability bounds and assessment of soundness

The assessments in earlier chapters have obtained lower bounds on the range of applicability. It is natural to also look for corresponding upper bounds for the same entailment criteria. Ideally, one would like to know that a certain chronicle family is both a lower bound and an upper bound, so that no additional improvement of the result is possible.

For practical use one may be content with entailment methods which are sound but not complete, so that $S(\Upsilon) \supseteq \Sigma_{\mathbf{IA}}(\Upsilon)$ without necessarily having equality. This means that some intended conclusions are lost, but that may be acceptable since in any practical application there are several other reasons why admissible conclusions may be lost or late for their intended usage. The assessment of correct applicability should be extended, therefore, with an assessment of the range in which a given entailment method is merely sound. Ideally, one would also like to have some information as to how much $S(\Upsilon)$ differs from $\Sigma_{\mathbf{IA}}(\Upsilon)$.

This chapter contains results on both of these topics.

13.1 Direct upper bounds

The following is the simplest and most obvious definition of an upper bound for the range of applicability of an entailment method:

Definition. *A family* $\mathcal{Z} \subseteq \mathcal{K}\text{-}\mathbf{IA}$ *of chronicles is a* **direct upper bound** *for the range of applicability of an entailment method* $\langle \mathbf{G}, S \rangle$ *iff* $\Sigma_{\mathbf{IA}}(\Upsilon) \neq S(\mathbf{G}(\Upsilon))$ *for any* $\Upsilon \in \mathcal{K}\text{-}\mathbf{IA} - \mathcal{Z}$. ⋈

One kind of direct upper bound is obtained at once, namely using pointwise consistency-retaining chronicles:

Proposition 13.1. $\mathcal{K}r\text{-}\mathbf{IA}$ *is a direct upper bound for the entailment methods PCM, OCM, PGM, GMON, and TMOC.*

Proof These entailment methods are consistency-preserving. Therefore, for any chronicle Υ that is not consistency-retaining they will select a non-empty set of models. □

Unfortunately, this is about as much as one can say with respect to direct upper bounds, since for most of the possible restrictions there are a number of cases where an entailment method, so to speak, accidentally obtains the intended models. One such example was discussed at the end of subsection 9.3.1.

13.2 Principles for upper bounds for range of applicability

In order to arrive at more useful definitions for upper bounds, we shall discuss general principles and concrete examples together in this section.

13.2.1 Merely ontological constraints

As a concrete example, consider first the method CMOC (chronological minimization of occlusion and change), which has been proved correct for $\mathcal{K}$-**IAe**.

Proposition 13.2. *If* $\Upsilon = \langle \mathcal{O}, \Pi, \text{SCD}, \text{OBS} \rangle \in \mathcal{K}\text{-}\mathbf{IA} - \mathcal{K}\text{-}\mathbf{IAe}$, *then there is some modified, safely sequential chronicle* $\Upsilon' = \langle \mathcal{O}, \Pi, \text{SCD}', \text{OBS}' \rangle$ *such that* $\Sigma_{\mathbf{IA}}(\Upsilon') \neq S_{cmoc}(\Upsilon')$.

Proof In this case, there is some action designator E, some states r and r', and some trajectory $v \in \text{Trajs}(E, r)$ of length k such that no trajectory $v' \in \text{Trajs}(E, r')$ has length k. If there are several r with this property, for the same E, k, and r', then select an r which minimizes the set of features where r and r' differ. Construct SCD' as $[1, k+1]E$ and OBS' as $[0]\phi$, where ϕ is a formula that precisely characterizes the state r'. Then Υ' does not have any intended models according to $\mathcal{K}$-**IA**. However, the interpretation $\langle M, R \rangle$ where $R(0) = r'$, $R(1) = r$, and R then continues according to v is a preferred CMOC model. □

Of course, if $\Upsilon \in \mathcal{K}\text{-}\mathbf{IA} - \mathcal{K}\text{-}\mathbf{IAe}$ then $\Upsilon' \in \mathcal{K}\text{-}\mathbf{IA} - \mathcal{K}\text{-}\mathbf{IAe}$ as well, since the ontological characterization is only based on the properties of the world, which is specified by the action laws Π. Notice that k is a metavariable so, for example, if $k = 5$ then the single member of SCD is $[1, 6]E$.

The meaning of Proposition 13.2 is that if one uses the entailment method CMOC for the prediction system of a robot and this robot is set to operate in an equidurational world, then it will obtain the correct conclusions in all the chronicles that can arise (given enough time to complete the computation), and conversely, if it is set to operate in a non-equidurational world then there are some scenarios in that world where CMOC will obtain the wrong set of conclusions. Therefore, this result provides an indirect upper bound on the range of applicability; a bound which will be useful for applications where the action laws are held fixed whereas schedule and observations vary. An indirect upper bound is also useful for identifying cases where the assessment can not be further improved. This seems to be a reasonable pattern for upper-bound assessments, and it can be used for several other of the entailment criteria.

In the particular case of CMOC, the lower bound and the upper bound that we have obtained are equal.

I shall continue to use the notation for compatibility-type subcharacteristics, for example [CAMC] and [CAMC$^+$], which was defined in Section 11.7. I shall also write $\mathcal{K}$-**IA**(1-λ) for $\mathcal{K}$-**IA** $-$ $\mathcal{K}$-**IA**λ, where λ is an arbitrary sequence of subcharacteristics for **A**. For example $\mathcal{K}$-**IA**(1-e[PCM]) stands for $\mathcal{K}$-**IA** $-$ $\mathcal{K}$-**IA**e[PCM].

Now consider the method CAMC, which has been proved correct for $\mathcal{K}$-**IA**[CAMC].

Proposition 13.3. *If* $\Upsilon = \langle \mathcal{O}, \Pi, \text{SCD}, \text{OBS} \rangle \in \mathcal{K}$-**IA**(1-[CAMC])*, then there is some modified, safely sequential chronicle* $\Upsilon' = \langle \mathcal{O}, \Pi, \text{SCD}', \text{OBS}' \rangle$ *such that* $\Sigma_{\text{IA}}(\Upsilon') \neq S_{camc}(\Upsilon')$.

Proof There is some action designator E and some state r such that $Min(\ll_{camc}, W) \neq W$ for
$$W = \{r \triangleright v \mid v \in \text{Trajs}(E, r)\}.$$
Choose SCD$'$ with $[0, t_1]E$ as its only member, and OBS$'$ as $[0]\phi$, where ϕ completely specifies r. The result follows immediately. □

Similarly for PCMF, the proven lower bound is $\mathcal{K}$-**IA**e[PCM], which combines the complications of the two previous criteria. The upper bound is obtained by combining the two previous proofs with some small modifications.

Proposition 13.4. *If* $\Upsilon = \langle \mathcal{O}, \Pi, \text{SCD}, \text{OBS} \rangle \in \mathcal{K}$-**IA**(1-e[PCM])*, then there is some modified, safely sequential chronicle* $\Upsilon' = \langle \mathcal{O}, \Pi, \text{SCD}', \text{OBS}' \rangle$ *such that* $\Sigma_{\text{IA}}(\Upsilon') \neq S_{pcmf}(\Upsilon')$.

Proof Since $\Upsilon \in \mathcal{K}$-**IA**(1-e[PCM]), it is either not equidurational or it does not satisfy PCM compatibility. In the former case, the proof of Proposition 13.2 applies. In the latter case, there is some action designator E, some state r, and some positive integer k such that $Min(\ll_{pcm}, W) \neq W$ for
$$W = \{r \triangleright v \mid v \in \text{Trajs}(E, r) \wedge |v| = k\}.$$
Choose SCD$'$ with $[0, k]E$ as its only member and OBS$'$ as $[0]\phi$, where ϕ completely characterizes r. The result follows immediately. □

13.2.2 Ontological constraints in an epistemological range

For PCM, the proven lower bound is $\mathcal{K}$p-**IA**e[PCM], which contains both ontological and epistemological constraints. The ontological constraint can be analyzed in similar terms as for the previous cases:

Proposition 13.5. *If* $\Upsilon = \langle \mathcal{O}, \Pi, \text{SCD}, \text{OBS} \rangle \in \mathcal{K}\text{p-IA}(1\text{-e}[\text{PCM}])$, *then there is some modified, safely sequential chronicle* $\Upsilon' = \langle \mathcal{O}, \Pi, \text{SCD}', \text{OBS}' \rangle$ *which also has the epistemological property* p, *and for which* $\Sigma_{\text{IA}}(\Upsilon') \neq S_{pcm}(\Upsilon')$.

Notice that without the qualification 'which also has the epistemological property p', this proposition would be uninteresting, since then the inequality could be explained by the choice of an observation set that violates the epistemological assumption.

Proof The proof for Proposition 13.4 applies for this case also. □

Therefore, as long as we stay within the epistemological constraint $\mathcal{K}$p-**IA**, the assessment $\mathcal{K}$p-**IAe**[PCM] is both lower and upper bound. But what about upper bounds for the epistemological constraint?

13.2.3 Upper bounds on the epistemological constraints

Looking back at the summary of assessments at the end of Chapter 11, we obtain a list of what epistemological constraints have actually been used. The weakest of those constraints is "r", which has already been identified as a direct upper bound. For PCM, we obtained the assessment $\mathcal{K}$p-**IAe**[PCM], that is, PCM-compatible chronicles in $\mathcal{K}$p-**IAe**. This epistemological property is not a direct upper bound; it is easy to find single examples of chronicles outside the bound where PCM still gives intended results, for example chronicles containing observations of the form $[t]\phi$, where $t > 0$ but ϕ is a tautology.

One obvious upper bound is given by $\mathcal{K}$r-**IAe**[PCM] since we know that $\mathcal{K}$p-**IAe** $\subseteq$ $\mathcal{K}$r-**IAe**. The relationships between the various constraints are illustrated by Figure 13.1. The entire rectangle in the figure represents the $\mathcal{K}$-**IA** family of chronicles; the horizontal axis represents the choice of schedule and observations; and the vertical axis represents the choice of action laws. Thus, the ontological subcategorization follows directly the vertical axis. The epistemological properties follow, in some cases, the horizontal axis (for example $\mathcal{K}$p). In other cases, they cut diagonally across the rectangle (for example $\mathcal{K}$r).

The lower left rectangle stands for $\mathcal{K}$p-**IAe**[PCM], which is the class in which PCM is guaranteed correct. If one extends this rectangle towards the right over the vertically striped area one obtains $\mathcal{K}$r-**IAe**[PCM], which is the region where PCM is at least possibly correct. Further to the right are chronicles which do not satisfy $\mathcal{K}$r, and for which PCM certainly obtains the wrong set of models.

Similarly, the upper left rectangle represents the area where the ontological upper bound is in force. This is suggested by the horizontal striping,

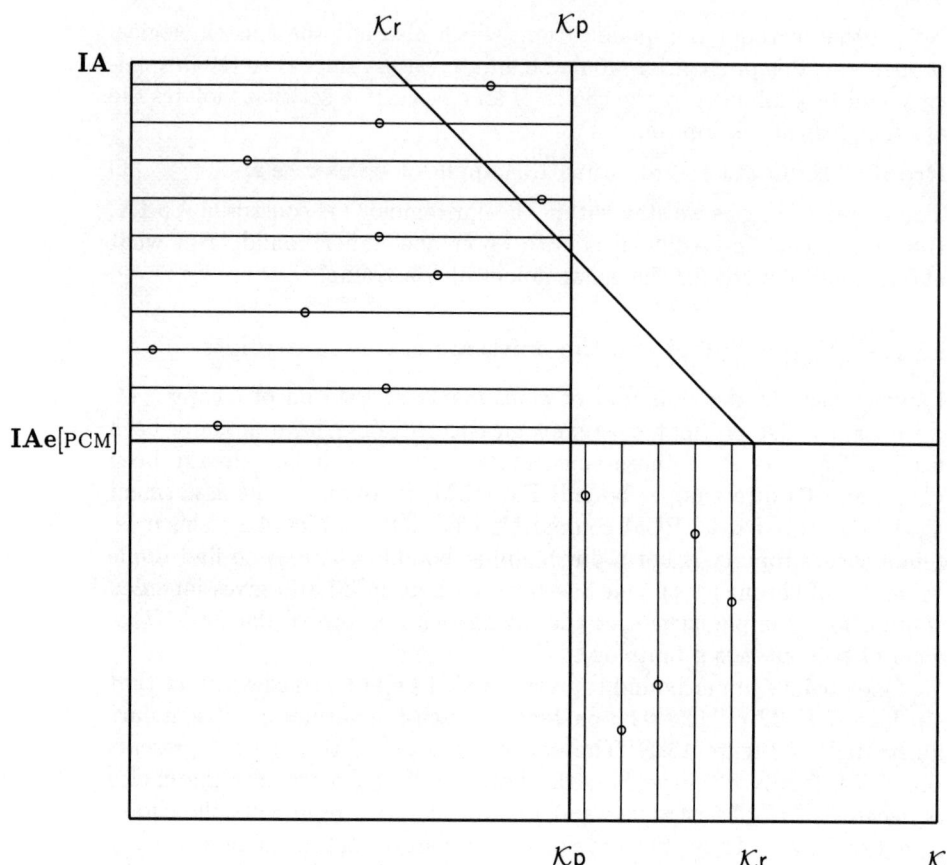

Fig. 13.1. Maximal assessments for PCM

as follows: For every chronicle Υ in $\mathcal{K}$p-**IA** $-$ $\mathcal{K}$p-**IAe**[PCM] (arbitrary position on any of the horizontal lines), there is some chronicle Υ' with the same action laws and, therefore, the same world (other position on the same line), and for which the entailment condition obtains the wrong result (dot on the line).

It would be tempting to conclude, from these assessments, that $\mathcal{K}$r-**IAe**[PCM] is an indirect upper bound on the range of applicability for PCM. Unfortunately, this is not true, because the proof of Proposition 13.4 depends crucially on the use of a chronicle which belongs to $\mathcal{K}$p-**IAe**[PCM] but not to $\mathcal{K}$r-**IAe**[PCM]. This is also illustrated by the figure, and suggests a certain care in the formal definition and use of indirect upper bounds.

With respect to upper bounds for PCM, the remaining question is what to do with the vertically striped area between $\mathcal{K}$p and $\mathcal{K}$r. One possibility would be that the $\mathcal{K}$p restriction is unnecessarily strong, and that a weaker restriction with a reasonable definition could be found. That does not seem to be within reach, however. A natural strategy would therefore be to use the same pattern as was found for the ontological restriction, but in a symmetrical fashion. The following definitions are obtained.

Definition. *Let $\langle \mathbf{G}, S \rangle$ be an entailment method and let $\mathcal{K}\xi$-Ψ be a chronicle family formed using the ontological family Ψ and the epistemological properties ξ. Then:*

- *$\mathcal{K}\xi$-Ψ is an **ontological upper bound** iff the following condition holds. If $\Upsilon = \langle \mathcal{O}, \Pi, \text{SCD}, \text{OBS} \rangle \in \mathcal{K}\xi$-**IA** $-$ $\mathcal{K}\xi$-Ψ, then there is some modified chronicle $\Upsilon' = \langle \mathcal{O}, \Pi, \text{SCD}', \text{OBS}' \rangle$ that also has the epistemological property ξ, and such that $\Sigma_{\mathbf{IA}}(\Upsilon') \neq S(\mathbf{G}(\Upsilon'))$.*

- *$\mathcal{K}\xi$-Ψ is an **epistemological upper bound** iff the following condition holds. If $\Upsilon = \langle \mathcal{O}, \Pi, \text{SCD}, \text{OBS} \rangle \in \mathcal{K}$-$\Psi - \mathcal{K}\xi$-$\Psi$, then there is some modified chronicle $\Upsilon' = \langle \mathcal{O}, \Pi', \text{SCD}, \text{OBS} \rangle \in \mathcal{K}$-$\Psi - \mathcal{K}\xi$-$\Psi$ (so that the world described by Π' is also in Ψ), such that $\Sigma_{\mathbf{IA}}(\Upsilon') \neq S(\mathbf{G}(\Upsilon'))$.*

*These two kinds of upper bounds are jointly called **indirect upper bounds**. Furthermore, $\mathcal{K}\xi$-Ψ is an **ontologically maximal assessment** (an **epistemologically maximal assessment**) for $\langle \mathbf{G}, S \rangle$ iff the method is correct in $\mathcal{K}\xi$-Ψ, and $\mathcal{K}\xi$-Ψ is an ontological (epistemological) upper bound for the method.* ⋈

Figure 13.2 illustrates these definitions with the simplifying assumption that the epistemological constraints only refer to the SCD and OBS components of the chronicle.

The horizontally striped rectangle to the upper left indicates how the ontologically maximal assessment works, using the same representation as for the previous diagram. The vertically striped rectangle on the lower right

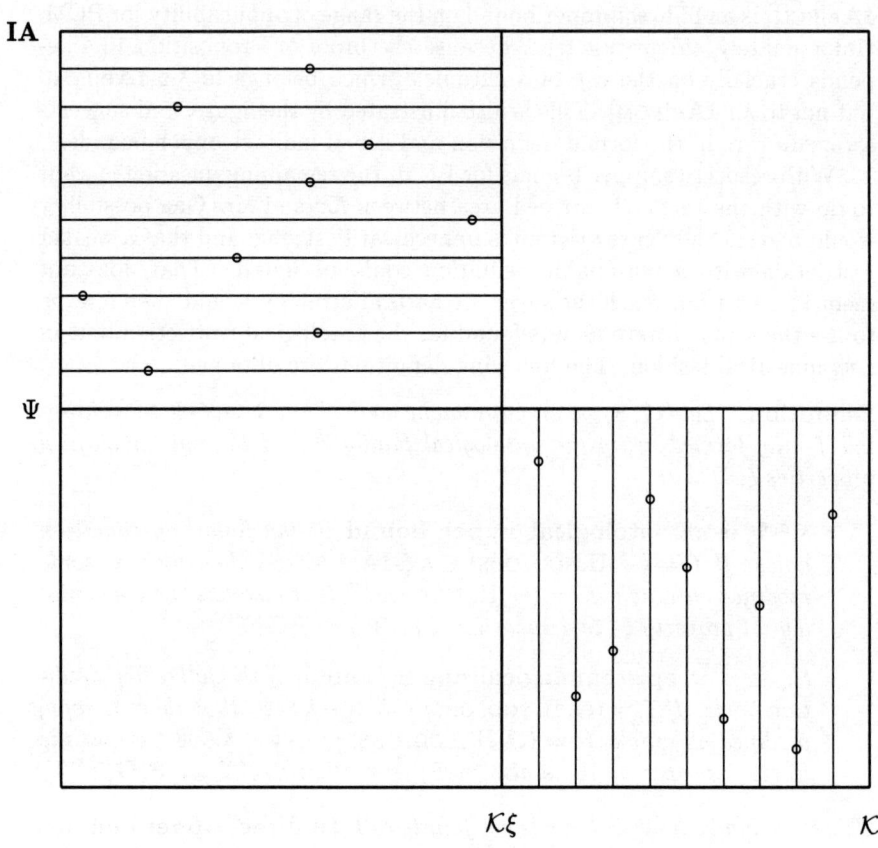

Fig. 13.2. Maximal assessments

indicates, in a similar way, how the epistemologically maximal assessment works. For every chronicle Υ (arbitrary position on any of the vertical lines) there is some chronicle Υ' with the same schedule and observations but not necessarily the same world (other position on the same vertical line) which satisfies the same epistemological properties, and where the entailment condition obtains the wrong result (dot on the line).

Therefore, although some chronicles Υ in the striped areas of the diagram may satisfy $S(\Upsilon) = \Sigma_{\mathbf{IA}}(\Upsilon)$, it is not possible to extend the assessed region, represented by the lower-left rectangle, either upward while retaining its right side, or rightward while retaining its upper side.

13.2.4 Summary of the upper-bound results in this section

In this section, we have developed definitions and upper bounds in parallel, but it is now appropriate to summarize the upper-bound results that have been obtained in this process:

CMOC has the assessment $\mathcal{K}$-**IAe**; CAMC has $\mathcal{K}$-**IA**[CAMC] as its assessment; PCMF has the assessment $\mathcal{K}$-**IAe**[PCM]. These assessments are ontologically maximal and (trivially) epistemologically maximal.

PCM has the assessment $\mathcal{K}$p-**IAe**[PCM], which is ontologically maximal but not epistemologically maximal. It also has $\mathcal{K}$r-**IA** as a direct upper bound.

13.3 Additional upper bounds on range of applicability

Now that the general principles have been established, we can proceed to analyzing the remaining entailment methods. In all cases, the upper bounds are only concerned with what applies within $\mathcal{K}$-**IA**, of course – it is only there that we have a definition of the set of intended models in the first place. For those entailment functions where the range of applicability is already $\mathcal{K}$-**IA** there is nothing left to do, therefore.

13.3.1 Original chronological minimization

In Chapter 9 we established two different assessments for OCM, which also illustrated the tradeoff between ontological and epistemological restrictions. Let us now use each of them as a starting point for upper bounds.

One proven range of applicability for OCM is $\mathcal{K}$sp-**IAde**[OCM]. This assessment is not ontologically maximal, as shown by the following counterexample.

Example. Consider an IDS world with the two propositional features p and q, and a single action type called A_1 that non-deterministically toggles the value of either p or q, but not both of them, in one timestep. For given schedule SCD, initial observations IOBS, and valuation M the intended models are those with all possible toggle combinations in $M[\text{SCD}]$, and none of these is OCM-preferred over any other. ⋈

However, if we want to use it for finding a broader range of applicability, this example is disappointing. It is even sensitive to the addition of an action in $\mathcal{K}$-**IAde**[OCM], as the following example shows.

Example. Consider the same IDS world as in the previous example, but with the addition of an action type A_2 which toggles q in one timestep if p is true, and otherwise does nothing. Then a schedule containing an occurrence of A_2 after an occurrence of A_1 will clearly OCM-prefer some intended models over others. ⋈

The assessment is not epistemologically maximal either, since the formula $[t]\phi$ with a tautology ϕ is harmless here as well. Similarly as for PCM, $\mathcal{K}$r-**IA** is a direct upper bound.

The other proven range of applicability for OCM is $\mathcal{K}$p-**IAe**[OCM$^+$], which actually is more similar in form to the assessment for PCM, but seems to be a 'worse' estimate from the point of view of its usage. Here we obtain, in a similar way as for Proposition 13.4,

Proposition 13.6. *If* $\Upsilon = \langle \mathcal{O}, \Pi, \text{SCD}, \text{OBS}\rangle \in \mathcal{K}\text{p-}\mathbf{IA}(\text{1-e}[\text{OCM}^+])$, *then there is some modified, safely sequential chronicle* $\Upsilon' = \langle \mathcal{O}, \Pi, \text{SCD}', \text{OBS}'\rangle \in \mathcal{K}\text{p-}\mathbf{IA}$ *such that* $\Sigma_{\mathbf{IA}}(\Upsilon') \neq S_{ocm}(\Upsilon')$.

Proof Since $\Upsilon \in \mathcal{K}$-**IA**(1-e[OCM$^+$]), it is either not equidurational or not OCM-favorable. In the former case, the proof of Proposition 13.2 applies. In the latter case, there is some action designator E, some states r and r', some positive integer k, and some trajectories $v \in \text{Trajs}(E, r)$ and $v' \in \text{Trajs}(E, r')$ both of length k such that $r \triangleright v \ll_{ocm} r' \triangleright v'$. Choose SCD$'$ with $[0, k]E$ as its only member, and OBS$'$ as $[0]\phi \vee \phi'$ where ϕ and ϕ' completely characterize r and r', respectively. The result follows immediately. □

This range of applicability for OCM is therefore ontologically maximal but not epistemologically maximal.

13.3.2 Prototypical global minimization

PGM has been assessed to be correct for chronicles that are PGM-perfect and pointwise consistency-retaining. The latter requirement is already a direct upper bound, and the example in Chapter 9, subsection 9.2.4, showed why it is needed.

The requirement of PGM-perfection is a very strong one, and it would be nice if it could be relaxed. Unfortunately, this is not so easy. The following example shows that it is not possible to obtain PGM-favorability as an assessment.

Example. Consider a world with the similarity type
$$\{f_1 : \{\texttt{G}, \texttt{Y}, \texttt{R}\}, p_2 : \{\texttt{T}, \texttt{F}\}\}$$
and two actions A_1 and A_2. A_1 is always performed with duration 1, and A_2 with duration 2, so they are equidurational. A_1 changes the value of f_1 as follows:
$$\texttt{G} \to \texttt{R}, \texttt{R} \to \texttt{G}, \texttt{Y} \to \texttt{R}.$$
The action $[t, t+2] A_2$ has the following effect. If $[t] f_1 \hat{=} \texttt{GR}$ then $[t, t+1] Rev$ else $[t+1, t+2] Rev$, where Rev is the action of reversing the value of p_2. This world is in the quite narrow family **IdAtue**, therefore, and it is PGM-favorable but not PGM-perfect. Already for this world, PGM gives some unintended models. Consider the schedule $\{\ulcorner[4,5] A_1\urcorner, \ulcorner[10,12] A_2\urcorner\}$. In every intended model, the change of value in p_2 must take place over $[10, 11]$, therefore, and the changeset of intended models is always $\{\langle f_1, 5\rangle, \langle p_2, 11\rangle\}$. However, there is also a model where f_1 has a phantom change to $\texttt{Y}$, for example at time 8, whereby the changeset becomes $\{\langle f_1, 5\rangle, \langle f_1, 8\rangle, \langle p_2, 12\rangle\}$. This model is also selected. ⋈

The following concept will be useful for the determination of the upper bound.

Definition. *A trajectory-semantics world is said to be* **PGM-semiperfect** *iff it satisfies the following requirement. For any action designator E, if $v \in \texttt{Trajs}(E, r)$ and $v' \in \texttt{Trajs}(E, r')$ are two trajectories of equal length and $r \neq r'$, then changeset$(r \triangleright v) = $ changeset$(r' \triangleright v')$. Also, if $v \in \texttt{Trajs}(E, r)$ and $v' \in \texttt{Trajs}(E, r)$ are two trajectories of equal length, then neither of changeset$(r \triangleright v)$ and changeset$(r' \triangleright v')$ is a subset of the other.* ⋈

The concept of PGM-semiperfect will be represented in shorthand by [PGM$^\times$]. It satisfies
$$[\text{PGM}^\star] \subseteq [\text{PGM}^\times] \subseteq [\text{PGM}],$$
and the definition is also a hybrid between those for PGM-compatible and PGM-perfect. Neither of [PGM$^+$] and [PGM$^\times$] is included in the other one.

Proposition 13.7. $\mathcal{K}r$-**IA**[PGM$^\times$] *is an ontological upper bound for PGM.*

Proof Let $\Upsilon \notin \mathcal{K}$-**IA**[PGM$^\times$]. We consider the two possible cases according to the definition of semiperfect.

1. There exists an action designator E, an integer k, two different states r and r', and two trajectories v and v' of length k such that $v \in \texttt{Trajs}(E, r)$, $v' \in \texttt{Trajs}(E, r')$, and where changeset$(r \triangleright v) \neq$ changeset$(r' \triangleright v')$. It is no restriction to assume that changeset$(r \triangleright v)$ is minimal for the chosen E and k. This means, in particular, that changeset$(r' \triangleright v')$ is not a subset of changeset$(r \triangleright v)$. Construct Υ' with the same action laws, the single observation $[0]\varphi'$ where φ precisely characterizes r', and the single action statement $[1, k+1] E$. It is clear that Υ' is pointwise consistency-retaining, and that $R(0) = R(1) = r'$ in all intended models. Now construct a model where $R(0) = r'$, $R(1) = r$, and the development continues with

the trajectory v. It is clear that this model is also minimal with respect to $\ll_{pgm}$.

2. There exists an action designator E, an integer k, a state r, and two trajectories v and v' both in $\text{Trajs}(E,r)$ and of length k, and where $changeset(r \triangleright v) \subset changeset(r \triangleright v')$. Construct the chronicle where $[0]\varphi$ is the only observation, φ characterizes r exactly, and the schedule has $[0,k]E$ as its only member. It is clear that this chronicle is pointwise consistency-retaining, and that $\ll_{pgm}$ will prefer one of its intended models over another one. This concludes the proof. □

In summary, $\mathcal{K}r\text{-}\mathbf{IA}$ is a direct upper bound on the applicability of PGM. Within $\mathcal{K}r\text{-}\mathbf{IA}$, the range of applicability for PGM has PGM-perfection as a lower bound and PGM-semiperfection as an upper bound. The difference between perfection and semiperfection is therefore the remaining distance between the lower and the upper bound of the range of applicability.

For PGMF we have obtained $\mathcal{K}\text{-}\mathbf{IAe}[\text{PGM}^\star]$ as an assessment, and we obtain a related family as an upper bound:

Proposition 13.8. $\mathcal{K}\text{-}\mathbf{IAe}[\text{PGM}^+]$ *is an ontological upper bound for PGMF.*

Proof Suppose $\Upsilon \in \mathcal{K}\text{-}\mathbf{IA}(1\text{-}\mathrm{e}[\text{PGM}^+])$. If it is not equidurational then the proof of Proposition 13.2 applies. If it is not PGM-favorable, on the other hand, then there exists an action designator E, an integer k, two states r and r', and two trajectories v and v' of length k such that $v \in \text{Trajs}(E,r)$, $v' \in \text{Trajs}(E,r')$, and $r \triangleright v \ll_{pgm} r' \triangleright v'$. It is no restriction to assume that $r \triangleright v$ is $\ll_{pgm}$-minimal for the chosen E and k. Construct Υ' with the same action laws, the single observation $[0]\varphi \vee \varphi'$ where φ and φ' precisely characterize r and r', respectively, and the single action statement $[0,k]E$. It is clear that Υ' is pointwise consistency-retaining, and that $\ll_{pgm}$ prefers one intended model over another one. □

13.3.3 Global minimization of occlusion with nochange premises

Global minimization of occlusion with nochange premises (GMON) was defined by
$$S_{gmon}(\Upsilon) = gmon([\![\Upsilon]\!] \cap [\![\text{NCH}_\nu]\!]),$$
and was assessed to be correct for $\mathcal{K}r\text{-}\mathbf{IA}[\text{GMON}^\star]$, that is, the family of GMON-perfect and pointwise consistency-retaining chronicles. Its preference relation was defined by
$$\langle M,R,X \rangle \ll_{gmon} \langle M',R',X' \rangle \quad \text{iff} \quad M = M' \wedge X \prec X'.$$
We obtain

Proposition 13.9. $\mathcal{K}r\text{-}\mathbf{IA}[\text{GMON}^+]$ *is an ontologically maximal assessment for GMON.*

Proof The proof is similar to the proof for Proposition 13.7. Consider a chronicle Υ which is not in $\mathcal{K}r\text{-}\mathbf{IA}[\text{GMON}^+]$. By definition, there exists some

E, r, r', k, and trajectories $v \in \text{Trajs}(E,r)$ and $v' \in \text{Trajs}(E,r')$ both of length k, and where $\text{Infl}(E,r) \subset \text{Infl}(E,r')$. Construct the chronicle where $[0](\varphi \vee \varphi')$ is the only observation, φ and φ' characterize r and r' exactly, and the schedule has $[0,k]E$ as its only member. Clearly, this chronicle is in $\mathcal{K}r$-**IA**. Since the occlusion predicate X is directly derived from the function Infl, according to the construction of FTNF, it follows that some intended models are GMON-preferred over some others. This concludes the proof. □

Proposition 13.10. $\mathcal{K}$-**IA**[GMON$^+$] *is an ontologically maximal assessment for GMONF.*

Proof The proof for Proposition 13.9 applies for this case as well. □

Therefore, for GMONF we have equality between the lower bound and the upper bound.

13.4 Summary of upper-bound results

The assessment results for lower and upper bound of range of applicability for these entailment methods are summarized in Table 13.1. For each entailment method it indicates the assessment that has been obtained here (column 2), and whether that assessment is epistemologically maximal, and whether it is ontologically maximal (on successive lines in column 4). Column 3 indicates whether the method requires the transformation $\mathbf{G}_{seqx}$ to simply executable schedules. Column 5 indicates an indirect ontological upper bound in those cases where the assessment itself is not ontologically maximal, and the last column indicates a direct upper bound when relevant.

Checking to what extent these results give a complete analysis, we notice that there are only two cases where the assessment is not epistemologically maximal (OCM and PCM), and in those two cases we can use $\mathcal{K}r$ as an epistemological upper bound. With respect to the ontological bounds, it is only for PGM/PGMF, GMON, and one of the assessments of OCM that the ontological upper bound differs from the assessment.

For OCM and PCM there may, in principle, be an exchangability between the ontological and the epistemological constraints, as illustrated by the figures: if one of them is weakened then possibly the other one can be strengthened. For PGM and GMON, such a tradeoff is in principle possible towards a stronger epistemological restriction, but not towards a stronger ontological restriction since $\mathcal{K}r$-**IA** is a direct upper bound.

Ideally one would like to have equality between assessment (lower bound) and a single indirect upper bound. This goal has now been reached for several of the twelve entailment methods, but a few gaps remain still.

Table 13.1. Summary of assessments and upper bounds

Entailment method	Assessment	Exec sched	Emax Omax	Upper on-tol bound	Direct upp bound
PGM	$\mathcal{K}$r-$\mathbf{IA}$[PGM*]	No	Yes / No	$\mathcal{K}$r-$\mathbf{IA}$[PGM$^\times$]	$\mathcal{K}$r-$\mathbf{IA}$
PGMF	$\mathcal{K}$-$\mathbf{IAe}$[PGM*]	No	Yes / No	$\mathcal{K}$-$\mathbf{IAe}$[PGM$^+$]	
OCM	$\mathcal{K}$p-$\mathbf{IAe}$[OCM$^+$]	No	No / Yes	–	$\mathcal{K}$r-$\mathbf{IA}$
	$\mathcal{K}$sp-$\mathbf{IAde}$[OCM]	No	No / No	?	
PCM	$\mathcal{K}$p-$\mathbf{IAe}$[PCM]	No	No / Yes	–	$\mathcal{K}$r-$\mathbf{IA}$
PCMF	$\mathcal{K}$-$\mathbf{IAe}$[PCM]	No	Yes / Yes	–	
CAMC	$\mathcal{K}$-$\mathbf{IA}$[CAMC]	Yes	Yes / Yes	–	
CMOC	$\mathcal{K}$-$\mathbf{IAe}$	(Yes)	Yes / Yes	–	
CAMOC	$\mathcal{K}$-$\mathbf{IA}$	Yes	Yes / Yes	–	
PMON	$\mathcal{K}$-$\mathbf{IA}$	No	Yes / Yes	–	
GMON	$\mathcal{K}$r-$\mathbf{IA}$[GMON*]	No	Yes / No	$\mathcal{K}$r-$\mathbf{IA}$[GMON$^+$]	$\mathcal{K}$r-$\mathbf{IA}$
GMONF	$\mathcal{K}$-$\mathbf{IA}$[GMON$^+$]	No	Yes / Yes	–	
TMOC	$\mathcal{K}$r-$\mathbf{IA}$	No	Yes / Yes	–	$\mathcal{K}$r-$\mathbf{IA}$

13.5 Assessments of soundness

The definition of correct applicability in Chapter 9, Section 9.2, has an obvious counterpart in the definition of soundness:

Definition. *An entailment method $\langle G, S \rangle$ where S operates on DFL-1 interpretations is said to be* **sound** *for the ontological family $\mathcal{Z} \subseteq \mathcal{K}$-IA iff $\Sigma_{IA}(\Upsilon) \subseteq S(G(\Upsilon))$ for any chronicle Υ in $\mathcal{Z}$.* ⋈

Therefore, the range of soundness must always be $\supseteq$ the range of correct applicability. In particular, for CAMOC and PMON, soundness is not an issue since they are correctly applicable throughout $\mathcal{K}$-IA anyway. For the other methods, most of the results are already available in the form of separate lemmas or parts of proofs in earlier chapters. Notice that a $\mathcal{K}r$ restriction on the range of applicability can always be removed for the range of soundness, since lack of consistency retention can only have the effect that the set of selected models is nonempty when the set of intended models is empty. For simplicity, we restrict the soundness assessments to simply executable chronicles.

CMOC is correctly applicable in $\mathcal{K}$-IAe. According to the proof of Proposition 11.3, CMOC is sound throughout $\mathcal{K}$-IA, that is, the equidurationality requirement is needed for completeness but not for soundness.

CAMC is correctly applicable for CAMC-compatible chronicles in $\mathcal{K}$-IA. We proved in Proposition 10.5 that $S_{camc}(\Upsilon) \subseteq \Sigma_{IA}(\Upsilon)$, that is, the opposite of what is required for soundness. In combination with Proposition 13.3, this shows that CAMC is not usable in CAMC-incompatible worlds.

The remaining entailment methods require separate proofs for the soundness results.

Lemma 13.11. *If $\Upsilon \in \mathcal{K}$-IA is PCM-compatible then $\text{Ci}(S_{cmoc}(\Upsilon)) \subseteq S_{pcmf}(\Upsilon)$.*

Proof It is sufficient to prove the result for $\Upsilon \in \mathcal{K}p$-IA[PCM], since both PCMF and CMOC are defined using filtering. Let $I = \langle M, R \rangle \ll_{pcm} I' = \langle M, R' \rangle$ be two members of DFL1:$[\![\Upsilon]\!]$, and let X and X' be chosen so that $\langle M, R, X \rangle$ and $\langle M, R', X' \rangle$ are members of DFL2:$[\![\Upsilon]\!]$ with minimal X. Let t be the timepoint of preferential comparison between I and I' according to $\ll_{pcm}$. It follows that $R_{0:\theta t} = R'_{0:\theta t}$, and therefore, that $X_{0:t} = X'_{0:t}$. It also follows that $breakset(I, t) \subset breakset(I', t)$. Let F be the set of those features f for which $X(f, t)$ is true. By definition, $brs(I, t) = breakset(I, t) \cap \bar{F}$, and similarly for I'. Suppose $brs(I, t) \not\subset brs(I', t)$. It follows $brs(I, t) = brs(I', t)$ and $breakset(I, t) \cap F \subset breakset(I', t) \cap F$. However, this contradicts the assumption that Υ is PCM-compatible. From $brs(I, t) \subset brs(I', t)$ we obtain $I \ll_{cmoc} I'$. The claim follows immediately. □

From this lemma and the soundness result for CMOC it follows that PCMF is sound for $\mathcal{K}$-**IA**[PCM]. Proposition 9.17 then shows that PCM is also sound for $\mathcal{K}$-**IA**[PCM]. For PGM we obtain:

Proposition 13.12. *PGMF is sound for $\mathcal{K}$-**IA**[PGM$^+$].*

Proof It is sufficient to prove the result for $\Upsilon \in \mathcal{K}\text{p-}\mathbf{IA}[\text{PGM}^+]$, since both PGMF and $\Sigma_{\mathbf{IA}}$ are defined using filtering. Suppose that $\Upsilon \in \mathcal{K}\text{p-}\mathbf{IA}$ is PGM-favorable. If PGMF is not sound for Υ, then there must be some member $I = \langle M, R \rangle$ of $[\Upsilon]$ and $I' = \langle M', R' \rangle$ of $\Sigma_{\mathbf{IA}}(\Upsilon)$ for which $I \ll_{pgm} I'$, that is, $changeset(I) \subset changeset(I')$. It follows $M = M'$, so $M[\text{SCD}]$ is the same for both. Let $\langle f, t \rangle$ be a member of $changeset(I') - changeset(I)$. Since the intended model I' has a change between θt and t, it follows that $M[\text{SCD}]$ has some member $[s_i, t_i] E_i$ such that $s_i < t \leq t_i$. It follows that $R(s_i) = R'(s_i)$. Then $changeset(R_{s_i:t_i}) \subset changeset(R'_{s_i:t_i})$ without possibility of equality because of the presence of $\langle f, t \rangle$. Let v and v' be the trajectories used for performing the action $[s_i, t_i] E_i$ in R and in R'. We obtain

$$changeset(R(s_i) \triangleright v) \subseteq changeset(R_{s_i:t_i}) \subset changeset(R'_{s_i:t_i}) =$$
$$= changeset(R(s_i) \triangleright v),$$

which is inconsistent with the assumption of PGM-compatibility. □

It follows from Proposition 9.17 that PGM is likewise sound for $\mathcal{K}$-**IA**[PGM$^+$]. It seems likely that a similar result can be obtained for GMON and GMONF.

The assessment results for soundness are summarized in Table 13.2. A question-mark indicates that no result has been obtained. One can observe a uniform pattern: those constraints for correctness that are directly derived from the preference relation (compatibility, favorability, perfection) are retained for soundness although a perfection requirement is weakened throughout to the corresponding favorability requirement. All other restrictions for correctness vanish for soundness.

13.6 Summary of assessment criteria and assessment results

The assessment results in the present volume can be summarized as follows.

Four main types of restriction have been identified for chronicles. There are two epistemological restrictions, namely initial observations and pointwise consistency-retention. There are also two types of ontological restriction, namely equidurationality and the various requirements of compatibility, favorability, etc.

All the twelve entailment methods have assessments for their range of applicability in terms of these four restriction types, although OCM also has another assessment in terms of a third epistemological restriction. The assessments include both a lower bound (the proven range of applicability),

Table 13.2. Summary of assessments of soundness

Entailment criterion	Correctness	Soundness
PGM	$\mathcal{K}r$-**IA**[PGM*]	$\mathcal{K}$-**IA**[PGM$^+$]
PGMF	$\mathcal{K}$-**IAe**[PGM*]	$\mathcal{K}$-**IA**[PGM$^+$]
OCM	$\mathcal{K}p$-**IAe**[OCM$^+$]	?
PCM	$\mathcal{K}p$-**IAe**[PCM]	$\mathcal{K}$-**IA**[PCM]
PCMF	$\mathcal{K}$-**IAe**[PCM]	$\mathcal{K}$-**IA**[PCM]
CAMC	$\mathcal{K}$-**IA**[CAMC]	$\mathcal{K}$-**IA**[CAMC]
CMOC	$\mathcal{K}$-**IAe**	$\mathcal{K}$-**IA**
CAMOC	$\mathcal{K}$-**IA**	$\mathcal{K}$-**IA**
PMON	$\mathcal{K}$-**IA**	$\mathcal{K}$-**IA**
GMON	$\mathcal{K}r$-**IA**[GMON*]	?
GMONF	$\mathcal{K}$-**IA**[GMON$^+$]	?
TMOC	$\mathcal{K}r$-**IA**	$\mathcal{K}$-**IA**

indirect upper bounds, and in some cases a direct upper bound. In most cases, the lower bound and the ontological upper bound are equal or close.

Systematic techniques have been found which serve to eliminate these restrictions. These techniques are modifications on the entailment methods which can be used alone or in combination. Such techniques include filtering for eliminating the requirement of initial observations, occlusion for eliminating (or in some cases, weakening) the compatibility restrictions, and chronological assignment of valuation for eliminating the requirement of equidurationality.

Among the four requirement types, compatibility and the corresponding use of occlusion are essential in the sense that without them one loses not only correctness but even soundness. When the other requirements and corresponding techniques are ignored, then completeness is lost but soundness is retained.

14
Future directions

Chapter 3, subsection 3.1.5, proposed a long range strategy for the use of the systematic methodology. The idea was to proceed stepwise in the taxonomy of scenario descriptions which is defined by the ontological and epistemological characteristics. One should address successively larger and more difficult families in that taxonomy; at each step one defines the set of intended models for a given scenario description as the current variant of the function Σ, and uses it as the basis for assessment of entailment methods. The present volume has applied the proposed methodology to the chronicle family $\mathcal{K}$-**IA** using the function $\Sigma_{\mathbf{IA}}$. Continuations of this work may address additional topics in the framework of $\Sigma_{\mathbf{IA}}$, or proceed to other families in the taxonomy.

Topics for continued work using $\Sigma_{\mathbf{IA}}$ or simple variations of it

Although we now have assessments for a number of non-monotonic logics of actions and change, viewed as entailment methods, there are several others that have been proposed during the last decade of research. An obvious next step is to obtain assessments of their respective ranges of expressivity and correct applicability. One may expect that the underlying semantics that has been used here will be relevant for them as well.

Several additional research problems can be addressed with merely small changes to the $\Sigma_{\mathbf{IA}}$ underlying semantics. The material in Chapter 12 suggests an approach to hierarchically defined actions, where action laws for higher level actions are expressed in terms of lower level actions. In this context, it may be useful to relax the requirement that all elements in all trajectories of a given $\text{Trajs}(E, r)$ must be defined over the same set of features.

The generalization to concurrent actions is also a natural next step. One may expect that, as usual, the case of non-interacting actions is relatively simple, and that substantial problems arise if concurrent actions can interact throughout their durations. Finally, one approach to the so-called 'ramification problem' may be obtained by keeping the same underlying semantics as has been used here, and to merely consider the domain constraints as a way of separating some of the information in the action laws into a formula of its own. This may be motivated if it improves the legibility and understandability of the action laws, and if it means that the same constraint can be stated just once instead of being repeated in several action laws.

Extensions using a different underlying semantics

The taxonomy of scenario descriptions provides a list of phenomena in common-sense reasoning which one would like to address. In the systematic methodology, one must first find a definition of intended models that both fits the needs and is possible to work with. The following are some possibilities.

- Graded possibility levels in the function Trajs. Even a crude distinction between "normal" and "unusual" trajectories may be used as the basis for addressing surprises and qualification. A more general solution would be to associate a probability distribution with Trajs(E, r) for every E and r, or to consider it as a fuzzy set.

- As an alternative or a complement of graded possibility levels for trajectories, one may also consider the use of graded plausibility levels for states.

- An abstraction of the ego (an 'ego model') which ascribes beliefs, goals, and intentions to the ego. This may provide an underlying semantics for a modal logic, where the current ego can be characterized in those categories. The ego-world game must be redefined accordingly.

- Relaxation of the assumption of complete knowledge about actions, in order to have a more realistic assumption from both the cognitive-robotics and the common-sense point of view.

Research has been started in all of these areas but the topics are almost unexhaustible.

Lower levels of aggregation

In Chapter 3, section 3.4, we discussed the possibility of deriving world descriptions in the trajectory semantics from other world descriptions on lower levels of aggregation, and ultimately from the level of standard physics. This suggests a complex and very challenging topic for interdisciplinary research. Research on (so-called) qualitative reasoning addresses this range of problems, but there would also be room for other approaches. The amendment of $\mathcal{K}$-**IA** to allow for continuous time and continuously changing feature-values, called $\mathcal{K}$-**RA** above, is only meaningful if it is related to those lower levels of aggregation.

Perspective on non-monotonic logics

Non-monotonic logics were originally proposed as modifications of standard logic which would make it adequate for reasoning about actions and change, as well as for some other aspects of common-sense reasoning. It is interesting, therefore, to compare the non-monotonic logics that work well for actions and change with those that are actively studied in contemporary research on non-monotonic logic.

There are two striking differences. The first one regards how the logics are formulated. In this book we have studied several entailment methods which were defined in terms of a common base logic. Each entailment method is defined in terms of a few kinds of operations, namely

- organizing the set of premises by partitions;
- obtaining the set of classical models for a set of formulae;
- syntactic transformations on sets of formulae;
- imposing a preference relation on a set of models;
- ordinary set-theoretic operations on sets of formulae and of models.

By contrast, the research literature on non-monotonic logics focuses on a small number of 'standard' approaches, in particular circumscription and default logic, each with an increasing number of variants. The entailment methods that have been defined in the present book can actually be re-expressed in terms of circumscription, but doing so does not seem to offer any advantage for assessment purposes. The proofs in this book fit naturally with the present type of definitions for entailment methods, and a translation into circumscription would probably be more of a hindrance than a help as far as the assessment proofs are concerned. This does not exclude, of course, that translations into 'standard' non-monotonic logics may be useful for the purpose of realization and implementation.

The other difference regards how proposed non-monotonic logics are motivated. The classical research paradigm for the frame problem was driven by the assumption that one could find a logic of common-sense reasoning which would be able to produce common-sense conclusions from *bona fide* common-sense scenario descriptions. Although it was recognized from the start that such a logic would have some non-classical properties, such as non-monotonicity, there was still an expectation that only one or a few basic principles would have to be added, for example, minimization of abnormality, and that those new principles could be formulated in reasonably concise form. Reasoning about actions and change was perceived as one important part of such a logic of common-sense reasoning.

This vision is now in serious doubt. The example-based methodology, which has been used for a large part of the research, has been useful for finding problems but not for analyzing them with precision. Results that have been obtained so far using a systematic methodology suggest that fairly complex logical operations are needed to obtain any level of generality. Even for the basic case, where strict inertia (persistence) is assumed, a reasonable range of applicability can only be obtained after a choice between some alternatives, none of which is structurally simple. One may use nontrivial syntactic operations on the axioms, as in explanation closure; nontrivial preference relations on models, as in the CAMOC method; or nontrivial combinations of syntactic and semantic techniques, as in the PMON method. In view of the complexity of the most general methods, one can also suspect that some of the simpler methods, having a more limited range of applicability, may continue to be competitive by having better computational properties.

The broader range of frame problems includes surprises, ramification, and qualification. Eventually, we will also need to address topics that have been little studied so far, such as cause-and-effect chains, continuous change, and delayed effects of actions. It does not seem likely that the entailment methods that have been investigated for strict inertia will be correct for these more general cases as well, so additional preference mechanisms will most likely have to be introduced. Perhaps the eventual logic of common-sense reasoning, rather than being unified, concise, and elegant, will have the character of a Swiss army knife and contain one tool for each purpose. However, unlike the Swiss army knife, one will use several tools at the same time and not one tool at a time.

To the extent that applications require the use of non-monotonic logics with complex entailment methods, a change of emphasis may be forthcoming for non-monotonic logic research as well. We have been used to a theory-driven research process, where one develops mathematically deep results for formally elegant logics and assumes that those logics will also be put to practical use. Now, maybe, it will be natural to first identify which variants of the logic are adequate for which classes of applications. Research on the formal and computational properties of non-monotonic logics could then obtain an external input with respect to relevance which simply has not been available to it before.

The role of logic in AI

The encounter between AI and knowledge representation on one hand, and formal logic on the other, started more than thirty years ago. It has already produced resolution theorem-proving, logic programming, and non-monotonic logics, to mention only the most obvious of the interaction results. The methodology in these branches of research has been relatively close to the traditional concepts and methods of logic. This applies, in particular, for the rapidly increasing literature on non-monotonic logics, which often seems to be quite remote from those AI problems that originally inspired its emergence. We have witnessed one more occurrence of a common phenomenon in science, namely that theory is becoming an end it itself.

The present work has demonstrated and used a methodology which, although logic-based, is quite different from the traditional methodology in logic. It is closer to the application, in the sense that it uses abstract versions of a number of concepts and phenomena that are of central importance for artificial intelligence, as well as for neighboring disciplines. For example, the trajectory concept and the ego-world game, which are characteristic of the underlying semantics, represent essential aspects of cognitive robotics. Since the trajectory semantics is a straight-forward generalization of state-transition semantics, it can also be related to traditional methods in control engineering. More generally, I take the view that when logic is applied it must be used for speaking about something, and then it is important to define with precision those phenomena that one is going to speak about. That is exactly why it was important to define the concept of inhabited dynamical systems and to develop their underlying semantics.

I believe that this approach to the use of logic and the role of theory can be applied in other areas besides for reasoning about actions and change in inhabited dynamical systems. If the present book has shown a way of bridging the gap between theory and practice in AI, then it may have made a contribution that goes beyond its proper topic.

Appendix A
Term index

Since the topic of the present book is fairly complex, it has been necessary to introduce and use a relatively large number of technical terms with precise definitions. Although the terminology can hopefully be grasped in a direct reading of the book, the following term index is offered as an aid when the precise meaning of a term has to be looked up again. The index provides a reference to those subsection(s) where the term is introduced or formally defined, and sometimes a short explanatory text. All references are to section/subsection numbers.

A.1 Abbreviations

The abbreviations for entailment methods (PCM etc.) are listed in table A.1, with an indication of where the method is defined, and where its lower bound(s) and upper bound(s) are obtained. Other abbreviations are as follows.

EFL, LFL, DFL: Variants of base logics. See Chapters 4, 5, and 6 respectively.

IDS: Inhabited dynamical system. See Chapter 1.

FTNF: Full trajectory normal form. See Chapter 8.

SDS: A scenario example, see Chapter 2, Section 2.1.

YSS, HTS, SMM, FCS, RTS, SCS, TCS, FAS: Scenario examples. See Chapter 7, Section 7.2.

Table A.1. Entailment methods

Entailment criterion	Definition	Lower bound	Upper bound	Full name
PGM	9.1.5	9.2.4	13.1, 13.3.2	Prototypical global minimization
PGMF	9.4.1	9.4.2	13.3.2	PGM with filtering
OCM	9.1.4	9.2.3	13.1, 13.3.1	Original chronological minimization
PCM	9.1.3	9.2.2	13.1, 13.2.2	Prototypical chronological minimization
PCMF	9.4.1	9.4.2	13.2.1	PCM with filtering
CAMC	10.1.2	10.3.2	13.2.1	Chronological assignment and minimization of change
CMOC	11.3.1	11.6.2	13.2.1	Chronological minimization of occlusion and change
CAMOC	11.3.2	11.6.1	(NA)	Chronological assignment and minimization of occlusion and change
PMON	11.4.3	11.6.3	(NA)	Pointwise minimization of occlusion with nochange premises
GMON	11.4.2	11.6.3	13.1, 13.3.3	Global minimization of occlusion with nochange premises
GMONF	11.4.2	11.6.3	13.3.3	GMON with filtering
TMOC	11.5	11.6.4	13.1	Two-stage minimization of occlusion and change

A.2 Technical terms

For each technical term, this list indicates the section(s) or subsection(s) where it has been introduced (italic number) or formally defined (boldface number). A cross-reference or a brief review of the definition is given for many of the entries. When common-sense words are used (such as 'now' or 'world') the reference is of course only to the technical usage. When a word is italicized in the explanation it means that it also has an entry in this index.

abbreviations in formal syntax: [**2.6.3**, 6.3.1]

abstraction: [1.4.4: footnote] A simplified description of an object system, often expressed in logical or other mathematical terms. Cf. *model*.

accessible timepoint: [8.3.1]

action, action class: [1.1, 1.2.2] An *occurrence* which is performed by an *agent* in such a way that it is initiated by the *ego*, but terminates when the *vehicle* considers it to be done.

action designator: [] See *occurrence designator*.

action law: [2.1.3]

action rewrite function: [8.2.5]

action statement: [8.2.1]

activity: [1.1, 1.2.2] An *occurrence* which is performed by an *agent* in such a way that it is both initiated and terminated by the *ego*. Cf. *action*.

agent: [1.1, 1.2.1] (IDS *)

agglomeration: [**12.1.2**] Set of *actions* that occur in immediate succession.

aggregation: [12.1.4, **12.1.5**] An aggregated action symbol is one which is introduced in order to replace a *composite* action expression.

alternative results of actions: [1.4.2, 7.3.1] Also an ontological characteristic, denoted by **A**.

ambiguity: [11.1.1]

assessment: [3.1.2]

axiom: [2.6.2] Self-evident truth, in the context of a given ontology. Compare *premise*.

axiomatic time: [6.1.2]

axiom schema: [4.4.4]

base logic: [**2.6.3**] Defined by a syntax and a semantics, not having underlying semantics.

branching time: [6.6] See also *Herbrand time*.

breakpoint: [**1.3.2**]

candidate trajectory set: [**8.4.2**]

characteristic: [] See *ontological* * and *epistemological* *.

chronicle: [2.1.6, **2.3.1, 8.2.4**]

chronicle completion: [8.3.5] See also *scenario completion*.

chronometric fluent: [6.1.7]

classical model set: [**8.3.4**] Model set identified by classical (Tarski) semantics, in contradistinction to the *intended model set* obtained from the *underlying semantics*.

codesignation algorithm: [**5.3.2**]

coherence verification: [**2.4.3**]

compatible: [**9.2.2, 9.2.3, 10.3.1, 11.7.3**] PCM-compatible etc.

complete development: [**1.3.2, 6.6.3**]

complete history: [**1.1.2**]

complete inertia: [1.4.2] Ontological characteristic, denoted by **I** in the case of discrete time and **R** in the case of continuous time.

complete tree: [**6.6.3**]

composite: [12.1.1] Term applied to an *occurrence expression* or to any formula containing an occurrence expression as a part.

concurrency: [1.4.2] Ontological characteristic, denoted by **C**.

condition: [12.1.1] Expression used after *if* in a *conditional expression*.

conditional action: [1.4.2] Context sensitive.

conditional expression: [12.1.1] Formed using if ... then ... else.

conditional reassignment formula: [6.5.3]

configuration: [1.1]

consistency-preserving, consistency-retaining: [9.2.4] See also *pointwise* *.

Technical terms

consistent for: [10.2.1] (Valuation M consistent for schedule SCD).

constant symbol: [1.3.2, 5.2.2, 6.3.1] Constant in the sense of logic; the same entity is often called a variable in other terminology.

core syntax: [**2.6.3**, 6.3.1]

correctly applicable: [**9.2**, 11.2.3]

correct extension of valuation: [10.2.1]

correct revision: [**1.3.3, 6.6.3**]

correct time-restriction: [**9.2.1**]

corresponding selection function: [4.4.3]

corresponding action rewrite function: [8.2.3]

corresponding world: [**3.2.4, 6.6.3, 10.2.3**]

current action set: [1.3.1, **1.3.2, 6.6.3**]

default formula: [4.4.3]

default inference relation: [**4.4.3**]

default rule: [4.4.3]

delayed effects: [1.4.2] Ontological characteristic, denoted by **L**.

denotation: [5.2.3, 5.2.4]

dependency between features: [1.4.2] Ontological characteristic, denoted by **D**.

describable: [7.3.1] For a set of *trajectories*, if some formula in a given language characterizes the set precisely.

designator (epistemological): [**2.2.2**]

designator (ontological): [**1.4.1**]

determistic action: [1.4.2]

development: [1.3.1, **1.3.2, 6.6.3**] A tuple consisting of a *history*, a *valuation*, and other components such as a set of *occurrences*.

direct model selection function: [9.1.2]

direct upper bound: [**13.1**]

discrete fluent logic: [6.1.1]

discrete time structure: [6.1.8]

domain: [2.6.1, 5.2.1, 6.3.1] 1. Set of objects of a given type. 2. The set of values (called *items*) that a feature can have. 3. (k-th domain) the actual domain (value domain) of the feature symbol f_k in a vocabulary. – For 'domain' in the sense of 'application domain' I use the term *world*.

domain closure premise: [4.1.2]

domain-specific scenario: [2.1.4]

dynamical system: [1.1]

ego: [1.2.1, **1.3.3, 6.6.3**] (IDS *): reasoning system, knowledge system.

elementary feature expression: [6.3.1]

elementary fixed formula: [6.3.1]

elementary fluent formula: [6.3.1, 11.1.1]

elementary formula: [4.2.2, 5.2.2]

ending time: [10.2.2]

entailment method: [**2.6.3, 9.1.1**]

entailment relation: [**2.4.1**]

epistemological assumption: [2.1.3, 3.1.2]

epistemological characteristic, property, designator: [**2.2.2**]

epistemological upper bound, epistemologically maximal assessment: [**13.2.3**]

equivalent world description: [**3.2.4**]

essential selection function: [9.1.2]

event: [1.2.2] An *occurrence* which characterizes some aspect of a *development* in the IDS *reality*.

example based methodology: [3.1.1]

executed model set: [**10.2.3**]

expand, expansion: [6.5.2] Cf. *abbreviation*.

explanation: [**2.4.4**]

explicit time: [6.2]

extension (I): [2.4.2] Formula corresponding to a subset of the intended models for a chronicle.

extension (II), extension candidate, extension relation: [4.4.3] Extension in certain types of non-monotonic logic, such as default logic.

extension (III): [10.2.2] A superset of a partial valuation.

favorable: [**9.2.3, 9.2.4, 11.6.3, 11.7.3**] OCM-favorable, etc.

feature: [1.2.2, 5.2.2] 1. A discrete-valued state variable in an *image*. 2. A rigid feature expression.

feature assignment: [5.2.4] Synonym: *state*.

feature expression: [5.2.2, 6.3.1]

feature symbol: [4.2.1, 4.2.2]

Ferryboat connection scenario: [7.2.4]

filter: [4.4.3] (Semantic filter)

filtering: [9.1.2] (Formed using filtering)

filtered preferential entailment: [**9.4**]

finite development: [**1.3.2**]

finite game: [] See *game*.

finite history: [**1.1.2**]

finite integer time: [**6.1.8**]

fluent: [1.2.2]

fluent formula: [6.3.1, 6.3.2] A formula whose value is a function of the time indicated by the formula's context.

formed using filtering: [9.1.1]

formed using minimization: [9.1.1]

free domain semantics: [5.2.1]

free-domain model set: [5.2.5]

full trajectory normal form: [8.1]

Furniture assembly scenario: [7.2.8]

game: [**1.3.4, 6.6.3**]

good valuation: [8.3.1]

good initial valuation: [10.2.1]

harmless: [**9.1.1**]

Herbrand time: [6.1.3, **6.1.8**, 6.6]

Hiding turkey scenario: [7.2.2]

hierarchical occurrences: [1.4.2] Ontological characteristic, denoted by **H**.

history: [**1.1.2, 1.3.2**, 6.4.1, **6.6.3**] Mapping from *timepoints* to *states*, or equivalently from *timepoints* times *features* to *items*.

hybrid system semantics: [3.2.2]

image: [1.1.1] A domain of descriptors for a system, providing less information than the state of the system itself.

image level: [1.1.1]

immediate predecessor function for timepoints: [6.1.8] Cf. *precedes relation*

implementation: [2.6.4]

indirect upper bound: [**13.2.3**]

inert, inertia: [1.1.3] 1. Features do not change unless there is a reason why they must or may do so. 2. All actions map a stable state of the IDS world to a new stable state.

inference operation: [**2.5.1**] (In a reasoning problem)

infinite development: [**1.3.2**]

inhabited dynamical system: [1.1, **1.3.5**] An inhabited dynamical system is one containing one or more *agents* which can influence the system's subsequent state by performing *actions* or *activities*.

initial observation: [8.3.1]

integer time: [**6.1.8**]

intended model: [3.1.2] Model specified by *underlying semantics*.

interpretation: [4.2.3, 5.2.1, 5.2.4, 6.4.1]

interval in branching time, **interval set**: [**6.6.2**]

item: [4.2.2] Member of the *value domain* of a feature.

knowledge level, knowledge system: [] See *ego*.

lack alternatives: [] See *alternative results*

lexical domain, lexical domain semantics: [4.1.2, 5.2.1]

lexical model set: [5.2.5]

linear time: [**6.1.8**]

Technical terms 305

locality: [] See *strict locality*.

logic: [2.1, **2.6.3**] Defined by a *base logic* and optionally an *entailment method*.

logical formula: [2.6.2, 4.2.2, 5.2.2, 5.4.1, 6.3.1, 6.3.2] The main type of formula in the logical language at hand. Used instead of the common term "well-formed formula".

loop expression: [12.1.1]

main language, main syntax: [**2.6.3**, 6.5.1] In contradistinction to *side language*.

material event: [1.2.2]

material level, system: [1.1.1, 1.2.1]

memory: [1.4.2] Ontological characteristic, denoted by **M**.

minimization (formed using *): [9.1.2]

model: [1.4.4: footnote] This terms is used as in logic, not as in engineering: a model for a formula is an interpretation or structure where the formula is true. This is a generalization of the ordinary mathematical concept of a 'solution' (for an equation etc.). A 'model' in the engineering sense (a simplified description of an object system, often expressed in mathematical terms) is here called an *abstraction*.

model set: [2.3.2, **8.3.2, 12.1.2**]

model selection function: [**9.1.1**, 12.2]

monitor: [1.2.3]

monotonic premise integration function: [4.4.4]

next executable: [10.2.2]

nochange premises: [11.4.1]

non-deterministic: [1.4.2, 6.1.9]

normal form for action laws: [] See *trajectory normal form*.

normality in features: [1.4.2] Ontological characteristic, denoted by **N**.

now: [1.3.2]

object, object domain: [1.1, 2.1.3, 4.1.1, 5.1.1, 5.2.1]

object expression: [5.2.2, 6.3.1]

object identification: [**2.4.3**, 5.3] See also *identification*.

object name: [4.1.2, 5.1.1]

objective: [1.2.3]

observation: [2.1.3, 8.2.2] 1. Member of premise partition OBS. 2. Formula with certain syntax, appropriate to be used as observation[1].

occlusion: [11.1.1, 11.2]

occurrence: [1.2.2, 1.3.2, 6.5.2] 1. A phenomenon defined to hold or not to hold over an interval of time. Either of an *action*, an *activity*, or an *event*. 2. A tuple $\langle s, E, t \rangle$ in $\mathcal{T} \times \mathcal{E} \times \mathcal{T}$ where $s < t$. 3. A variable-free syntactic expression $[s,t]E$ where $s < t$.

occurrence definition: [6.5.2]

occurrence designator: [1.3.1, 6.5.2]

occurrence expression: [6.5.2, 12.1.1]

occurrence formula: [6.5.2]

occurrence language: [6.5.2]

occurrence symbol: [6.5.2]

occurrence vocabulary: [6.5.2]

ontological designator: [**1.4.1**]

ontological characteristic/family: [1.4.1]

ontological subcharacteristic/subfamily: [1.4.3]

ontological taxonomy: [1.4.1, 3.1.2]

ontological upper bound, ontolog. maximal assessment: [**13.2.3**]

oracle features: [11.1.1, 11.7]

origo: [6.1.3]

partition: [2.1.3]

past action set: [1.3.1, **1.3.2**, **6.6.3**]

path in tree: [**6.6.2**]

perception, perception function: [1.1, **1.1.2**]

perceived inert: [1.1.3]

perfect: [**9.2.4, 11.6.3, 11.7.3**] PGM-perfect etc.

perform: [**10.2.3**]

Technical terms

persistence: [1.1.3] As applied to persistence over time, here called *inertia*. (To distinguish from other uses of the term persistence in logic).

plan: [1.2.2]

plan construction: [**2.4.3**]

pointwise consistency-retaining: [**9.2.4**]

postdiction: [] See *chronicle completion*.

precedes relation: [6.1.8] See also *immediate predecessor function*.

predeterminate ego, model set: [**8.3.2**]

prediction: [**2.4.3**] See also *chronicle completion*.

preference relation, preferred: [**4.4.3**, 9.1.2, 10.1, 11.3]

premise: [2.6.2] Three types of premises are primarily used: *axioms*, *laws*, and *observations*.

premise integration: [4.4.4]

preserve: [**9.2**]

pretransformation: [**9.1.1**]

process: [1.2.2]

progressive: [**8.4.2**]

Quine quotes: [4.3.2]

random (action): [1.4.2]

reality: [1.2.1] (IDS *)

realize: [**12.1.2**]

realization: [2.6.4]

reasoning problem: [**2.5.1**]

reasoning system: [] See *ego*.

reassignment expression/formula: [6.5.3]

reduction: [4.4] Mapping a set of models to one of its subsets, in order to *remove* models that do not satisfy the underlying semantics, and to *retain* models that do.

reference image: [1.2.3]

reformulation function: [**10.1.3**]

reification: [6.2]

remove model: [4.4.1]

restricted model: [**8.4.1**]

resulting development: [**1.3.4**, **6.6.3**]

retain model: [4.4.1]

rigid: [5.2.2, 12.1.1]

rigid feature expression: [6.3.1]

rigid occurrence formula: [6.5.2]

Russian turkey scenario: [7.2.5]

safely sequential: [**9.1.1**]

scenario: [1.1, **1.3.6**]

scenario completion: [**2.4.2**]

scenario description: [1.1, 1.3.6, 2.1.3, **2.3.1**]

schedule: [1.2.2, 2.1.3, 8.2.2]

schedule statement: [8.2.1]

schema (axiom schema): [4.4.4]

selected models: [**9.1.1**]

selection function: [**4.4.1, 9.1.1**] 1. A function from a set of models to a subset thereof. 2. A function from a chronicle to a set of models, usually a subset of its *classical model set*.

sequential schedule: [8.3.1, 8.5.1] See also *safely sequential*.

sequential expression: [12.1.1] Expression for a sequence of actions, formed using the ; operator.

set expression: [5.4.1]

set relationship: [5.4.1]

side language: [**2.6.3**, 6.5.1]

signal feature: [**12.1.3**] Auxiliary feature used to indicate the end of one cycle in the execution of a *loop expression*.

similarity type: [4.2.2, 5.2.1]

simply executable: [**10.2.1**]

situation: [6.1.5]

skeptical selection function: [4.4.3]

solution to reasoning problem: [**2.5.1**]

stable period: [6.1.6]

standard time: [6.1.2]

Stanford murder mystery: [7.2.3]

state: [1.1, 1.3.2, 5.2.4, 6.1.4] A *feature assignment*.

statement: [8.2.1]

state-transition semantics: [3.2.2]

Stolen car scenario: [7.2.6]

strict locality: [7.3.1] An action can only depend on, and influence, features of objects that occur as some of its arguments.

strict inertia: [1.1.4, 1.4.2]

structure: [5.2.1, 5.2.4, 6.4.1] A pair of an *object domain* and an *interpretation*. Also, as in *time structure, feature-value structure*, a tuple consisting of one or a few *domains* and operations on them.

subjective: [1.2.3]

sub-language: [**2.6.3**]

surprises: [1.4.2] Ontological characteristic, denoted by **S**.

syntax neutral: [4.4.4]

system: [1.1]

systematic methodology: [3.1.2]

temporal gap: [6.1.6]

tense syntax: [6.2]

Ticketed car scenario: [7.2.7]

time-bound objects: [1.4.2] Ontological characteristic, denoted by **T**.

time domain, time-point: [6.1.8]

time-point expression: [6.3.1]

time-stamp: [6.1.7]

time structure: [] See *discrete time structure*.

timing statement: [8.2.1]

trajectory: [**3.2.3**] A sequence of partial states over those features which are affected by an action, used for defining the underlying semantics of the action.

trajectory semantics: [**3.2.3**]

trajectory semantics ego: [**3.2.5**, **6.6.3**]

trajectory semantics world: [**3.2.4**, **6.6.3**]

tree in branching time: [**6.6.2**]

truly inert: [1.1.3]

underlying semantics: [3.1.2] Formal construct which defines the *intended models* using a *game* between an IDS *ego* and an IDS *world*.

uniform world description: [**3.2.4**]

unique names premise: [4.1.2]

valuation: [**1.3.2**, 5.2.4, 6.4.1] A function that assigns values to constant symbols, used as a part of an *interpretation*.

value of formula: [5.2.3, 5.2.4]

value domain: [4.1.1, 4.1.2, 5.1.2] Members are called *items*.

vehicle: [1.2.1] (IDS vehicle): A system for *perception* and actuation.

vocabulary: [5.2.1, 6.3.1] A pair of a *similarity type* and an *object domain*, for use in *lexical-domain semantics*.

weakened entailment relation: [**2.4.1**]

weakly inert: [1.1.4]

well-formed formula: [] See *logical formula*.

widening of world description: [**3.2.4**]

world: [1.2.1, **1.3.3**, 2.6.1, **6.6.3**] (IDS world).

world description: [**3.2.3**]

Yale shooting scenario: [7.2.1]

Appendix B
Notation

The following tables are intended as a reference for the use of symbols and of the various letters in the available typestyles. For symbols, the conventional notation of set theory and logic is of course used, as listed in table B.6. Tables B.1 through B.5 specify the notation that is specific for the present book. Tables after B.6 specify the usage of single-letter symbols, for one typestyle at a time.

Underlining of a function symbol is used in meta-level syntactic expressions and means to locally evaluate the function. See the definitions of Quine quotes in Chapter 4, subsection 4.3.2.

Table B.1. Infix symbols introduced in this book

Symbol	Introduced in subsection	Usage
$\models$	2.4.1	Entails (for scenario descriptions).
$\cong$	2.4.1	Semantically equivalent (for scenario descriptions).
$\|\diamond$	2.4.1	Consistent with (for scenario descriptions).
$\rightsquigarrow$	2.5.1	Inference operation in reasoning problem.
$\ll$	2.5.1	Preference relation on solutions for reasoning problem
	4.4.3	Preference order on formulae or interpretations.
	9.2.2	Preference order on trajectories.
$\oplus$	3.2.3	Override operator on functions.
$\triangleright$	3.2.4, 9.2.2, 12.1.4	Extension of history by trajectory.
$\hat{=}$	4.2.1	Assignment of value to feature. (Assignment here means "has the value" and not "is changed to have the value").
$\prec$	4.2.3	Ordering on truth-values where $F \prec T$.
	11.2.2	Ordering on occlusion predicates.
$\equiv$	4.3.2	Equality with implicit Quine-quoting of the arguments.
$:$	4.3.2	Maplet ($\mapsto$) with implicit Quine-quoting of the left argument.
$\|\diamond$	4.3.5	$\Gamma \|\diamond \alpha$ means α is satisfied in some model of Γ.
$\simeq$	4.3.5	$\alpha \simeq \beta$ means α and β are logically equivalent.
$\iff$	4.3.5	Meta level equivalence.
$\approx_\zeta, \approx_w$	4.4.1, 4.4.3, 9.1.2	Selective entailment for a selection function ζ or a criterion called w.
$\div$	5.4.1	$f \div \mathcal{X}$ set of objects with given feature value.
$\Uparrow$	5.4.1	$W \Uparrow W'$ the sets are disjoint.
$<$	6.1.8	Later-time ordering in time domain (may be total or partial order).
$\Rightarrow$	6.5.2	Action law (expressed as an abbreviation rule).
$:=$	6.5.3, 11.2.1	Reassignment of value to feature.
$\asymp, \smile$	6.6.1	Equivalence relations in branching time.
$\diamond$	6.6.2	Extension of branching history with timestep without action.
$/$	7.3.1, 8.1.2	r/O restriction of a partial state.
$\dagger$	12.1.4	$r \dagger F$ restriction of a partial state.

Table B.2. Prefix and subscript symbols and constants.

Symbol	Introduced in subsection	Usage
$n_\mathcal{B}$	1.3.2	Largest member of $\mathcal{B}$.
$\mathcal{T}_\mathcal{B}$	1.3.2	$[0, n_\mathcal{B}]$.
$\emptyset$	4.2.2	The empty set of items, in logic formulae.
#	5.1.1	#1, #2,... are object names.
¶	5.1.2	Represents the unknown item.
α_y^x	5.2.4	Substitution, replacing all free occurrences of y by x.
$\Box$	6.3.2	Always.
$R_{0:t}$	1.1.2, 8.3.1	Restriction of R to the time interval $[0, t]$.
R_B	8.5.2	Restriction of R to $[B]$ where B is a subtree of the tree where R is defined.
$M_{0:t}$	10.1.2	Restriction of temporal maplets in M to those where value is in $[0, t]$.
$M[\alpha]$	8.3.1, 10.2.1	Substitution $\alpha_t^{M[t]}$ for every time-point constant t, and subsequent simplification.
DFL1:, DFL2:	11.2.2	Choice of logic variant relative to which the operator $[.]$ is to be used.

Table B.3. Constant identifiers

Symbol	Introduced in subsection	Usage
LAW	2.1.3	Set of action laws in a chronicle. Compare table B.4, 9.2.2.
SPR	2.1.3	Set of surprise patterns.
Diw	2.1.3, 2.3.1	Combination of designator of ontological assumptions and set of action laws.
SCD	2.3.1	Set of change statements in a chronicle.
OBS	2.3.1	Set of observations in a chronicle.
PRE	2.3.1	Preconditions.
IOBS	8.3.1	Set of observations of initial state in a chronicle.
SXS	10.2.1	Simply executable schedule.
NCH_ν	11.4.1	Set of nochange premises.

Table B.4. Prefix identifiers

Symbol	Introduced in subsection	Usage
Perc	1.1.2	Perception function.
Mod	2.3.2, 2.4.1, 6.6.3, 8.3.2, 12.1.2	Model set or intended developments.
Infl	3.2.3, 12.1.4	Set of influenced features.
Rstat	3.2.3	Set of resulting states from an action.
Trajs	3.2.4, 12.1.4	Set of trajectories.
neg	4.2.3	$neg(\text{T}) = \text{F}, neg(\text{F}) = \text{T}$.
val	4.2.3, 5.2.4, 6.4.1, 12.1.2	Value of a formula or other expression.
$mval$	6.4.1, 11.2.2	Value of formula or expression at timepoint.
Dom	5.2.1	Domain in a vocabulary.
den	5.2.4, 6.4.1, 11.2.2	Denotation of an expression.
$result$	6.1.5	Forms timepoints in Herbrand time.
$args$	7.3.1	Set of objects occurring in arguments.
$Pdur$	7.3.3	Set of possible durations.
$last$	8.1.2	Last element of a sequence.
$resl$	8.1.2	Auxiliary function for FTNF.
$assign$	8.1.2	Auxiliary function for FTNF.
$change$	8.1.2	Auxiliary function for FTNF.
pt	8.1.2	Auxiliary function for FTNF.
At	8.3.1	Set of accessible timepoints.
Mod^t	8.3.2	Intended models up to time t.
Ci	8.3.3, 11.2.2	Coerce interpretation.
Cats	8.4.2	Candidate trajectory set.
$breakset$	9.1.3	Used in defintion of prototypical chronological minimization.
$changeset$	9.1.5	Used in defintion of prototypical global minimization.
Wdraw	10.1.3	Withdraw information from interpretation
$Xmod^u$	10.2.4	Intended executed models up to and including an occurrence that starts at time u.
brs	11.3.1	Used in definition of chronological minimization with occlusion.
chs	11.5	Used in definition of two-stage minimization.
$esig$	12.1.3	Auxiliary function in the definition of Π^c and Π^{cw}.
$widen$	12.1.4	Add members to all partial states in a trajectory.
$chaw$	12.1.6	Auxiliary function for FTNF.
pw	12.1.6	Auxiliary function for FTNF.
atr	12.1.6	Auxiliary function for FTNF.

Table B.5. Bracket symbols

Symbol	Introduced in subsection	Usage
[...]		Range of a quantifier.
	4.3.2	[...] abbreviates ('...').
$M[...]$		Substitution, see table B.2.
[EM], [EM$^+$], [EM*]	11.7	EM-compatible, EM-favorable, EM-perfect for entailment methods EM.
[EM$^\times$]	13.3.2	EM-semiperfect.
[...]	6.3.2	$[s,t]$ is the closed (time) interval between s and t. Similarly (s,t), $[s,t)$, $(s,t]$.
	6.6.2	$[B]$ is the set of timepoints in the tree B.
◁...▷	6.6.2	History extension for branching time.
[[...]]	4.3.1, 5.2.5, 8.3.6	$[\![\Gamma]\!]$ is the classical model set for the set Γ of formulae. Variants for lexical or free semantics.
	8.3.4	$[\![\Upsilon]\!]$ is the classical model set for the chronicle Υ.
$[\![...]\!]^t$	8.4.1	Restricted models
$[\![...]\!]^c, [\![...]\!]^{cw}$	12.1.3	Like $[\![\Upsilon]\!]$ but using aggregation of composite action statements.
⌜...⌝	4.3.2	Quine quotes. Metavariables and underlined function symbols are evaluated inside the quotes.
{...}		A set.
|...|		Cardinality of a set, length of a sequence.
⟨...⟩		A tuple.
(...)		All other standard uses of parentheses.

Table B.6. Standard symbols

Symbol	Usage
$\mathrm{T}, \mathrm{F}, \neg, \wedge, \vee, \rightarrow, \leftrightarrow$	Propositional connectives.
$\forall, \exists$	Quantifiers.
$\cup, \cap, \subset, \subseteq, \supset, \in, -, \emptyset$	Set algebra. $\subset$ stands for "subset and not equal"; $-$ for set difference.
$\overline{}$	Overbar for complement.
$\rightarrow$	$W \rightarrow U$ is the set of all mappings from W to U.
$\mapsto$	$x \mapsto y$ is a maplet saying that x maps to y. A mapping is considered as a set of maplets.
$\models$	Semantic entailment. (Nonmonotonic entailment is represented as $\mathrel{\vert\!\approx}$).
$\bigwedge$	Conjunction over a set of formulae.
$+, -, *$	Arithmetics. Overloaded use in 6.6.1.
$<, \leq$	Arithmetics. See also 6.1.8, 6.3.2, and 6.6.1.
$\times$	Cartesian product. $A \times B \times C$ is $\times(A, B, C)$.
$\infty, -\infty$	Plus and minus infinity. In logic formulae: considered as abbreviation formers (see 6.3.2).
min, max	Minimum or maximum w.r.t. a total order.
$Min(\ll, W)$	The subset of W consisting of those members which are minimal w.r.t. the ordering $\ll$. See 4.4.3.
lim	Limit to ∞ (see 1.3.4).

Table B.7. Small italic letters

Letter	Introduced in subsection	Usage
a		
b	1.3.3	Breakpoint (timepoint in $\mathcal{B}$).
c		
d	4.2.3	Meta-variable for items (members of feature domains): d_i.
e		
f	3.2.2, 4.2.1	Feature symbol, feature: f_i.
g		
h		
i		Integer offset.
j		Integer offset.
k		Integer offset or range.
l		
m		
n	1.3.1 ff.	Timepoint, especially for 'now'.
o	5.2.2	Object, object variable.
p	4.2.2	Propositional feature symbol p_i.
q	11.1.1	Oracle feature.
r	1.3.2, 5.2.4	Feature assignment = state.
s	5.2.3	Structure $s = \langle \mathcal{O}, I \rangle$.
s,t	6.3.1	Timepoint (general, variable).
u		Timepoint (general).
v	3.2.4	Trajectory.
w		
x		
y		
z		

Table B.8. Capital italic letters

Letter	Introduced in subsection	Usage
A	6.5.2	Occurrence symbol.
B	6.6.2	Tree (in branching time).
C	6.6.1	Set of pairs $\langle m, E \rangle$.
D	5.2.2	Set of items.
	4.4.3	Subset of Δ in the definition of maximal consistency selection.
E	1.3.2	Metavariable for rigid elementary occurrence expression (= occurrence).
F	4.4.3	Filter (predicate on interpretations).
G	12.1.2	Agglomeration (set of occurrences).
H		
I	4.2.3	Interpretation.
J	1.3.3	Development.
K	10.1.3	Set of constant symbols.
L		
M	1.3.2, 5.2.4	Valuation: mapping from constant symbols to values.
N	6.1.8, 6.6.1	Next-timepoint operation.
O	7.3.1	Set of object names.
P		
Q		
R	1.3.2, 6.4.1	History: mapping from timepoints to states.
S	3.1.2, 9.1.1	Model selection function.
T		
U		
V		
W	3.3.5	Set of interpretations.
X	11.2.2	Occlusion component in interpretations.
Y		
Z	4.4.3	Extension relation.

Table B.9. Script letters

Letter	Introduced in subsection	Usage
$\mathcal{A}$	1.3.2	Set of completed occurrences in a development.
$\mathcal{B}$	1.3.2	Domain of breakpoints in a development.
$\mathcal{C}$	1.3.2	Set of incompleted occurrences in a development.
$\mathcal{D}$	4.2.2	Feature domain $\mathcal{D}_i$.
$\mathcal{E}$	1.3.2	Domain of occurrence designators.
$\mathcal{F}$	4.3.1	Domain of features $\mathcal{F}_\sigma$.
$\mathcal{G}$		
$\mathcal{H}$		
$\mathcal{I}$	2.2.2	Epistemological characteristics of illusion (intended).
$\mathcal{J}$	1.3.2	Domain of finite developments.
$\mathcal{J}^*$	1.3.2	Domain of developments, finite or infinite.
$\mathcal{K}$	2.2.3	Epistemological characteristics of omniscience.
$\mathcal{L}$	4.3.1	Language of logical formulae $\mathcal{L}_\sigma$.
$\mathcal{M}$	2.2.2	Epistemological characteristics of misperception (intended).
$\mathcal{N}$	2.2.3	Empty epistemological assumption.
$\mathcal{O}$	5.2.1	Domain of objects.
$\mathcal{P}$	2.2.3	Epistemological characteristics allowing composite actions ('programs').
$\mathcal{Q}$	2.2.3	Epistemological characteristics of qualified knowledge.
$\mathcal{R}$	1.1.2	Domain of (image-level) states.
$\mathcal{R}_M$	1.1.2	Domain of material-level states.
$\mathcal{S}$	4.3.1	Domain of interpretations $\mathcal{S}_\sigma$.
$\mathcal{T}$	1.1.2, 6.1.8	Time domain.
$\mathcal{U}$	2.2.2	Epistemological characteristics of uninformedness (intended).
$\mathcal{V}$		
$\mathcal{W}$		
$\mathcal{X}, \mathcal{Y}$	4.2.1	Meta-variable for set of feature values ($f \hat{=} \mathcal{X}$).
$\mathcal{Z}$	2.2.2	Variable for epistemological designator.
	13.1	Chronicle family.

Table B.10. Greek letters. These are often used as syntactic metavariables or for other syntax-related purposes.

Letter	Introduced in subsection	Usage
α		Metavariable for logical formulae.
β		Metavariable for logical formulae.
γ		Metavariable for logical formulae (especially members of Γ).
δ	4.4.3	Default inference relation.
ε	6.5.2	Metavariable for occurrence expressions.
ζ	4.3.1, 4.4.1	Selection function on sets of models.
η	6.6.2	$\eta(B)$ Set of interval endpoints in a tree.
θ	6.1.8	Previous timepoint.
κ		
λ		Lambda-abstraction.
	13.2.1	Set of ontological subcharacteristics.
ν	5.2.1	Vocabulary.
ξ	13.2.3	Variable for epistemological property.
π	6.5.2	Occurrence vocabulary.
σ	4.2.3	Similarity type.
$\varsigma, \tau, \upsilon$	6.3.1	Metavariables for timepoint expressions.
ϕ	6.3.1	Metavariable for fluent formula.
φ	12.1.1	Metavariable for condition.
ψ	8.1.3	Component of FTNF formula characterizing the inner structure in a set of trajectories.
ω	6.3.1	Metavariable for object expressions.
Γ		Set of formulae.
Δ	4.4.3	Set of formulae (default, or addition to given set).
Θ	6.1.8	Origo time.
Λ	6.3.1	Previous-time operator on feature expressions.
Ξ	2.5.1	Set of chronicles (scenario descriptions) from which solution(s) is to be taken.
Π	8.1.2, 12.1.3, 12.1.6	Action laws, viewed as a syntactic transformation on action statements.
$\Sigma_{IA}, \Sigma_{IA}^t$	8.3, 8.3.3, 8.4.1	Set of intended models.
$\hat{\Sigma}_{IA}^u$	10.2.4	Set of finite progressive models.
Υ	2.3.1	Chronicle, scenario description.
Φ		
Ψ	2.1.3, 13.2.3	Variable for ontological family or descriptor.
Ω	5.1.1	Object universe.

Appendix C
References to related work

As I described in the preface, the early stages of work on this book involved a quite detailed survey of the existing literature in the present field. However, as the book developed, I decided to postpone the assessment oriented survey to volume II. This was not only in order to keep volume I at reasonable size, but also because the topics of surprises and ramification could not be covered in the present volume. Many of the currently proposed approaches intend to deal with surprises and/or ramification, in addition to strict inertia. It would therefore not do them justice to assess their range of applicability only with respect to $\mathcal{K}$-**IA**.

Meanwhile, I have also co-authored a historical overview of many of the main approaches in the field of reasoning about actions and change [74]. That survey describes the background and the main ideas in several of the approaches, but without any attempt at applicability assessment. Since it will be published almost concurrently with the present book, it did not seem meaningful to dublicate such a survey chapter here.

Consequently, the bibliography in the present volume is relatively short, and consists only of the immediate references from the actual text in the book. It is by no means a bibliography of all the relevant literature on the book's topic. Possibly the bibliography in volume II will be more complete.

Many of the referenced articles have been published in major AI conference proceedings. The publishers and distributors) for those series of proceedings are as follows. For the *International Joint Conference on Artificial Intelligence* and the *International Conference on Knowledge Representation and Reasoning*: Morgan Kaufmann, San Francisco, Calif. For the *National (U.S.) Conference on Artificial Intelligence*: The MIT Press.

Bibliography

1. James Allen, Henry Kautz, Richard Pelavin, and Josh Tenenberg. *Reasoning about Plans*. Morgan-Kaufmann, 1991.

2. James F. Allen. Towards a general theory of action and time. *Artificial Intelligence*, **23**(2):123–154, 1984.

3. Christer Bäckström and Inger Klein. Parallel non-binary planning in polynomial time. In *International Joint Conference on Artificial Intelligence*, pp. 268–273, 1991.

4. Christer Bäckström and Bernhard Nebel. Complexity results for SAS$^+$ planning. In *International Joint Conference on Artificial Intelligence*, pp. 1430–1435, 1993.

5. Andrew B. Baker. A simple solution to the Yale shooting problem. In *International Conference on Knowledge Representation and Reasoning*, pp. 11–20, 1989.

6. Andrew B. Baker. Nonmonotonic reasoning in the framework of the situation calculus. *Artificial Intelligence*, **49**:5–23, 1991.

7. Chitta Baral and Michael Gelfond. Representing concurrent actions in extended logic programming. In *International Joint Conference on Artificial Intelligence*, pp. 866–871, 1993.

8. Avron Barr and Edward A. Feigenbaum. *Handbook of Artificial Intelligence*. Addison-Wesley, 1981.

9. John Bell. Pragmatic logics. In *International Conference on Knowledge Representation and Reasoning*, pp. 50–60, 1991.

10. Suzanne Biundo and Richard Waldinger. *Proceedings of the International Dagstuhl Seminar on Deductive Approaches to Plan Generation and Plan Recognition*. German Research Institute for Artificial Intelligence, Saarbrücken, 1993.

11. Geneviève Bossu and Pierre Siegel. Saturation, nonmonotonic reasoning and the closed-world assumption. *Artificial Intelligence*, **25**:13–63, 1985.

12. Gerhard Brewka. *Nonmonotonic reasoning: logical foundations of commonsense*. Cambridge University Press, 1991.

13. David Chapman. Planning for conjunctive goals. *Artificial Intelligence*, **32**:333–377, 1987.

14. K. Clark. Negation as failure. In H. Gallaire and J. Minker, editors, *Logics and Databases*, pp. 293–322. Plenum Press, 1978.

15. James M. Crawford and David Etherington. Formalizing reasoning about change: A qualitative reasoning approach (preliminary report). In *National (US) Conference on Artificial Intelligence*, 1992.

16. Ernest Davis. *Representation of Commonsense Knowledge*. Morgan-Kaufmann, 1990.

17. Thomas Dean and Michael P. Wellman. *Planning and Control.* Morgan-Kaufmann, 1991.

18. R. Dechter, I. Mieri, and J. Pearl. Temporal constraint networks. *Artificial Intelligence*, **49**:61–95, 1991.

19. Patrick Doherty. NML3 – a three-valued cumulative non-monotonic formalism. In *Proceedings of the European Workshop on Logics in A.I. (JELIA)*, pp. 196–211. Springer Verlag, Lecture Notes in Artificial Intelligence, 1990.

20. Patrick Doherty. Reasoning about actions and change using occlusion. In *European Conference on Artificial Intelligence*, pp. 401–405, 1994.

21. Patrick Doherty and Witold Łukaszewicz. FONML-3 – a first-order non-monotonic logic with explicit defaults. In *European Conference on Artificial Intelligence*, pp. 294–298, 1992.

22. Patrick Doherty and Witold Łukaszewicz. NML3 - a nonmonotonic logic with explicit defaults. *Journal of Applied Non-Classical Logics*, **2**(1):9–48, 1992.

23. Patrick Doherty and Witold Łukaszewicz. Circumscribing features and fluents. In *International Conference on Temporal Logic*, pp. 82–100, 1994.

24. David W. Etherington. *Reasoning with incomplete information*. Pitman, 1988.

25. Kenneth D. Forbus. Introducing actions into qualitative simulation. In *International Joint Conference on Artificial Intelligence*, pp. 1273–1278, 1989.

26. Michael Gelfond and Vladimir Lifschitz. Representing actions in extended logic programs. In *International Conference on Logic Programming*, pp. 559–573, 1992.

27. Michael R. Genesereth and Nils J. Nilsson. *Logical Foundations of Artificial Intelligence*. Morgan-Kaufmann, 1987.

28. Matthew L. Ginsberg and David E. Smith. Reasoning about actions I: A possible worlds approach. In Frank Brown, editor, *Proceedings of the 1987 Workshop on the Frame Problem in Artificial Intelligence*, pp. 233–258. Morgan Kaufmann, Lawrence, Kansas, 1987.

29. Matthew L. Ginsberg and David E. Smith. Reasoning about actions II: The qualification problem. In Frank Brown, editor, *Proceedings of the 1987 Workshop on the Frame Problem in Artificial Intelligence*, pp. 259–288. Morgan Kaufmann, Lawrence, Kansas, 1987.

30. Scott D. Goodwin and André Trudel. Persistence in continuous first order temporal logics. Technical Report TR 89-24, Department of Computer Science, University of Alberta, October 1989.

31. Andrew R. Haas. The case for domain-specific frame axioms. In Frank Brown, editor, *Proceedings of the 1987 Workshop on the Frame Problem in Artificial Intelligence*. Morgan Kaufmann, 1987.

32. Steve Hanks and Drew McDermott. Default reasoning, non-monotonic logics, and the frame problem. In *National (US) Conference on Artificial Intelligence*, pp. 328–333, 1986.

33. Steve Hanks and Drew McDermott. Nonmonotonic logics and temporal projection. *Artificial Intelligence*, **33**(3):379–412, 1987.

34. Brian Haugh. Simple causal minimizations for temporal persistence and projection. In *National (US) Conference on Artificial Intelligence*, pp. 218–223, 1987.

35. Lars Karlsson. DFLEAK, planning with features and fluents. Master's thesis, Linköping University, Sweden, 1993.

36. G. Neelakantan Kartha. Soundness and completeness theorems for three formalizations of action. In *International Joint Conference on Artificial Intelligence*, pp. 724–729, 1993.

37. G. Neelakantan Kartha and Vladimir Lifschitz. Actions with indirect effects (preliminary report). In *International Conference on Knowledge Representation and Reasoning*, pp. 341–350, 1994.

38. Henry Kautz. The logic of persistence. In *National (US) Conference on Artificial Intelligence*, pp. 401–405, 1986.

39. Vladimir Lifschitz. Computing circumscription. In *Proceedings of the Ninth International Joint Conference on Artificial Intelligence*, pp. 121–127. Morgan Kaufmann, 1985.

40. Vladimir Lifschitz. Formal theories of action. In F. Brown, editor, *Proceedings of the 1987 Workshop on the Frame Problem in Artificial Intelligence*, Lawrence, Kansas, 1987. Morgan Kaufmann.

41. Vladimir Lifschitz. Benchmark problems for formal nonmonotonic reasoning, version 2.00. In Michael Reinfrank, Johan de Kleer, Matthew L. Ginsberg, and Erik Sandewall, editors, *Non-Monotonic Reasoning. Proceedings of the Second International Workshop*, number 346 in Lecture Notes in Artificial Intelligence, pp. 202–219. Springer Verlag, 1988.

42. Vladimir Lifschitz. Frames in the space of situations. *Artificial Intelligence*, **46**:365–376, 1990.

43. Vladimir Lifschitz. Toward a metatheory of action. In *International Conference on Knowledge Representation and Reasoning*, pp. 376–386, 1991.

44. Fangzhen Lin and Ray Reiter. How to progress a database (and why) I. Logical foundations. In *International Conference on Knowledge Representation and Reasoning*, pp. 425–436, 1994.

45. Fangzhen Lin and Yoav Shoham. Provably correct theories of action (preliminary report). In *National (US) Conference on Artificial Intelligence*, pp. 349–354, 1991.

46. Sten Lindström. A semantic approach to nonmonotonic reasoning: inference and choice. In *Presented at 9th International Congress of Logic, Methodology and Philosophy of Science*, 1991.

47. David Makinson. General theory of cumulative inference. In M. Reinfrank, J. de Kleer, M. L. Ginsberg, and E. Sandewall, editors, *Non-Monotonic Reasoning. Proceedings of the 2nd International Workshop*, pp. 27–41. Springer Verlag, 1989.

48. John McCarthy. Circumscription - a form of non-monotonic reasoning. *Artificial Intelligence*, **13**:27–39, 1980.

49. John McCarthy. Applications of circumscription to formalizing common sense knowledge. In *Proceedings of the Nonmonotonic Reasoning Workshop*, pp. 295–324, October 1984.

50. Drew V. McDermott. A temporal logic for reasoning about processes and plans. *Cognitive Science*, **6**:101–155, 1982.

51. Leora Morgenstern and Lynn Andrea Stein. Why things go wrong: a formal theory of causal reasoning. In *National (US) Conference on Artificial Intelligence*, pp. 518–523, 1988.

52. Bernhard Nebel and Christer Bäckström. On the computational complexity of temporal projection, planning, and plan validation. *Artificial Intelligence*, **66**(1):125–160, 1994.

53. Allen Newell. The knowledge level. *Artificial Intelligence*, **18**(1):87–127, 1982.

54. Edwin P.D. Pednault. ADL: Exploring the middle ground between STRIPS and the situation calculus. In *International Conference on Knowledge Representation and Reasoning*, pp. 324–332, 1989.

55. Tommy Persson and Lennart Staflin. A causation theory for a logic of continuous change. In *European Conference on Artificial Intelligence*, pp. 497–502, 1990.

56. David Poole. What the lottery paradox tells us about default reasoning. In *International Conference on Knowledge Representation and Reasoning*, pp. 333–340, 1989.

57. W.V. Quine. *Mathematical Logic. Revised edition.* Harvard University Press, 1981.

58. Han Reichgelt. Semantics for reified temporal logic. In J. Hallam and C. Mellish, editors, *Advances in Artificial Intelligence*, pp. 49–61. Wiley and Sons, 1987.

59. Ray Reiter. Equality and domain closure in first order data bases. *Journal of the ACM*, **27**:81–132, 1980.

60. Ray Reiter. A logic for default reasoning. *Artificial Intelligence*, **13**:81–132, 1980.

61. Ray Reiter. A theory of diagnosis from first principles. *Artificial Intelligence*, **32**:57–95, 1987.

62. Ray Reiter. The frame problem in the situation calculus: a simple solution (sometimes) and a completeness result for goal regression. In Vladimir Lifschitz, editor, *Artificial Intelligence and Mathematical Theory of Computation*, pp. 359–380. Academic Press, 1991.

63. Hans Rott. Belief contraction in the context of the general theory of rational choice. *Journal of Symbolic Logic*, **58**(4):1426–1450, December 1993.

64. Piotr Rychlik. The generalized theory of model preference. In *National (US) Conference on Artificial Intelligence*, pp. 615–620, 1990.

65. Earl Sacerdoti. The non-linear nature of plans. In *International Joint Conference on Artificial Intelligence*, pp. 206–214, 1975.

66. Earl D. Sacerdoti. *A Structure for Plans and Behavior.* Elsevier, 1980.

67. Erik Sandewall. An approach to the frame problem, and its implementation. In *Machine Intelligence, Vol. 7*, pp. 195–204. Edinburgh University Press, 1972.

68. Erik Sandewall. The pipelining transformation on manufacturing cells with robots. In *International Joint Conference on Artificial Intelligence*, pp. 1055–1062, 1987.

69. Erik Sandewall. Non-monotonic entailment for reasoning about time and action, part I: Sequential actions. Technical Report LiTH-IDA-R-88-27 and LAIC-IDA-88-TR13, IDA, 1988.

70. Erik Sandewall. A decision procedure for a theory of actions and plans. In Zbigniew Ras, editor, *Methodologies for Intelligent Systems, IV*. North-Holland, 1989.

71. Erik Sandewall. Filter preferential entailment for the logic of action in almost continuous worlds. In *International Joint Conference on Artificial Intelligence*, pp. 894–899, 1989.

72. Erik Sandewall. Features and fluents. Review version of 1991. Technical Report LiTH-IDA-R-91-29, IDA, 1991.

73. Erik Sandewall and Ralph Rönnquist. A representation of action structures. In *National (US) Conference on Artificial Intelligence*, pp. 89–97, 1986.

74. Erik Sandewall and Yoav Shoham. Nonmonotonic temporal reasoning. In Dov Gabbay, editor, *Handbook of Artificial Intelligence and Logic Programming (to appear)*. Oxford University Press, 1994.

75. Lenhart Schubert. Monotonic solution of the frame problem in situation calculus. In Henry E. Kyburg, Ronald P. Loui, and Greg N. Carlson, editors, *Knowledge Representation and Defeasible Reasoning*, pp. 23–67. Kluwer, 1990.

76. Yoav Shoham. Reified temporal logics: Semantical and ontological considerations. In *European Conference on Artificial Intelligence*, pp. 390–397, 1986.

77. Yoav Shoham. *Reasoning about Change.* MIT Press, 1988.

78. Léa Sombé. *Reasoning under Incomplete Information in Artificial Intelligence.* Wiley, 1991.

79. J.M. Spivey. *The Z notation.* Prentice-Hall, 1987.

80. Michael Thielscher. An analysis of systematic approaches to reasoning about actions and change. In P. Jorrand, editor, *International Conference on Artificial Intelligence: Methodology, Systems, Applications (AIMSA)*, Sofia, Bulgaria, 1994. World Scientific Publishing Co.

81. David S. Touretzky, John F. Horty, and Richmond H. Thomason. A clash of intuitions: The current state of nonmonotonic multiple inheritance systems. In *International Joint Conference on Artificial Intelligence*, pp. 476–482, 1987.

82. Marianne Winslett. Reasoning about actions using a possible models approach. In *National (US) Conference on Artificial Intelligence*, pp. 89–93, 1988.